APPLIED OPTIMIZATION WITH MATLAB® PROGRAMMING

Second Edition

APPLIED OPTIMIZATION WITH MATLAB® PROGRAMMING

Second Edition

P. Venkataraman

Rochester Institute of Technology

WILEY

JOHN WILEY & SONS, INC.

Library of Congress Cataloging-in-Publication Data:

Venkataraman, P.
  Applied optimization with MATLAB programming / P. Venkataraman.—2nd ed.
    p. cm.
  Includes index.
  ISBN 978-0-470-08488-5 (cloth)
  1. Mathematical optimization—Data processing. 2. MATLAB. I. Title.
  QA402.5.V42 2009
  519.3—dc22

                            2008044564

Printed in the United States of America

10 9 8 7 6 5 4

For the peace, harmony, and the coexistence of the people of this world

CONTENTS

PREFACE TO THE SECOND EDITION

The intervening years between the first and the second edition of the book have seen an explosion in application of optimization techniques to the design of products, allotment of resources, search for innovative solutions, and development of efficient process and planning. The optimization methods have migrated beyond engineering practice through a synergistic development of a multidisciplinary framework for product development. The costs of design, resources, production, quality assurance, distribution, and promotion continue to affect the success of the enterprise. Competitive pressures require the enterprise to embrace optimization in all of its endeavors to be successful. At the same time, there has been an explosion in the use of MATLAB® in academia and in the industry.

In recent years there is a lot of excitement in applying optimization to larger and newer problems. New problem areas such as particle swarm optimization and stochastic optimization are attracting much interest. The optimization techniques themselves have not changed significantly. Large, inexpensive computer resources and relative easy software development environments provide the opportunity to customize algorithms and techniques for particular problems. This is an important advantage for the serious design problems are described by nonlinear relations that cannot be easily characterized and their solutions are difficult to predict or obtain.

Optimization practice is mostly through digital computation. The astonishing power and speed of personal computation today allows all designers to embed optimization ideas in their practice. The use of optimization in professional practice still requires the same three pre-requisites identified in the first edition. First, the design must be defined mathematically in a form so that the search for the solution can take place. Second, the knowledge of the optimization technique is necessary so that if required, techniques can be modified to yield a solution. Third, knowledge of computer programming will make the intervention in the second item easier. There are many optimization software packages that do not require any programming knowledge. These are very suitable for problems that do not require creative handling. However, standard applications do not guarantee a solution when dealing with difficult design issues. In such situations, the knowledge of the optimization techniques will allow progress through better parameter selections for navigating the design space. Many computer-aided design and analysis programs provide an opportunity to develop an optimum design through parametric modeling using just a few clicks. Such applications have limited opportunities for user intervention or problem redefinition.

There are several excellent books on the subject of traditional optimization. The few released recently are newer editions that take advantage of software programs that have shallow learning curves. The first edition of *Applied Optimization with* MATLAB® *Programming* recognized the importance of MATLAB as an incredible resource for technical computing, especially as a nimble vehicle for customized exploration, particularly in the educational arena. This connection was a good choice, since MATLAB is now recognized by the industry and by academia, and is a serious resource for advanced engineering degrees. MATLAB allows the practice of professional design optimization through its Optimization Toolbox. Another significant provider of enhanced optimization tools in the MATLAB environment is TOMLAB©.

All of the books on optimization cover the subject of optimization in depth, which is necessary because the mathematical models are nonlinear and require special techniques that are usually not part of any core curriculum. There are also increasing resources on the internet for learning and understanding optimization. In addition there are software systems that are easy to master, that handle mathematics, programming, and graphics effortlessly but are available at some cost. MATLAB, MathCAD©, Mathematica©, Maple©, and Excel© are some of these systems that students and practitioners will have been exposed to. The serious advent of free open source software through OCTAVE, Python, Perl, TCL, etc provides a large set of resources that can be used to develop education materials and accessed by those who cannot afford to buy one of the stable platforms.

Applied Optimization with MATLAB® *Programming* uses MATLAB to illustrate and implement the various techniques of optimization. The use of graphics to empower visualization for explaining concepts continues to remain an important element in the code accompanying the book. In the past six years, the use of MATLAB has grown significantly in undergraduate engineering education. The book still espouses the two objectives of the first edition—teach MATLAB and teach optimization. Symbolic, numerical, and graphic programming skills are developed. These then are used to translate the optimization algorithms into applied numerical techniques. To balance the development of programming skills and the presentation of optimization concepts, programming issues are incrementally explored and implemented. Investment of time in developing programming skills is the most effective way to imbibe them. The author is not aware of shortcuts. The student is expected to independently program, make mistakes, debug errors, and incorporate improvements as part of his learning experience delivered through this book.

The book contains two elements that are new. First, a new technique called the *Scan and Zoom* method is introduced. It is a simple but effective method that mimics how you would normally eyeball the optimal solution, by scanning the design space and homing in on the region that you believe holds the minimum. In the limited examples in the book its performance appears inversely proportional to its simplicity. The method works for unconstrained, constrained, discrete, and global optimization problems. It is more effective in identifying the neighborhood of the optimum. It is also based on the objective function and does not require derivative computation. The second item is Chapter 11, which is an example of the use of optimization techniques in an unexpected area. It is called *Hybrid Mathematics—An Application of Optimization*, and illustrates through examples of how optimization can be used to develop analytical or explicit solutions to differential systems and data fitting problems.

Almost all of the numerical techniques are supported by MATLAB code and are available for download through as the companion web site www.wiley.com/go/venkat2e. These files are necessary as the book will essentially be incomplete without them. In comparison with the first edition, the code is more structured, streamlined and standardized. There is a lot of use of wrapper functions to deal with specific examples. There are links to other resources available at the site, including a complete course on MATLAB programming, examples of the use of MATLAB for undergraduate engineering math, and its use in the various core courses in mechanical engineering. The site should also contain updated information about *Applied Optimization with MATLAB® Programming*. The code that is part of the book is basically used to illuminate the techniques that are explained in the book. They have been only tested with respect to the specific information they provide in the book. Anything else is the responsibility of the user. The code available through the book is expected to provide a seed through which the student can develop and explore new ideas and problems. The reader must be careful and cautious in trying to extend it other problem. They should be useful for simple problems and will serve excellently as a test bed for exploring extensions of the techniques. They may also provide the avenue for original algorithms. Serious problems deserve to be explored through professional software mentioned earlier. Several examples are included in the book that contradict conventional expectations of performance of standard methods in optimization.

An important element of *Applied Optimization with MATLAB® Programming* is its small size for the number of topics it addresses. One reason is that the code is available outside the book. Another reason is that the topics are developed through simple ideas making the book relatively easy to read. Sophisticated mathematical ideas are illustrated through example rather than serious mathematical development, retaining the important concepts without being burdensome. Traditional continuous design optimization techniques do take up a significant portion of the book. Continuous problems generally allow the algorithms to be mathematically developed in a relatively easier fashion. These methods could be the core for original algorithms developed by readers for their special needs. Once again discrete and global optimization techniques are included to illustrate the very different algorithms that arise in the search for particular solutions. A special chapter, Chapter 11, on new applications of optimization ideas is included to demonstrate that the methods used here can transcend traditional uses.

The book can be used as a formal text on design optimization and also by the self learner. The preferred setting requires access to hands-on computation and availability of MATLAB software. A formal classroom setting without computational experience is also possible. In this case the algorithms can be presented and the numerical results illustrated. The target audience is the senior/graduate students in various disciplines, particularly engineering, as are industry professionals who would like to know more about the subject of optimization. Optimization techniques are an important tool to solve design problems in all professional areas of study.

Many academic programs include exposure to MATLAB as part of the curriculum prior to the course on Optimization. In those instances all of the topics can be covered in a single term. For programs where this book will also provide a formal exposure to MATLAB programming, it is likely that only continuous optimization problems can be covered effectively in a single term. A final design project that embraces an open-ended design is an effective mechanism to instill the ideas developed in this course. The best experience is for students to *research*, *discover*, *formulate*, *solve*, and *evaluate* real problems in design optimization.

This book could not have been developed without the work published by scores of authors and researchers in the area of design optimization over the years. The purpose of this book, since the first edition, is to bring the reader up to speed in generating and translating his own ideas into promising numerical techniques and deploying them through MATLAB. The book includes all of the mathematical ideas necessary to present the subject of optimization. The focus is on applications and the recognition that optimization techniques can be used in a wide variety of situations. Sometimes practical discussion and numerical experiments are substituted for higher-order mathematical analysis and rigor. A few of the algorithms are simplified so that they can be better appreciated. Some of the examples are explored through different techniques to illustrate the impact of different algorithms and approaches.

The book was made possible through the encouragement and support from Robert Argentieri, Executive Editor of the Professional/Trade Engineering Program, John Wiley & Sons, Inc., and Daniel Magers, his Editorial Assistant. Gilda Gaeta, from John Wiley was patient and remarkable as the Production Editor. Thanks are also due to MathWorks, Inc., for continuing to provide resources through their book program. Then, there are all those who read the first edition and were encouraging and helpful in offering suggestions for improvement. I am depending on them once again for their assistance.

I would also take the time to express my thanks to my family comprising of Jayanti, my wife, Archana, my daughter, Vinayak, my son, and Comet, the family dog who have been kind and understanding during the development of the manuscript. Their patience and understanding was particulary useful during the time when a trivial omission caused the code to respond in unexpected ways for several days at a time.

The author welcomes comments, criticisms, and suggestions for improvement at all times.

P. VENKATARAMAN

Rochester, New York
10/29/2008

1

INTRODUCTION

Optimization is, in essence, a search for the best *objective* when operating within a set of constraints. For example you want the *lightest* car that can at least deliver standard performance and possess good safety features while being powered a hybrid engine. Maybe you want to generate *maximum* shareholder return on investment while *meeting* the 10 percent sales growth and holding inventory at the *minimum*. Sometimes you want to *minimize* commuting time by choosing alternate routes during peak traffic periods. You can come up with several examples of this kind from everyday experience. What you have done is loosely define an *optimization* problem that needs a solution. The procedure by which you will establish a solution to the above examples uses *optimization techniques*. A second read through these sentences will suggest you are really talking about designs that are qualified by the words *lightest, maximum, minimize*. All of these, and many others like it can be described by the word **optimum**. In addition, these designs must be within a *certain envelope* or satisfy certain conditions or must be limited, if they are to be acceptable. These are the *constraints* on the design. With this basic definition you can identify problems of *design optimization*. Most of the book is about formally expressing these kinds of problems and looking at several techniques that can be employed to establish the solution.

   Optimization is an essential part of design activity in all major disciplines. These disciplines are not restricted to engineering. In product development, competition demands producing economically relevant products with embedded quality. Today, globalization demands that additional dimensions such as location, language, and expertise must also merit consideration as new constraints in the development process. Improved production and design tools coupled with inexpensive computational resources have made optimization an important part of the process. Even in the absence of a tangible product, **optimization ideas**

1

provide the ability to define and explore problems while focusing on solutions that subscribe to some measure of usefulness. Generally, the use of the word *optimization* implies the best result under the circumstances. This includes the particular set of constraints on the development resources, current knowledge, market conditions, and so on. The ability to make the *best choice* is a perpetual desire among us all. The techniques that are used in optimization are also used for obtaining solutions to nonlinear problems in many disciplines—so the subject has an attraction to a wider audience from many fields.

In this book, optimization is often associated with **design**, be it a product, service or a strategy. Aerospace design was among the earliest disciplines to embrace optimization in a significant way driven by a natural need to lower the tremendous cost associated with carrying unnecessary weight in aerospace vehicles. Minimum mass structures are the norm. It forms part of the psyche of every aerospace designer. Just recently, Boeing introduced the *Dreamliner* to the world. The aircraft uses more than 50% plastic in its structure, replacing metals, without compromising on safety. Today, an emerging discipline is multidisciplinary optimization (MDO) of greater significance to aerospace designers, where structural design and aerodynamics are combined to produce an optimal vehicle shape. A good example is the Boeing blended wing-body design. Saving on fuel through trajectory design was another problem that suggested itself. Very soon, the entire engineering community could recognize the need to define solutions based on merit.

Recognizing the desire for optimization and actually implementing it has taken some time to mature. In the past, optimization was usually attempted only in those situations where there were significant penalties for generic designs. The application of optimization demanded large computational resources. In the nascent years of digital computation, these were available only to large national laboratories and programs. These resources were necessary to handle the nonlinear problems that are associated with engineering optimization. As a result of these constraints most of the everyday products were designed without regard to optimization.

Today, it is inconceivable that current replacement products, such as the car, the house, the desk, the pencil, and so on, are not designed optimally in some sense or another. The most obvious contemporary example of the use of optimization manifests in the same streamlined bubblelike look in all cars from all manufacturers. Minimizing drag coefficient, particularly through software, improves fuel consumption. This is an easier option for meeting fuel consumption standards without requiring a change in the engine performance, which has proved very stubborn.

Today, you would definitely explore procedures to optimize your investments by tailoring your portfolio. You would optimize your business travel time by appropriately choosing your destinations. You can optimize your commuting time by choosing your time and route. You can optimize your necessary expenditure for living by choosing your day and store for shopping. You can buy software that will optimize your connection to the Internet. You can have affordable access to resources that will allow you to perform all these various optimizations. Seeing the variety of problems that need to generate optimum solutions, the study of optimization is actually more of a tool that can be applied to a variety of disciplines. If so, all of the optimization problems from many disciplines should be described in a common way.

The partnership between *design* and *optimization* activity is still more often found in *engineering*. This book recognizes that connection. Much of the problems used for illustrations and practice are from engineering, primarily mechanical, civil, and aerospace design. Nevertheless, the study of optimization, particularly **applied optimization**, is not an exclusive property of any specific discipline. It involves the discovery and design of solutions through *appropriate techniques* associated with the *formulation of the problem in a specific manner*. This can be done for example, in economics, chemistry, and business management. A Google search on *"optimization of"*, at the time of this writing, resulted in 128 million hits, the first six of them related to optimization in digital circuits, trading system, immunomagnetic separation, object-oriented programming, risk measures, and chemical process.

1.1 OPTIMIZATION FUNDAMENTALS

Optimization was described as the process of *search* for the solution that is more useful than several others. Qualitatively, this assertion implicitly recognizes the necessity of choosing among alternatives. This book deals with optimization in a quantitative way. This means that an outcome of applying optimization techniques to the problem, design, or service must yield numbers that will define our solution—in other words, numbers or values that will characterize the particular design or service. Quantitative description of the solution requires a quantitative description of the problem itself. This description is called a **mathematical model**. The design, its characterization, and its circumstances must be expressed mathematically prior to the application of the optimization methods. In many situations, coming up with a mathematical model will prove to be very challenging. The development of a suitable mathematical model presupposes knowledge of content in the particular design area that the optimization problem is being formulated. Consider the design activity in the following cases:

1. New consumer research, with deference to the obesity problem among the general population, suggests that people should drink no more than about 0.25 liter of soda pop at a time. The fabrication cost of the redesigned soda can is proportional to the surface area, and can be estimated at $1.00 per square centimeter of the material used. A circular cross-section is the most plausible, given current tooling available for manufacture. For aesthetic reason, the height must be at least twice the diameter. Studies indicate that holding comfort requires a diameter between 5 and 8 cm. Create a design that will cost the least.

2. Design a cantilevered beam, of minimum mass, carrying a point load F at the end of the beam of length L. The cross-section of the beam will be

in the shape of the letter I (referred to as an I-beam). The beam should be sufficiently strong in bending and shear. There is also a limit on its deflection.

3. MyPC Company has decided to invest $12 million in acquiring several new component placement machines to manufacture different kinds of motherboards for a new generation of personal computers. Three models of these machines are under consideration. Total number of operators available is 100 because of the local labor market. A floor-space constraint needs to be satisfied because of the different dimensions of these machines. Additional information relating to each of the machines is given in Table 1.1. The company wishes to determine how many of each kind is appropriate to maximize the number of boards manufactured per day.

4. The first-order differential equation $(t + 1)dy/dt - (t + 2)y = 0$, subject to the initial condition $y(0) = 1$, can be solved analytically using the power series method. Set up an alternate procedure using optimization.

This list represents four problems that will be used to define the elements of **problem formulation**. Each problem requires information from the specific area or discipline it refers to. To recognize or design these problems assumes that the designer is conversant with the particular subject matter. Such kinds of problem are quite common, and they are expressed in far greater detail than formulated here. The problems are kept simple to focus on optimization issues. The last example is included to show that optimization ideas and techniques can migrate very well to solve standard problems in a nontraditional manner. In fact, with creative formulation, you should be able to solve most problems through optimization. Optimization also provides an opportunity to establish useful values for designs that are unique or one of a kind.

1.1.1 Elements of Problem Formulation

In this section, we will introduce the formal elements of the optimization problem. Please keep in mind that optimization presupposes the knowledge of the design rules for the specific problem, primarily the ability to describe the design in mathematical terms. For engineering problems this means that the designer is aware of the relevant physics from the particular area, including all of the

Table 1.1 Component Placement Machines

Machine Model	Board Types	Boards/ Hour	Operators/ Shift	Operable Hours/Day	Cost/ Machine
A	10	55	1	18	400,000
B	20	50	2	18	600,000
C	18	50	2	21	700,000

expressions and techniques used in mathematical analysis of the design. The terms in optimization include **design variables, design parameters**, and **design functions**. Traditional design practice—that is, design without regard to optimization, includes all of these elements, except it is not necessary to formally recognize them as such. It is also a good idea to recognize that optimization is a procedure for searching the best design among candidates, each of which can produce an acceptable product.

Design Variables: Design variables are entities that *define* a particular design. The values of a complete set of these variables will establish a specific design. In the search for the optimal design, the values of these entities will change over a prescribed range, hence the tag *variables*. The number of these design variables used to be very significant in the early days of optimization, with the recommendation that the set be as small a possible. This prohibition was related to available computational resources and is no longer a limitation in applied optimization. The type of these variables, *continuous*, or *discrete*, or *integer*, or *mixed*, is important in identifying and setting up the quantitative optimal design problem and the optimization procedure. It is crucial that this choice capture the essence of the object being designed and at the same time provide a quantitative characterization of the design problem. We will develop the elements of optimization assuming the variables are continuous. In applied mathematical terminology, design variables serve as the **unknowns** of the problem being solved. Using an analogy from the area of system dynamics and control theory, they are equivalent to defining the **state of the system**—in this case, the **state of design**. Typically, design variables can be associated with describing the size of the object, like length and height. In other cases, they may represent the number of items. The choice of design variables is the responsibility of the designer guided by intuition, expertise, and knowledge. There is a fundamental requirement to be met by this set of design variables. They must be **linearly independent**. This means that you cannot establish the value of one of the design variables viathe values of the remaining variables through basic arithmetic (*scaling or addition*) operations. For example, in a design having a rectangular cross-section, you cannot have three variables representing the length, height, and area. If the first two are prescribed, the third is automatically established. In complex designs, these relationships may not be very apparent. Nevertheless, the choice of the set of design variables must meet the criteria of linear independence for applying the techniques of optimization. Many of these techniques are borrowed from linear algebra, where this property is necessary for a solution. From a practical perspective, the property of linear independence identifies a *minimum set of variables* that can completely describe the design. This is significant because the effort in obtaining the solution varies as an integer power of the number of variables, and this power is typically greater than two.

The *set* of design variables is identified as the **design vector**. This vector will be column vector in this book. In fact, all vectors are column vectors in this book, unless indicated otherwise. The length of this vector, which is n, is the number of design variables in the problem. The design variables can express different dimensional quantities in the problem, but in the *mathematical model*, they are distinguished by the lowercase x for the variable and the uppercase X for the vector. All the techniques of optimization in the book are based on the **generic mathematical model**. The subscript on x, for example, x_3 represents the third design variable, which may be the height of an object in the characterization of the product. This abstract model is only necessary for mathematical convenience. This book will refer to the design variables in one of four ways:

1. $[X]$: (the square parenthesis defines a vector) the vector of design variables.
2. X or x (without subscripts): referring to the vector again, omitting the square brackets for convenience if appropriate.
3. $[x_1, x_2, \ldots, x_n]^t$: indicating the vector through its elements. Note the transposition symbol ^{t} to identify it as a column vector.
4. x_i, $i = 1, 2, \ldots, n$: referring to all the elements of the design vector.

The above notational convenience is extended to all vectors in the book. Once again, vectors refer to a collection or a related set of values.

Design Parameters: In this book, *design parameters* identify constants that will not change as different designs are generated and compared during optimization. Expressing the design and its properties requires more than design variables. Please be aware that many texts use the term *design parameters* to represent the *design variables* we defined earlier and do not formally recognize design parameters as defined here. The principal reason is that parameters have no role to play in determining the optimal design. They are significant in the discussion of modeling issues. The book will draw attention to these issues when the occasion arises. Examples of parameters include material property, applied loads, and choice of shape. The parameters in the generic mathematical model are recognized similar to the design vector, except that we use the character p. Therefore $[P]$, P, p, $[p_1, p_2, \ldots, p_q]$ represent the parameters of the problem. Note the length of the parameter vector is q. It is important to restate that except in the discussion of modeling, the parameters will not be explicitly referred, as they are primarily predetermined constants in the evaluation of the design.

Design Functions: The *design functions* will define meaningful information about the design. They are evaluated using the design variables and design parameters discussed earlier. They establish the ***mathematical model*** of the design problem. These functions can represent design objective(s) and constraints. The *design objective*, as its name implies, drives the search for the optimal design. The satisfaction of the *constraints* establishes the validity of the design. The designer is responsible for identifying the objective and constraints. "Minimize the mass of the structure" will translate to an **objective function**. The stress in the material must be less than the yield strength will translate to a **constraint**

function. In many problems, it is possible for the same function to switch roles to provide different design scenarios.

Objective Function(s): This is a more specific name for the design objective function. The traditional design optimization problem is defined using a single objective function. The format of this statement is usually to minimize or maximize some design function. This function must depend explicitly or implicitly, on the design variables. In the literature, this problem is expressed exclusively, without loss of generality, as a **minimum or minimization** problem. A **maximum** problem can be recast as a **minimization** problem using the negative or the reciprocal of the function used for the objective function in a maximization problem. In the first example introduced earlier, the objective is to minimize cost. Therefore, the design function representing cost will be the *objective function*. In the second case, the objective is to minimize mass. In the third case, the objective is to maximize machine utilization. The area of single objective design is considered mature today. Today, much of the work in applied optimization is directed at expanding applications to practical problems. In many cases, this has involved creative use of the solution techniques. In the generic mathematical model, the objective function is represented by the symbol f. To indicate its dependence on the design variables it is frequently expressed as $f(x_1, x_2, \ldots, x_n)$. A more concise representation is $f(X)$. Single objective problems have only one function, denoted by f. It is a scalar (not a vector). Note, although the objective function (and the other functions too) depends on P (parameter vector), it is not explicitly included in the format since it does not vary during the search for solution. We should recognize that if the parameter value changes, then the optimization problem must be solved again. In other words, the solution will be valid only for a defined parameter vector.

Multi-objective and **multi-disciplinary designs** are important developments today. Multi-objective design, or **multiple objective design**, refers to using several different design functions as objectives to drive the search for optimal design. Essentially, they permit a noticeably larger set of design solutions by permitting the mathematical model to represent design functions from several disciplines. Generally, they are expected to be conflicting objectives. They could also be cooperating objectives. The current approach to the solution of these problems involves standard optimization procedures applied to single reconstructed objective optimization problems based on the different multiple objectives. A popular approach is to use a suitably weighted linear combination of the multiple objectives. A practical limitation with this approach is the choice of weights used in the model. This approach has not been embraced widely. An alternative approach of recognizing a premier objective and solving it as a single objective problem with additional constraints based on the remaining objective functions can usually generate a good solution. In multi-objective problems the *objective function* will be a vector.

Between the first and the second edition of this book, **multi-disciplinary optimization** (MDO) problems were being seriously addressed. They are predominantly addressed in aerospace industry, where people are reconciling different requirements from structural design and aerodynamic design together. One of the requirements is to identify design variables that will define the structure as well as directly affect the aerodynamic analysis leading to new mathematical models. MDO optimization problems tend to be very large to accommodate aerodynamic analysis of reasonable fidelity. External aerodynamic problems are quite finicky and are difficult to deal with even without optimization. Another feature of the coupling between structure and aerodynamics is that objective function tends to have a large number of local minimums. In such cases, global optimization techniques, which are notoriously slow, must be part of the solution process. This is an exciting area, and definitely not for the resource poor. Although it raises some new issues of coupling, it has not altered the standard techniques of single objective optimization. This book will focus primarily on **single objective optimization**.

Constraint Functions: Design functions will be dependent on the design variables and parameters. A well-described optimization problem is expected to include several such functions that will ensure that the design will exist and interact well with its operating environment. Multiple constraint function can be represented as a vector of constraint functions. The format of these functions requires them to be compared to some numerical value that is established by design requirement, or the designer. This value remains constant during the optimization of the problem. The comparison is usually set up using the three standard *relational* operators: $=$, $\leq$, and $\geq$.

Consider our first example. Let $fun_1(X)$ represent the function that calculates the volume of the new soda can we are designing. The constraint on the design can be expressed as:

$$fun_1(X) = 250 \text{ cc}$$

In the second example, let $fun_2(X)$ be the function that calculates the deflection of the beam under the applied load. The constraint can be stated as:

$$fun_2(X) \leq 1 \text{ mm}$$

The constraint functions can be classified as *equality* constraints [like the one involving $fun_1(X)$ above] or *inequality* constraints [like the expression involving $fun_2(X)$ above].

Problems without constraints are termed as **unconstrained** problems. If constraints are present, then meeting them is *more paramount* than optimization. Constraint satisfaction is necessary for the design to be considered valid and acceptable. If constraints are not satisfied, then there is *no solution*. The design space enclosed by the constraints is called the *feasible domain* A **feasible** design is one in which all the constraints are satisfied. An **optimal** solution is one that has met the design objective. An optimal design *must* be feasible. Design space is described a few paragraphs later.

Equality Constraints: Equality constraints are mathematically neat and analytically easy to handle. Numerically, they require more effort to satisfy.

They are also more restrictive on the design as they limit the region from which the solution can be obtained. The symbol representing equality constraints in the abstract model is h. There may be more than one equality constraint in the design problem. A vector representation for equality constraints is introduced through the following representation. $[H]$, $[h_1, h_2, \ldots, h_l]$, and h_k: $k = 1, 2, \ldots, l$ are ways of identifying the equality constraints. The dependence on the design variables X is often omitted for convenience. Note the length of the vector is l. An important reason for distinguishing the equality and inequality constraints is that they are manipulated differently in the search for the optimal solution. The number of design variables in the problem, n, must be greater than the number of equality constraints, l, $(n > l)$, for a valid optimization problem. If n is equal to l, $(n = l)$, then the problem can be solved without any reference to the design objective. If $(n < l)$, then you have an over determined set of relations which could result in an inconsistent problem definition. The set of equality constraints must be **linearly independent** for a meaningful problem. Broadly, this implies you cannot obtain one of the constraints from elementary arithmetic operations on the remaining constraints. This is to ensure that the methods based on linear algebra will not fail. In the standard format for optimization problems, the equality constraints are written with a 0 on the right-hand side. This means that the equality constraint in the first example will be expressed as:

$$h_1(\mathbf{X}): \quad fun_1(X) - 250 = 0$$

In practical problems, equality constraints are rarely employed as they are less flexible compared to inequality constraints.

Inequality Constraints: Inequality constraints are more natural in problem formulation. They provide more choices for the design. The symbol representing inequality constraints in the abstract model is g. There may be more than one inequality constraint in the design problem. The various vector representation for inequality constraints are $[G]$, $[g_1, g_2, \ldots, g_m]$, and g_j: $j = 1, 2, \ldots, m$. m represents the number of inequality constraints. All design functions explicitly or implicitly depend on the design (or independent) variable X. g is used to describe both less than or equal to ($\leq$) constraints and greater than or equal to ($\geq$) constraints. The strictly greater than ($>$) and the strictly less than ($<$) are not used in optimization because the solutions are usually expected to lie at the constraint boundary ($g = 0$). In the standard format, all problems are expressed with the ($\leq$) relationship. Moreover, the right hand side of the $\leq$ sign is 0. The inequality constraint from the second example $fun_2(X)$ is set up as:

$$g_1(\mathbf{X}): \quad fun_2(X) - 1 \leq 0$$

In the case of inequality constraints a distinction is made whether the design variables lie on the constraint boundary ($g = 0$) or in the interior of the region bounded by the constraint ($g < 0$). If the design variables determine a solution on the boundary of the constraint the constraint acts like an equality constraint. In optimization terminology, this particular constraint is referred to as an **active constraint**. If the design variables determine a solution that does not lie on the constraint boundary, that is they lie inside the region of the constraints, they are considered **inactive constraints**. An inequality constraint can therefore be either *active* or *inactive*.

Side Constraints: Side constraints are necessary part of the solution techniques, especially numerical ones. It expresses the acceptable region for the design variable. Each design variable must be bound by numeric values for its lower and upper limit. The designer makes this choice based on anticipation of an acceptable design. The *design space*, the space that will be searched for optimal design, is the **Euclidean** or **Cartesian n-dimensional** space generated by the *n* independent design variables X. This is a generalization of the three dimensional physical space we are familiar with. For 10 design variables, it is a 10-dimensional space. This is not easy to imagine. It is also not easy to express this information through a figure or graph because of the limitation of the three-dimensional world. However, if the design variables are independent, then the *n* dimensional considerations are mere extrapolation of the three dimensional reality. Although we cannot geometrically define them we can deal with the numbers describing the design. The **side constraints** limit the search region, implying that only solutions that lie within a certain region will be acceptable. It defines an *n* dimensional rectangular region (*hyper cube*) from which the feasible and optimal solutions must be chosen. Later, we will see that the mathematical models in optimization are usually described by nonlinear relationships. The solutions to such problem cannot be predicted, as they are typically governed by the underlying numerical technique used to solve them. It is necessary to restrict the solutions to an acceptable region. The side constraints provide a ready mechanism for implementing this limit. Care must be taken that these limits are not imposed over zealously. There must be a sufficient space for the numerical techniques to conduct a robust search for the optimal design.

The Standard Format: These definitions allow us to assemble the generic *mathematical model* for optimization through the various design functions:

$$\text{Minimize} \quad f(x_1, x_2, \ldots, x_n) \tag{1.1}$$

$$\text{Subject to:} \quad h_1(x_1, x_2, \ldots, x_n) = 0$$

$$h_2(x_1, x_2, \ldots, x_n) = 0 \tag{1.2}$$

$$h_l(x_1, x_2, \ldots, x_n) = 0$$

$$g_1(x_1, x_2, \ldots, x_n) \leq 0$$

$$g_2(x_1, x_2, \ldots, x_n) \leq 0 \tag{1.3}$$

$$g_m(x_1, x_2, \ldots, x_n) \leq 0$$

$$x_i^l \leq x_i \leq x_i^u \qquad i = 1, 2, \ldots, n \qquad (1.4)$$

The same problem can be expressed concisely using this notation:

$$\text{Minimize} \quad f(x_1, x_2, \ldots x_n) \qquad (1.5)$$

$$\text{Subject to:} \quad h_k(x_1, x_2, \ldots, x_n = 0, k = 1, 2, \ldots, l \qquad (1.6)$$

$$g_j(x_1, x_2, \ldots, x_n) \leq 0, j = 1, 2, \ldots, m \qquad (1.7)$$

$$x_i^l \leq x_i \leq x_i^u \quad i = 1, 2, \ldots, n \qquad (1.8)$$

The generic model in vector notation (length of vector is shown in the definition):

$$\text{Minimize} \quad f(X), [X]_n \qquad (1.9)$$

$$\text{Subject to:} \quad [h(X)]_l = 0 \qquad (1.10)$$

$$[g(X)]_m \leq 0 \qquad (1.11)$$

$$X^{\text{low}} \leq X \leq X^{\text{up}} \qquad (1.12)$$

The previous mathematical model expresses the standard optimization problem in natural language as:

Minimize the *objective function f*, subject to *l equality constraints*, *m inequality constraints*, with the *n designx variables* lying between prescribed lower and upper limits.

The techniques in the book will apply to the problem described in the standard format, which refers to a single objective function. To solve any specific design problem, it is required to reformulate the problem in this manner so that the methods can be applied directly. Also in the book, the techniques are developed progressively, starting from the standard model without constraints. For example, the *unconstrained* problem will be explored first. The *equality* constraints are considered next followed by the *inequality* constraints, and finally, the complete model. This represents a natural progression, as prior information is used to develop additional conditions that need to be satisfied by the solution in these instances.

1.1.2 Mathematical Modeling

In this section, the four design problems introduced earlier will be translated to the *standard format* after first identifying the mathematical model. The second problem requires information from a course in mechanics and materials. This should be within the scope of most engineering students.

Example 1.1 New consumer research, with deference to the obesity problem among the general population, suggests that people should drink no more than about 0.25 liter of soda pop at a time. The fabrication cost of the redesigned soda can is proportional to the surface area, and can be estimated at $1.00 per square centimeter of the material used. A circular cross-section is the most plausible, given current tooling available for manufacture. For aesthetic reason, the height must be at least twice the diameter. Studies indicate that holding comfort requires a diameter between 5 and 8 cm. Create a design that will cost the least.

Figure 1.1 represents a sketch of the can. In most product designs, particularly in engineering, it is necessary to work with a figure. The diameter d and the height h are sufficient to describe the soda can. What about the thickness t of the material of the can? What are the assumptions for the design problem? Is t small enough to be ignored in the calculation of the volume of soda in the can? Another important assumption could be that the can will be made using a given stock roll. Another one is that the material required for the can will include only the cylindrical surface area and the area of the bottom. The top will be fitted with an end cap that will provide the mechanism by which the soda can be poured. The top is not part of this design problem. In the first attempt at developing the mathematical, we could start out by considering the quantities identified, including the thickness, as design variables:

$$\text{Design variables:} \quad d, h, t$$

Reviewing the statement of the design problem, one of the parameters is the cost of material per unit area that is given as $1.00 per square meter. Let us identify the cost of material per unit area as constant C. During the search for the optimal solution this quantity will be held at the given value. Note, if this value changes then our cost of the can will correspondingly change. This is what we mean by a **design parameter**. Typically, change in parameters will require a new solution to the optimization problem:

$$\text{Design parameter:} \quad C$$

The design functions will include the computation of the volume enclosed by the can and the surface area of the material used. The volume in the can is $\pi d^2 h/4$. The surface area is $\pi dh + \pi d^2/4$. The aesthetic constraint requires that

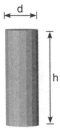

Figure 1.1 Example 1.1 — Design of a new beverage can.

$h \geq 2d$. The side constraints on the diameter are prescribed in the problem. For completeness, the side constraints on the other variables have to be prescribed by the designer. We can formally set up the optimization problem:

$$\text{Minimize} \quad f(d, h, t): \quad C(\pi dh + \pi d^2/4) \tag{1.13}$$

$$\text{Subject to:} \quad h_1(d, h, t)): \quad \pi d^2 h/4 - 250 = 0 \tag{1.14}$$

$$g_1(d, h, t): \quad 2d - h \leq 0 \tag{1.15}$$

$$5 \leq d \leq 8; \quad 4 \leq h \leq 20; \quad 0.001 \leq t \leq 0.01$$

In the mathematical model (1.13–1.15) of the optimization problem for the first example the values of the design variables are expected to be expressed in consistent units-centimeters. It is the responsibility of the designer to ensure correct and accurate problem formulation including dimensions and units. Note that the cost C was originally expressed as meter squared. Hence a scaling factor must be used in (1.13). We will call this C_1.

Intuitively, there is some concern with the problem as expressed by the Equations (1.13–1.15) even though the description is valid. How can the value of the design variable t be established? The variation in t does not affect the design. Changing the value of t does not change f, h_1, or g_1. Hence it *cannot be a design variable* (Note: The variation in t may affect the value of C, but we have already decided this value will not change during optimization). If this were a serious design example then the cans have to be designed for impact, stacking strength, and stresses occurring during transportation and handling. In that case t will probably be a critical design variable. This will require several additional structural constraints in the problem. Moreover, it is likely that development of these functions will not be a simple exercise. This could serve as an interesting extension to this problem for homework or project. The new mathematical model for the optimization problem after dropping t, and expressing $[d, h]$ as $[x_1, x_2]$ becomes:

$$\text{Minimize} \quad f(x_1, x_2): \quad C_1(\pi x_1 x_2 + \pi x_1^2/4) \tag{1.16}$$

$$\text{Subject to:} \quad h_1(x_1, x_2): \quad \pi x_1^2 x_2/4 - 250 = 0 \tag{1.17}$$

$$g_1(x_1, x_2): \quad 2x_1 - x_2 \leq 0 \tag{1.18}$$

$$5 \leq x_1 \leq 8; \quad 4 \leq x_2 \leq 20$$

The problem represented by Equations (1.16)–(1.18) is the mathematical model for the design problem expressed in the **standard format**. For this problem, simple geometrical relations were sufficient to set up the optimization problem.

Example 1.2 Design a cantilevered beam, of minimum mass, carrying a point load F at the end of the beam of length L. The cross-section of the beam will

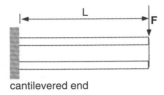

cantilevered end

Figure 1.2 Example 1.2 — Cantilever beam.

be in the shape of the letter I (referred to as an I-beam). The beam should be sufficiently strong in bending and shear. There is also a limit on its deflection.

Figure 1.2 represents the side view of the beam carrying the load. Figure 1.3 describes a typical cross-section of the beam. The I-shaped cross-section is symmetric. The symbols d, t_w, b_f, and t_f correspond to the depth of the beam, thickness of the web, width of the flange, and thickness of the flange, respectively. These quantities are sufficient to define the cross-section of the beam, which is uniform along the length. A handbook on mechanical engineering or a textbook on strength of materials can aid the development of the design functions. An important assumption for this problem is that we will be working within the elastic limit of the material, where there is a linear relationship between the stress and the strain. All the variables identified in Figure 1.3 will strongly affect the solution. They are good candidates for being design variables. The applied force F will also directly affect the problem. So will its location L. How about the material? Steel is definitely superior to copper for the beam. Should F, L, and the material properties be included as design variables?

Intuitively, the larger the value of F, the greater is the cross-sectional area necessary to handle it. Since the mass of the beam will be directly proportional to the area of cross-section, it can be easily concluded that if F were a design variable then it should be set at its lower limit, No techniques are required for this conclusion. If we know the optimum value of F without effort, then F is a good candidate for a parameter rather then a design variable. The larger the value of L the greater is the moment that F will generate about the cantilevered end. This will require an increase in the cross-section properties for strength of the structure. By the same reason, L is not a good choice for the design variable since

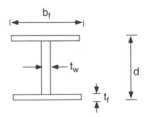

Figure 1.3 Example 1.2 — Cross-section shape.

it will be at its lowest value for an optimum design. To represent a material as a design variable we need to use its set of structural properties in the formulation of the design functions. Some useful material properties are: its specific weight, γ; the modulus of elasticity, E; its modulus of rigidity, G; its yield limit in tension and compression, σ_{yield}; its yield in shear, τ_{yield}; its ultimate strength in tension, σ_{ult}; in tension; its ultimate strength in shear, τ_{ult}. There are still more properties like coefficient of thermal expansion, conduction coefficient, and fatigue limit that will be useful is special problems. Optimization techniques will identify values of the design variables at the solution. If E is the variable used to represent the material design variable in a problem, then it is possible the solution can require a material that does not exist, that is a value of E still undiscovered, or a combination of material properties that are inconsistent with known ones. This for all intents and purposes will not generate a useful solution and it is certainly beyond the scope of the designer. Again, material as a parameter makes a lot of sense. To summarize, the optimization problem should be reinvestigated if F, L, or if the *material* changes. Concluding from this discussion on modeling, the following are defined:

Design parameters: $F, L, \{\gamma, E, G, \sigma_{yield}.\tau_{yield}, \sigma_{ult}, \tau_{ult}\}$

Design variables: d, t_w, b_f, t_f

In the development of the mathematical model, standard technical definitions are used. Much of the information can be obtained from a mechanical engineering handbook, or an elementary book on mechanics of materials. The first design function is the weight of the beam. This is the product of γ, L, and A_c, where A_c is the area of cross-section. The A_c of the cross-section of the beam is $(2b_f t_f + t_w(d - 2t_f))$. The maximum stress due to bending, σ_{bend}, can be calculated as the product of $FLd/2I_c$, where I_c, is the moment of inertia about the centroid of the cross-section along the axis parallel to the flange. The moment of inertia I_c, in terms of the design variables, is $(b_f d^3/12) - ((b_f - t_w)(d - 2t_f)^3/12)$. The maximum shear stress in the cross-section is expressed as $FQ_c/I_c t_w$, where Q_c is first moment of area about the centroid parallel to the flange. The first moment of area Q_c can be calculated as $(0.5b_f t_f(d - t_f) + 0.5t_w(d - t_f)^2)$. The maximum deflection (δ_L) of the beam will be at the end of the beam calculated by the expression $FL^3/3EI_c$. In the following, for ease of representation we will continue to use A_c, Q_c, and I_c instead of detailing their dependence on the design values. Associating x_1 with d, x_2 with t_w, x_3 with b_f, and x_4 with t_f, so that $X = [x_1, x_2, x_3, x_4]$ and the problem in standard format is as follows:

$$\text{Minimize} \quad f(X): \gamma L A_c \tag{1.19}$$

$$\text{Subject to:} \quad g_1(X): FLx_1/2I_c - \sigma_{yield} \leq 0 \tag{1.20a}$$

$$g_2(X): FQ_c/I_c x_2 - \tau_{yield} \leq 0 \tag{1.20b}$$

$$g_3(X): \ FL^3/3EI_c - \delta_{\max} \leq 0 \tag{1.20c}$$

$$0.01 \leq x_1 \leq 0.25; \quad 0.001 \leq x_2 \leq 0.05; \tag{1.20d}$$

$$0.01 \leq x_3 \leq 0.25; \quad 0.001 \leq x_4 \leq 0.05 \tag{1.20e}$$

The designer must ensure that the problem definition is also consistent with the unit system chosen for the parameters and variables. To solve this problem, F must be prescribed (10,000 N), L must be given (3 m). Material must be selected (steel : $\gamma = 7860$ kg/m^3; $\sigma_{\text{yield}} = 250E + 06$ N/m^2; $\tau_{\text{yield}} = 145E + 06$ N/m^2). Also the maximum deflection is prescribed ($\delta_{\max} = 0.005$ m).

This is an example with four design variables, three inequality constraints, and no equality constraints. Other versions of this problem can easily be formulated. It can be reduced to a two-variable problem if symmetry was imposed. Standard failure criteria with respect to combined stresses or principal stresses can also be included through additional functions. If the cantilevered end is bolted, then additional design functions regarding bolt failure needs to be examined.

Example 1.3 MyPC Company has decided to invest \$12 million in acquiring several new component placement machines to manufacture different kinds of motherboards for a new generation of personal computers. Three models of these machines are under consideration. Total number of operators available is 100 because of the local labor market. A floor-space constraint needs to be satisfied because of the different dimensions of these machines. Additional information relating to each of the machines is given in Table 1.1. The company wishes to determine how many of each kind is appropriate to maximize the number of boards manufactured per day.

It is difficult to use a figure in this problem to set up our mathematical model. The number of machines of each model needs to be determined. This will serve as our design variables. Let x_1 represent the number of component placement machines of model A. Similarly, x_2 will be associated with model B, and x_3 with model C.

Design variables: x_1, x_2, x_3

The values in the tables can be regarded as parameters. For this problem, it is more useful to work with the values directly and hence identifying them as parameters does not serve any useful purpose. The information in Table 1.1 is used to set up the design functions in terms of the design variables directly. An assumption is made that all machines are run for three shifts. The cost of acquisition of the machines is the sum of the cost per machine multiplied the number of machines (g_1). The machines must satisfy the floor space constraint (g_2). The constraint on the number of operators is three times the sum of the product of the number of machines of each model and the operator per shift (g_3). The utilization of each machine is the number of boards per hour times the number of hours the machine operates per day. The optimization problem can be assembled in the following form:

$$\text{Maximize} \quad f(X): 18^*55^*x_1 + 18^*50^*x_2 + 21^*50^*x_3 \qquad (1.21)$$

or

$$\text{Minimize} \quad f(X): \; -18^*55^*x_1 - 18^*50^*x_2 - 21^*50^*x_3$$

$$\text{Subject to:} \quad g_1(X): 400{,}000x_1 + 600{,}000x_2 + 700{,}000x_3 \le 12{,}000{,}000$$
$$(1.22a)$$

$$g_2(X): 3x_1 - x_2 + x_3 \le 30 \qquad (1.22b)$$

$$g_3(X): 3x_1 + 6x_2 + 6x_3 \le 100 \qquad (1.22c)$$

$$x_1 \ge 0; \; x_2 \ge 0; \; x_3 \ge 0 \qquad (1.22d)$$

Equations (1.21) and (1.22) express the mathematical model of the problem. Note in this problem that there is no product being designed. Here, a strategy for placing order for the number of machines is being determined. Equation (1.21) illustrates the translation of the maximization objective into an equivalent minimizing one. The inequality constraints in (1.22) are different from the previous two examples. Here, the right-hand side of the less than and equal to operator is nonzero. Similarly, the side constraints are bound on the lower side only. This is done *deliberately*. There is a significant difference between this problem and the previous two. All of the design functions here are *linear*. This problem is classified as a **linear programming problem**. The solution technique for this type of problem is very different than the solution to first two examples, which are recognized as **nonlinear programming problems**. Linear programming problems are essential in decision making in commerce and business. They are critically explored in those disciplines. Before moving on to the next section, the inequality constraint in (1.22a) stands out because of the large numbers in all of the terms. Dividing the constraint through by 1,000,000 will not change the problem. This is termed **scaling**, and is an important step in optimization.

Example 1.4 The first-order differential equation $(t + 1)dy/dt - (t + 2)y = 0$, subject to the initial condition $y(0) = 1$, can be solved analytically using the power series method. Set up an alternate procedure using optimization.

The differential equation and initial condition is rewritten as follows:

$$(t + 1)y' - (t + 2)y = 0; \quad 0 \le t \le 3 \qquad (1.23)$$

$$y(0) = 1 \qquad (1.24)$$

The solution, in terms of power series, can be expressed as

$$y(t) = \sum_{m=0}^{\infty} c_m t^m = c_0 + c_1 t + c_2 t^2 + \cdots \qquad (1.25)$$

The initial condition in (1.24) will determine the value of c_0 as 1. That leaves the determination of the remaining coefficients. These will be our design variables of the problem. Let us consider the series solution for this problem to 10 terms. We require *nine design variables*. We will also monitor the differential equation at 100 points in the interval of t, labeled as t_i. The solution, at each t_i:

$$y(t_i) = y_i = 1 + x_1 t_i + x_2 t_i^2 + \cdots + x_9 t_i^9; \quad i = 1..100 \tag{1.26a}$$

where the x_j are the design variables. The derivatives at each i can be expressed as

$$y'(t_i) = y_i' = x_1 + 2x_2 t_i + \cdots + 9x_9 t_i^8; \quad i = 1..100 \tag{1.26b}$$

At each t_i Equation (1.23) will measure the error in the series solution (if the term is nonzero).

$$E_i = (t_i + 1)y_i' - (t_i + 2)y_i$$

We will drive this error to zero by setting up our objective function as

$$f(X) = \sum_{i=1}^{100} E_i^2 \tag{1.27}$$

This can be stated as the minimum of the square error over the range of the variable indicated at the 100 points. This is an unconstrained optimization problem. The exact solution to the problem is

$$y(t) = (1 + t)e^t = 1 + 2t + \frac{3}{2}t^2 + \frac{2}{3}t^3 + \frac{5}{24}t^4 + \frac{1}{20}t^5 \cdots$$
$$+ \frac{7}{720}t^6 + \frac{1}{630}t^7 + \frac{1}{4480}t^8 + \frac{1}{36288}t^9 + \frac{1}{329891}t^{10} \tag{1.28}$$

As formulated, the optimization problem has nine design variables. You could do this to other classical methods of solution, too. If it works, you are numerically identifying an analytical solution. Another advantage is that it is a simple matter to identify a different example by this method, but the analytical procedure must be worked out in detail for each different example. Chapter 11 addresses using optimization in nontraditional ways to solve nonlinear differential equations using Bezier surfaces, using exactly this idea.

1.1.3 Nature of Solution

We will look at the nature of solutions to the optimization problem using the first three examples we have modeled. Examples 1.1 and 1.2 describe a nonlinear programming problem. This means that some of the design functions in the model are nonlinear. They contain terms that include the product of design variables, or the variables themselves raised to powers other than 1. Example 1.3, as noted earlier is a linear programming problem, which is very significant in decision sciences, but rare in product design. Its inclusion here, while necessary for completeness, is also important for understanding contemporary optimization techniques for non-linear programming problems. Example 1.4 is a nonlinear problem in nine variables because the terms are squared. It is quite simple to solve numerically, especially using the optimization toolbox. In this chapter it has served to showcase how optimization can provide an alternative to a classical method of solution to math problems. Examples 1.1 and 1.2 are dealt first. Between the two, we notice that Example 1.1 is quite simple in comparison to Example 1.2. Second, the four variables in Example 1.2 make it difficult to use illustration to establish some of the discussion. We will use Example 1.1 in the following discussion.

Solution to Example 1.1   The simplest determination of nonlinearity is through a graphical representation of the design functions involved in the problem. This can be done easily for one or two variables. If the function does not plot as a straight line or a plane, then it is nonlinear. Figure 1.4 represents the three-dimensional plot of Example 1. The figure is obtained using MATLAB. Chapter 2 will provide detailed instructions for drawing such plots for graphical optimization. The three-dimensional representation of Example 1.1 in Figure 1.4 does not really enhance our understanding of the problem or the solution. The region of the solution, which is at the intersections of the functions, is hidden and difficult to see. Figure 1.5, the contour plot, is an alternate representation of the problem. The solution will be clearer in this figure. The horizontal axis represents the diameter (d, x_1), The vertical axis represents the height (h, x_2). In Figure 1.5 the *equality*

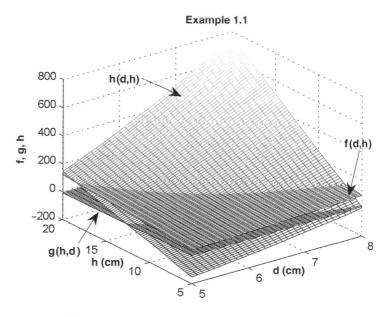

Figure 1.4   Graphical representation of Example 1.1.

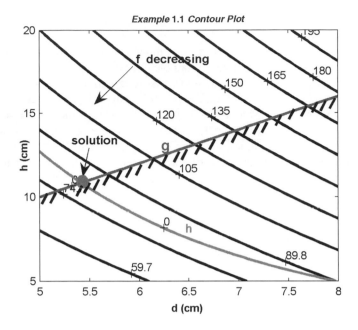

Figure 1.5 Contour plot of Example 1.1.

constraint $h(x_1, x_2)$ is marked appropriately as h on the figure. Since there is only one constraint the subscript is unnecessary. Any pair of values on this line will give a volume of 250 cc. The *inequality constraint* g is also identified. The pair of values on this line *exactly* satisfies the aesthetic requirement. The dashed lines on the side of the inequality constraint establish the *disallowed* region for the design variables. The constraints are drawn thicker for emphasis. The *feasible* region is on the other side of the dashed lines. The *objective function* is represented through several contours. Each contour is associated with a fixed value of the objective function and these values are shown on the figure. Since this is a minimum problem we are interested in the decreasing cost. The range of the two axes establishes the *side constraints*. The objective function f, and the equality constraint h are non-linear and therefore they are not drawn as straight lines. This is substantiated by the products of the two unknowns (design variables) in both these functions. The inequality constraint g is linear. In Equation (1.18) this is evident because the design variables appear by themselves without being raised to a power other than 1. As we develop the techniques for applying optimization, this distinction is important to keep in mind. It is also significant that graphical representation is typically restricted to two variables. For three variables, we need a fourth dimension to resolve the information while three-dimensional contour plots are not easy to illustrate. The power of imagination is necessary to overcome these hurdles.

The problem represented in Figure 1.5 is a graphical solution to the optimization problem. First, the *design space* is the region spanned by the side constraints—the boxed area. The *feasible* region is the space of acceptable designs. The presence of equality constraints complicates this discussion, since the solution must lie on the constraint. In this problem the feasible region is *on the equality constraint* above the intersection with the inequality constraint and limited by the y axis. The *optimal* solution is the feasible solution with the minimum value for the objective function. For this example, there are many feasible solutions (points on the feasible portion of the equality constraint) but only one optimal solution. There are *infinite* feasible solutions but a *unique* optimal solution. In this problem/figure, the smallest value of f is desired. The lowest value of f is about 82. Since this is in dollars, this can is quite expensive. Maybe there was an error in the cost. It should have been per square meter. Will it change the optimum solution? We will look at this and other answers later (you can run the code and see if it changes the solution). The solution to the optimization problem is at the intersection of the two constraints. While the value of f needs to be calculated, the optimal values of the design variables, read from the figure, are about 5.5 and 11 cm, respectively. From Figure 1.5, it can be seen that g is an **active** constraint.

Solution to Example 1.3   In contrast to Example 1.1, all the relationships in the mathematical model, expressed by Equations (1.21) and (1.22), are linear. The word equation is generally used to described the equivalence of two quantities on either side of the *equal* (=) sign. In optimization, we come across mostly *inequalities*. The word equation will be used to include both of these situations. In Example 1.1, the nature of solution was explained using Figure 1.5. In Example 1.3, the presence of three design variables denies this approach. Instead of designing a new problem, suppose that the company president has decided to buy five machines of the third type. This reduces the problem to two design variables allowing a graphical solution to be constructed. The mathematical model, with two design variables x_1, x_2, is reconstructed using the information that x_3 is now a parameter with the value of five. Substitute for x_3 in the model for Example 1.3, and cleaning up the first constraint:

$$\text{Maximize} \quad f(X): 990^* x_1 + 900^* x_2 + 5{,}250 \qquad (1.29)$$

$$\text{Subject to:} \quad g_1(X): 0.4x_1 + 0.6x_2 \le 8.5 \qquad (1.30a)$$

$$g_2(X): 3x_1 - x_2 \le 25 \qquad (1.30b)$$

$$g_3(X): 3x_1 + 6x_2 \le 70 \qquad (1.30c)$$

$$x_1 \ge 0; \quad x_2 \ge 0 \qquad (1.30d)$$

An interesting observation in this set of equations is that there are only two design variables but three constraints. This is a valid problem. If the constraints were equality constraints, then this would not be an acceptable problem definition, as it would violate the relationship between the number of variables and

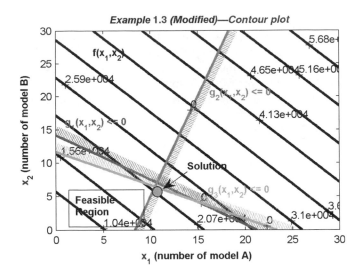

Figure 1.6 Contour plot of modified Example 1.3.

the number of equality constraints. Therefore, the *inequality* constraints are not related to the number of design variables used. This fact also applies to nonlinear constraints. Figure 1.6 is the plot of the functions (1.29) and (1.30). Choosing the first quadrant also satisfies the side constraints. The linearity of all the functions is indicated by the straight lines on the figure. Again, several contours of f are shown to aid the recognition of the solution. The figure was generated by MATLAB. A special piece code will generate the hash marks in MATLAB. In lien of hash marks, sometimes a thicker line is drawn parallel to the constraint ahead of the hash lines—or the forbidden region. The feasible region is enclosed by the two axes and the constraints g_2 and g_3. The objective is to increase the value of f, which is indicated by the contour levels, without leaving the feasible region. The solution can be spotted at the intersection of the constraints g_2 and g_3. The values of the design variables at the solution can be read off from the plot. We are not done yet. Before proceeding further it is necessary to acknowledge that a solution of 11.5 machines for model A is unacceptable. The solution must be adjusted to the nearest integer values. In actual practice, this adjustment has to be made without violating any of the constraints. The need for *integer variables* is an important consideration in design. Variables that are required to have only integer values belong to the type of variables called *discrete variables*. This book mostly considers *continuous variables*. Almost all of the mathematics an engineer encounters, especially academically, belong to the domain of continuous variables. The assumption of continuity provides very fast and efficient techniques for the search of constrained optimum solutions, as we will develop in this course. Many real-life designs, especially in the use of off-the-shelf materials and components determine discrete models. Discrete/integer programming is the area that addresses these problems. The nature of the methods searching the optimal solution in discrete programming is very different from those in continuous programming. Many of these methods are based on the scanning and replacement of the solutions through exhaustive search of the design space in the design region. In this book discrete problems, except in the Chapter 8, are handled as if they were continuous problems. The conversion to the discrete solution is the final part of the design effort and is usually performed by the designer outside of any continuous optimization technique we develop in this book. Chapter 8 discusses discrete optimization with examples.

Getting back to the *linear programming* problem we just discussed, it can be observed that only the boundary of the feasible region will effect the solution. Here is a way that will make it apparent. Note that the objective function contours are straight lines. A different objective function will be displayed by parallel lines with a different slope. Imagine these lines on the figure instead of the current objective function. Note the solution always appears to lie at the intersection of the constraints or the *corners*. Solution can never come from inside of the feasible region, unlike in *nonlinear programming* problem which accommodates solution on the boundary and from within the feasible region. Recognizing the nature of solutions is important in developing a search procedure. For example, in linear programming problems, a technique employed to search for the optimal design need only search the boundaries, particularly the intersection of the boundaries. Whereas in nonlinear problems, the search procedure cannot ignore interior values.

1.1.4 Characteristics of the Search Procedure

The search for the optimal solution will depend on the nature of the problem being solved. For nonlinear problems, it must consider the fact that the solution could lie inside the feasible region. Figure 1.7 will provide the context for discussing the search procedure for nonlinear problems. Similarly, the characteristics of the search process for linear problems will be explored using modified Example 1.3 and Figure 1.6.

Nonlinear Problems: In practice, solutions to nonlinear problems are obtained through *numerical techniques* that are applied iteratively. The methods or techniques for finding the solution to optimization problems are called *search methods*. These methods will require the following:

- Several *iterations* before the solution can be obtained.
- Each iteration or search is executed in a consistent manner where information from the previous iteration is utilized in the compute values in the present sequence.
- This consistent manner or process by which the search is carried out is called an *algorithm*. This term also refers to the translation of the particular search procedure into an ordered sequence of step by step actions.

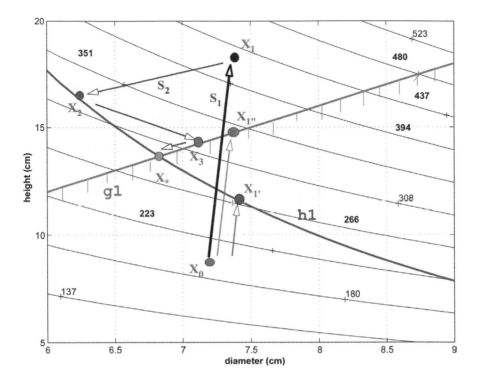

Figure 1.7   A typical search method for nonlinear problems.

- The algorithm is applied by converting these steps, through some computer language, into code that will execute on the computer. In this book, the algorithm is developed and translated into MATLAB code.

The search process is started typically by trying to guess the *initial design* solution. The designer can base this selection on his experience. On several occasions given the nonlinear nature of the problems, the success of optimization may hinge on his ability to choose a good initial solution. Any search method, even when used consistently, may fail to solve a particular problem. The degree and type of nonlinearity may frequently cause the method to fail, assuming there are no inconsistencies in problem formulation. Another method may yet solve the problem. This has led to the creation of several different techniques for optimization. Practical experience has also suggested that a *class* of problems respond well to certain algorithms.

These iterative methods are currently the only way to implement numerical solutions to *nonlinear* problems, irrespective of the discipline in which these problems appear or the nature of the problems they define. Nonlinear problems are plentiful in optimization, optimal control, structural and fluid mechanics, quantum physics, and astronomy. Although optimization techniques are applied in all of the noted disciplines, in traditional design optimization the problems are mainly characterized by nonlinear algebraic equations. In optimal control they can be nonlinear dynamics systems, which are described by non-linear differential equations. In structural mechanics, there is the possibility of nonlinear integral equations. Computational fluid dynamics deals mostly with nonlinear partial differential equations. The principal aim of search methods is to get closer to the solution with every iteration. This is referred to as *converging to the solution*. In optimization techniques, this could also mean determining a feasible solution.

The graphical representation of a problem similar to Example 1.1 can be seen in the background of Figure 1.7. It is embellished with several iterations of a *hypothetical* search method. The initial starting guess is the point marked X_0. The solution is at X_*. Remember, when solving the optimization problem such a figure *does not exist*. If it did, then we can easily obtain the solution by inspection. This statement is especially true if there are more than two variables. In the event there are two variables, there is no reason not to obtain a *graphical solution* (Chapter 2). The methods in optimization rely on the direct extrapolation of the concepts and ideas developed for two variables to any number of design variables. That is the reason we will use two variables to discuss and design most of the algorithm and methods. Using two variables also allow the concepts to be reinforced using geometry and graphics.

The design at the point X_0 is not feasible. It is not useful to discuss if it is an optimal solution. At least one more iteration is necessary to obtain the solution. Consider the next solution at X_1. Most methods in optimization will attempt to move to X_1 by first identifying a **search direction**, S_1. This is a **vector** (same dimension as design variable vector X) pointing from X_0 to X_1. The actual displacement from X_0 to X_1, call it ΔX_1 is some multiple of S_1, say αS_1. X_1 once again is not feasible though it as improvement on X_0 since it satisfies the inequality constraint g_1 (lying in the interior of the constraint). Alternately, using the same search direction S_1, another technique/method may choose for its next point, $X_{1'}$, at a distance βS_1. This point satisfies the constraint h_1 but violates g_1. A third strategy would be to choose X_1. This lies on the constraint boundary of g_1. Please note that all the three choices are not feasible. Which point is chosen is decided by the particular implementation of the algorithm. Assume point X_1 is the next point in this discussion.

Point X_1 now becomes the new point X_0. A new search direction S_2 is determined. Point X_2 is located. Another sequence of calculations locates X_3. The next iteration yields the solution. The essence of updating the design during successive iteration lies in two major computations, the **search direction** S_i and **stepsize** α. The change in the design vector, defined as ΔX, which represents the vector difference, for example, $(X_2 - X_1)$, is obtained as αS_i. For each iteration, the following sequence of activity can be identified:

Step 0: Choose X_0
Step 1: Identify S
          Determine α
          $\Delta X = \alpha S$

$$X_{new} = X_0 + \Delta X$$
Set $X_0 \leftarrow X_{new}$
Go to Step 0

This is the basic structure of a typical optimization algorithm. A complete one will indicate the manner in which S and α are computed. It will include tests, which will determine if the solution has been reached (convergence), or if the procedure needs to be suspended or restarted. The different techniques we will explore in this book are mostly different by the way S is established. Take a look at Figure 1.7 again. At X_0 there are infinite choices for the search direction. To reach the solution in reasonable time, an acceptable algorithm will try not to search along directions that do not improve the current design. There will be several ways to satisfy this condition.

Linear Programming Problem: It was previously mentioned that the solution to these problems would be on the boundary of the feasible region (also called *feasible domain*). In Figure 1.6, the axes and some of the constraints determine this region. This region is distinguished also by the points of intersection of the constraints among themselves, the origin, and the intersection of the *binding* (active) constraints with the axes. These are *vertices* or *corners* of the quadrilateral. Essentially the design improves by moving from one of these vertices to the next one through improving feasible designs. You have to acquire a *feasible* corner to start the iteration. This will be explored more fully in the chapter on linear programming. A reminder here once again that only with two variables can we actually see the geometric illustration of the technique, that is the selection of vertices as the solution converges.

So far, Chapter 1 has dealt with the introduction of the optimization problem. It was approached from an engineering design perspective, but the mathematical model, in abstract terms, is not sensitive to any particular discipline. A few design rules with respect to specific examples were included in the discussion. The standard mathematical model for optimization problem was established. Two broad classes of problems, linear and nonlinear were introduced. The geometrical characteristics of the solution for these problems were shown through the graphical representation of the model. An overview of the search techniques for obtaining the solution to these problems was also illustrated. It was also mentioned that the solutions were to be obtained *numerically*, primarily through iterative computation.

The numerical methods used in this area are iterative methods. Numerical solutions are obtained through running computer programs. These programs involve two components. The first is an algorithm that will establish the iterative set of calculations. The second is the translation of the algorithm into computer codes, using a *programming language*. This finished code is referred to as *software*. The early software in optimization ran on *mai frames*, or large computer systems, using legacy programs written in FORTRAN, Pascal, or C. Since the mid-1990s the standard computation environment has transformed to individual personal computers (PCs), running Windows, Macintosh, Linux, or other flavors of UNIX, operating systems. They are more powerful than old mainframe computers and are inexpensive to boot. Users interact with their PCs through a graphical user interface (GUI) avoiding time and tedious effort in learning the software and as well as setting up the problem. New *programming paradigms*, especially, *object-oriented programming*, has thrown up a lot of new and improved ways to develop software. Today engineering software is created by general programming languages such as C, C++, Java, Pascal, and Visual Basic. Many software vendors develop their own extensions to the standard language, providing a complete environment for specific deployment by the user. This makes it very convenient for users, who are not necessarily programmers, to get their job done in a reasonably short time.

One such vendor is Mathworks, Inc. with their flagship products MATLAB and Simulink. They also provide a complementary collection of Toolboxes, which deliver software for specific disciplines and applications. Between the first and the second edition of this book, the use of MATLAB has grown exponentially in academia and industry. It has also evolved from core mathematical areas to envelop almost every discipline. Of specific note is the *Optimization Toolbox*, which implements most of the concepts developed in the book. Knowledge of MATLAB is a skill that is in demand from those entering the workforce to work in analytical areas. In this book, MATLAB is used for numerical and graphical support. *No prior familiarity* with MATLAB is expected, though familiarity with the PC is required. Like all higher-level languages, effectiveness with this new programming resource comes with effort, frequency of use, original work, and a consistent manner of application. By the end of the book, the reader will become very familiar with the use of MATLAB to develop programs for optimization. The experience gained from this book will be substantial, and sufficient for the use of MATLAB in other settings. Nevertheless, an enormous part of MATLAB will not be used. It should be possible to explore those with the experience gained in this course. All of the code in this book is developed only after identifying a need for its implementation. New commands will be recognized and explained (commented on) when they appear for the first time. The code will be kept simple, as this book is also a means for understanding MATLAB through writing numerical techniques for optimization. The author welcomes suggestions from new users and those familiar with MATLAB who have considerable experience in programming, for ways to make the learning process more effective. The next part of this chapter deals with MATLAB, establishing its appropriateness for this book, as well as getting started with its use. The reader will need access to MATLAB to be able to use this book effectively.

1.2 INTRODUCTION TO MATLAB

MATLAB is introduced by Mathworks as the language for technical computing. Borrowing from the description in an earlier brochure, valid even now, MATLAB integrates computation, visualization, and programming in an easy-to-use environment where problems and solutions are expressed in familiar mathematical notation. The typical uses for MATLAB include the following:

- Math and computation
- Algorithm development
- Modeling, simulation, and prototyping
- Data acquisition, analysis, exploration, and visualization
- Scientific and engineering graphics
- Application development, including graphic user interface building

The use of MATLAB has exploded recently and so has its features. This continuous enhancement makes MATLAB an ideal vehicle in exploring problems in all disciplines that manipulate mathematical content. This ability is multiplied by application-specific solutions through *toolboxes*. To know more about the latest MATLAB, its features, the toolboxes, collection of user-supported archives, information about Usenet activities, other forums, information about books, and so on, please visit MATLAB at http://www.mathworks.com.

1.2.1 Why MATLAB?

An answer may have been necessary at the time of the first edition of this book. Today, with MATLAB being used in over 3,500 universities and over 1,000 research and industrial institutions across the globe, it would be a serious omission *not* to incorporate its support in learning about applied optimization. MATLAB is a standard tool for introductory and advanced courses in mathematics, engineering, and science in many universities around the world. In industry, it is a tool of choice for research, development and analysis. In this book, MATLAB's basic array element is exploited to manipulate vectors and matrices that are natural to the subject. In the next chapter, its powerful visualization features are used for graphical optimization. In the rest of the book the other built in features of MATLAB is utilized to translate algorithms into functioning code, in a fraction of the time needed in other languages such as C or FORTRAN. At the end, we will see some examples from its optimization toolbox.

Most books on optimization provide excellent coverage of the techniques and algorithms, and some provide a printed version of the code. Almost all new editions have moved away from legacy environments to embrace new software environments that are user friendly, with shallower learning curves. These software systems are feature rich and very resourceful in delivering various aspects of computation. MATLAB's code is compact, and it is reasonably intuitive. To make problems work requires only few lines of code compared to traditional computer languages. Instead of programming from the ground up, standard MATLAB pieces of code can be threaded together to develop the new algorithm. It avoids the standard separate compilation, link, and execution sequence by using an interpreter. Another important reason is code portability. The code written for the PC version of MATLAB does not have to be changed for the Macintosh or the Unix systems as they are text files. This will not be true if system dependent resources are used in the code. As an illustration of MATLAB's learning agility, both MATLAB and applied optimization are covered in this book, something very difficult to accomplish with say, FORTRAN and optimization, or C++ and optimization. MATLAB is a globally relevant product and can work with C and Java if desired.

1.2.2 MATLAB Installation Issues

The code in this book will work on MATLAB installed on an individual PC or Mac, a networked PC or Mac, or individual or networked UNIX workstations. MATLAB takes care of the issue of portability as user-created code is in ASCII text. The book assumes you have a working and appropriately licensed copy of MATLAB in the machine in front of you. This book uses MATLAB Version 7.2 (R2006a), as this was the current version when the manuscript was started. The version shipping during the first print is R2008a. The code was developed on a PC. During the development of the previous edition, the author had access to UNIX and Mac platforms. This time the development is based on a 32-bit Windows PC. New releases of MATLAB are supposed to be seamless across platforms in terms of user use and experience. If there are issues that are different, the author solicits feedback from the user. In terms of code itself, newer versions of MATLAB should not be a problem. There was however some significant changes in some commands with version 7. The same command can be run in a backwards-compatible form by including 'V6' as the first item in the Command. Installing MATLAB on your machine should not be a problem, as the software is accompanied by detailed installation instructions. If you are using a networked version, please consult your system administrator. While many of you may have been exposed to MATLAB, this book assumes that you are new to MATLAB. You are strongly urged to use the MATLAB help resources through its excellent documentation system, particularly the Help browser.

Start MATLAB from the start menu on the PC or through an icon on the desktop (Mac, UNIX, Linux users should find the program as a menu item in the appropriate menu). It is a good idea to click the various menus on the top menubar. You will see various menu sub-items that are some of the common commands you are likely to use. You can also customize how MATLAB will tile the various windows when it opens. My personal preference is using only the Command window and the Editor window in stand-alone mode instead of the default appearance, since it will be less distracting. If you are already a MATLAB user, you are welcome to use your favorite setup. You can change the way MATLAB opens by clicking through the menus *Desktop → Desktop Layout → Command Window Only*.

The MATLAB window, Figure 1.8, also called the **Command window**, is the means through which commands are issued to MATLAB during interactive use. Note that MATLAB command prompt is ≫. The second graphical window that will be very useful is the MATLAB M-file **Editor/Debugger window**, shown in Figure 1.9. This window is opened by clicking the *new* file icon in the Command

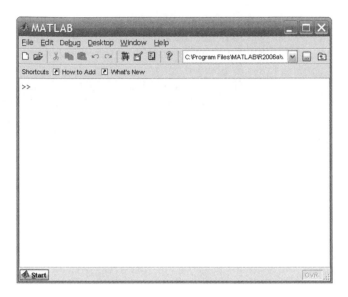

Figure 1.8   MATLAB Command window in the PC.

window under the *File* menu. This is the window where we will write the code. The editor uses color for enhancing code readability. It is also a **debugger;** it helps us find and correct errors in code. The Command window and the Editor window can spawn additional dialog boxes (if necessary) for setting additional features during the MATLAB session, like the path *browser* and *workspace browser*. These windows are very similar in other platforms.

Although this is not so much an installation issue, it would be appropriate to discuss the way the book plans to write and execute MATLAB code. Generally, there are two ways to work with MATLAB: *interactively* and through *scripts*. In interactive mode, the commands are entered and executed one at a time in the MATLAB Command window. Once the session is over and MATLAB exited, the set of commands that were used and what worked is stored in the **History window**. Large sequences of commands are usually difficult to keep track off in an interactive session. Another option is to store the sequence of commands in a text-file (ASCII), and execute the complete file in MATLAB. This is called a **script m-file** and can be considered a batch execution—a batch of commands. Most MATLAB files are called **m-files**. The *.m* is a file extension usually reserved for MATLAB. Any code changes are made through the editor, the changes to the file are saved, and the file is executed in MATLAB again. Executing MATLAB through script files is predominantly followed in the book. MATLAB allows you to revert to interactive mode any time by just typing in the Command window. In order to enhance the programming experience and expose you to more features of MATLAB, a lot of bookkeeping activities are done through code rather than the rich resources provided by newer versions of MATLAB. In a sense, interactive use

Figure 1.9 MATLAB editor/debugger window on the PC.

of MATLAB is eschewed so that code will run faster and the user is introduced to more programming.

1.2.3 Using MATLAB the First Time

MATLAB will be used interactively in this section. The author strongly recommends a hands-on approach to understanding MATLAB. This means typing the code yourself. This is essential to understand the syntax errors, the debugging, and rectifying of the code. The author is not aware of a single example where the reader learned programming without typing/entering code. Moreover, *your* intuitive understanding of MATLAB commands and programming is based on *your* practice and not the authors. In this book, MATLAB code segments are courier font with the boldface style used for emphasizing commands or other pieces of information. Anything else is recommendations, suggestions, or exercises. The script file name containing the commands is in italic.

Before We Start: Table 1.2 includes a few punctuation and special characters that you will use often. It is a good idea to become familiar with them. For those of you programming for the first time, the equal to sign (=) has a very special meaning in all programming languages, including MATLAB. It is called the **assignment operator**. The variable on the left-hand side of the sign is assigned the value of the right-hand side. The actual **equal-to** operation in MATLAB is usually accomplished by a double equal-to sign, which is (==).

Table 1.2 Some Punctuation and Special Characters

Punctuation/ Characters	What It Does
>>	MATLAB command prompt.
	You have to wait for the prompt before MATLAB will listen to you. It applies mostly for interactive sessions or when MATLAB is busy computing (see ^c).
%	All text to the right of the % is a comment and is ignored by MATLAB.
	Commenting your code seriously helps you remember what you were trying to achieve when you wrote the piece of code.
	Comments are used liberally in the code in the book. Without comments you will have difficulty understanding what you coded just a week ago.
%%	Together, they give you a bold comment.
;	A semicolon prevents information echoing to the screen.
	Removing it will cause information to be printed to the screen. This is useful when you want to debug the code.
,	A comma is used as a command separator and will cause information to be printed to the screen. You can also use a semicolon as a command separator.
:	A colon is used to establish a range of values. By itself, it implies all of the values.
...	A succession of three periods at the end of the line implies that the code will continue to the next line.
	You cannot continue variable name across lines. You cannot continue a comment on another line.
^c	Ctrl C (hold down Ctrl and C together) will stop MATLAB execution and return the control prompt.

Some Additional Features:

- MATLAB is **case sensitive**. An a is different than A. This is difficult for persons who are used to FORTRAN. All MATLAB built-in commands are in *lowercase*. Hence, you can define your own variables in uppercase so that they do not coincide with built-in functions or variable names.

- MATLAB does not need a **type definition** or a **dimension** statement to use and introduce variables. It automatically creates one on first encounter in code. The type is assigned in context

- Variable names start with a letter and contain up to any number of characters which can only include letters, digits, and underscore. Only the first 63 characters are used by MATLAB.

- MATLAB uses some built-in variable names and keywords. Avoid using built in variable names. For example the variable pi has the value of π or 3.14. It is probably a good idea to declare all your constants in the code.

- All numbers are stored internally using the long format specified by IEEE floating-point standard. These numbers have roughly a precision of 16 decimal digits (32-bit operating system). They range roughly between $10E-308$ and $10E+308$. However they are *displayed* differently, depending on the context.

- MATLAB uses conventional decimal notation for numbers using decimal points and leading signs optionally (e.g., **1, −9, +9.0, 0.001, 99.9999**).

- Scientific notation is expressed with the letter **e**, (e.g., **2.0e-03, 1.07e23, −1.732e+03**

- Imaginary numbers use either **i** or **j** as a suffix, for example, **1i, -3.14j, 3e5i**

Operators: The following are the arithmetic operators in MATLAB.

+	Addition (when adding matrices/arrays subscripts must match)
−	Subtraction (same as above)
*	Multiplication (the size of arrays must be consistent when multiplying them)
/	Division
∧	Power
'	Complex conjugate transpose (also array transpose)

In the case of arrays, each of these operators can be used with a **period** prefixed to the operator; for example, (.*) or (.∧) or (./). This has a special meaning only in MATLAB. It implies element-by-element operation. It is useful for quick computation. It has no relevance in mathematics or anywhere else. We will use it a lot in the next chapter for generating data for graphical optimization.

1.2.4 An Interactive Session

Start MATLAB. In the MATLAB Command window, there should be the MATLAB prompt followed by a blinking cursor. MATLAB is ready to go to work. Please hit return at the end of the line (or before the comment, as you do not need to type the comments in this exercise). Please do read the comments as they contain useful information. Please feel free to try your own variations and extend the exercise. In fact, to understand and reinforce the commands, it is recommended that you make up your own examples often in this exercise.

```
>> a = 1.0; b = 2.0; c = 3.0, d = 4.0; e = 5.0
>> % why did only c and e echo on the screen?

>> who % lists all the variables in the workspace
```

```
>> a                      % echoes the value stored in a
>> A = 1.5;      % another variable A
>> a, A          % echoes a and A - Note case matters
>> aA = a*A      % multiplying a and A and a new variable

>> one = a; two = b; three = c;        % new variables
>> % assigns the values of a, b, c to new variables

>> four = d; five = e; six = pi % value of pi is available
>> A1 = [a b c ; d e f]
>> % A1 is a 2 by 3 matrix
>> % space or comma separates columns
>> % semi-colon separates rows

>> A1(2,2)       % accesses the Matrix element on the second
>>               %      row and second column
>> size(A1)      % gives you the size of the matrix
>>                         % (row, columns)
>> AA1 = size(A1)      % What should happen here?
>> % from previous statement the size of A1
>> % contains two numbers organized as a row
>> % matrix. This size information is assigned to AA1
>> size(AA1)    % AA1 is a one by two matrix

>> B1 = A1'     % the transpose of matrix A1
>> % is assigned to B1. B1 is a three by two matrix
>> C1 = A1 * B1 % Since A1 and B1 are matrices this is a
>>              % matrix multiplication
>>              % Should this multiplication be allowed?

>> C2 = B1 * A1 % How about this?
>> C1 * C2      % What about this?
>>              % read the error message
>>              % it is quite informative

>> D1 = [1 2]' % D1 is a column vector
>> % by default row vectors are created

>> C3 = [C1 D1] % C1 is augmented by an extra column
>> C3 = [C3 ; C2(3,:)]
>> % means do the right hand side and overwrite the
>> % old information in C3 with the result of the right
>> % hand side calculation
>> % On the right you are adding a row to current
>> % matrix C3. This row has the value of
```

```
>> % the third row of C2 - Notice the procedure of
>> % identifying the third row. The colon represents all
>> % the column elements

>> C4 = C2 * C3        % permissible multiplication
>> % Note the presence of a scaling factor
>> % in the displayed output
>> C5 = C2 .* C3        % seems to multiply!
>> % Is there a difference between C4 and C5?
>> % The .* is represents the product of each element
>> % of C2 multiplied with the corresponding
>> % element of C3.  Such a multiplication is
>> % not defined mathematically

>> C6 = inverse(C2)    % find the inverse of C2
>> % apparently Inverse is not a command in Matlab
>> % if command name is known it is easy to obtain help
>> lookfor inverse      % this command will find all files
>> % where it comes across the word "inverse" in the
>> % initial comment lines
>> % The command we need appears to be INV which says
>> % Inverse of a Matrix
>> % The actual command is in lower case. To find out how
>> % to use it - Now
>> help inv      % shows how to use the command
>> inv(C2)       % inverse of C2

>> for i = 1:20
        f(i) = i^2;
        end
>> % This is an example of a for loop
>> % the index ranges from 1 to 20 in steps of 1(default)
>> % the loop must be terminated with "end"
>> % the prompt does not appear until "end" in entered

>> plot(sin(0.01*f),cos(0.03*f))
>> % plots an x-y plot in the Figure Window
>> % the Figure window is the third MATLAB window
>> % you will use often to display graphics
>> % it is also loaded with rich features
>> % that you can explore by yourself
>> % about the MATLAB command used above - plot
>> % all commands in MATLAB are between regular parenthesis
>> % first element of plot is the x information
>> % second element of plot is the y information
```

```
>> xlabel('sin(0.01*f)') % text appear in single quotes
>> % text is also called strings
>> ylabel('cos(0.03*f)')
>> legend ('Example')
>> title ('A Plot Example')
>> grid
>>      % The previous set of commands will create plot
>>      % label axes, write a legend, title and grid the plot

>> exit % finished with MATLAB
>> % you can exit using the exit button on the top right
```

This completes the first session with MATLAB. Additional commands and features will be encountered throughout the book. In this session, it is evident that MATLAB allows easy manipulation of matrices, definitely in relation to other programming languages. Plotting is not difficult either. These advantages are quite substantial in the subject of optimization.

This session introduced

- The MATLAB Command Window and Workspace
- Variable assignment
- Basic Matrix operations
- Accessing rows and columns
- Suppressing echoes
- **Who, inverse** commands
- .* multiplication
- Figure window
- Basic plotting commands

In the next session, we will use the m-file editor to gain additional experience in using MATLAB. In this case we will not be interactively using MATLAB.

1.2.5   Using the Editor

In this section, we will use the MATLAB editor to create and run a script file. A script file is a collection of commands executed in sequence by MATLAB. This is also called batch execution. The file has to be saved before execution. Normally, the editor is used to generate two kinds of MATLAB files. These files are termed as **script files** and **function files**. Both these files contain MATLAB commands like the ones we have used. However, the second type of files has a specified format. Both file types should have the extension .m. Though these files are ASCII text files, and can be generated in any text editor, the generic .m extension should be used because MATLAB searches for this extension. The MATLAB editor will append this extension automatically as the file is saved. This extension is unique to MATLAB.

This is different from the interactive session of the previous section where MATLAB responded to each command immediately. The script file is more useful when there are many commands that need to be executed to accomplish some objective, like running an optimization technique. It is important to remember that MATLAB allows you to switch back to interactive mode by just typing commands in the workspace window after the prompt. You can also work with the variables created by the script file.

The MATLAB editor uses color to identify MATLAB statements and elements. It also provides the current values of the variables (if/after they are available in the workspace) when the mouse is over the variable name in the editor. There are two ways to access the editor through the MATLAB Command window on the PCs. Start MATLAB. This will open a MATLAB Command or Workspace window. In this window, the editor can be started by using the menu or the toolbar. On the **File** menu, click on **New** and choose **M-file**. Alternately, click on the leftmost icon on the toolbar (the tooltip reads **New File**). The icon for the editor can also be placed on the desktop, in which case the editor can be started by double-clicking the icon. In this event, a MATLAB Command window need not be opened. The Editor provides its own window for entering the script statements.

The commands are the same as in the interactive session, except there is no MATLAB prompt prefixing the expressions. To execute these commands you will have to save them to a file. We will save the commands in a file called *script1.m*. The *.m* extension need not be typed if you are using the MATLAB editor. You can save the file using the **Save** or **Save As** command from most editors. This will open an explorer type window where you can save the file, including creating a directory to save it in. If you want to break this session in segments, you can save your current work and continue where you left off, by traversing to the directory where you stored it and opening it in the editor. You can also open the file by typing line commands in the Command window. You will need the full **path** to this file. We plan to save this file in the sub-directory *Ch1* of the directory *Opt_book*, which is off the main file directory *C*. The path for the file is therefore **C:\Opt_book\Ch1\script1.m**. Note, the path here is specified as a PC path description. In UNIX, usually the slashes are forward slashes. The reason we need this information is to inform MATLAB where to find the file. We do this in the MATLAB Command window using an **addpath** command to inform MATLAB of the location of the directory to search for the file:

```
>> addpath C:\Opt_book\Ch1\
```

The simpler way to do this is to open the Editor window, find the file script1.m by traversing the directory, and open it. When you try and run the file, MATLAB will ask you if you want to add the directory to the path, and you can accept the choice. This way you will not need to know about the *addpath* command. The script that was created in script1.m can be run by typing (note the extension is omitted).

```
>> script1
```

It can also be saved and run by clicking the icon that represents a *written page with a down arrow*, called the *RUN* icon, located at the right of the group of icons below the menu bar in the Editor window. If you just created the file and are clicking it for the first time a save file window will appear before the code is executed.

To understand the programming concepts in the script, particularly for users with limited programming experience, it is recommended to run the script after a block of statements have been written rather than typing the file in its entirety. You can run by clicking the RUN icon, so that the changes in the file are recorded and the changes are current. Another recommendation is to deliberately *mistype* some statements and attempt to debug the error generated by the MATLAB debugger during execution. Open the MATLAB editor and type the following code. The code is in courier font.

Creating the Script M-file: The following will be typed/saved in a file.

```
% script1.m - (second edition)
clear   % clear all values - clean start
clc     % position curser at the top of the screen
format compact % print single spacing on the screen
               % default is double spacing

%   example of using script
A1 = [1 2 3];  % a row vector
A2 = [4 5 6];  % another row vector
% the commands not terminated with semi-colon will display
% information on the screen
A = [A1; A2]    % a 2X3 matrix
B = [A1' A2']   % a 3x2 matrix
C = A*B         % matrix multiplication

%  press the run icon and save file as script1.m in
% a directory.  Look at the output in command window
% you should see A, B, C

%  you can use blank lines to make the code readable

% now recreate the matrix and perform matrix
% multiplication as in other programming languages
% example of for loop
```

```
for i = 1 : 3    % variable i ranges from 1 to 3 in steps
                 % of 1 (default)
    a1(1,i) = i;
end              % loops must be closed with end
a1               % print a1 on the screen

%  press the run icon and look at the output
%  in command window


for i = 6:-1:4   % note loop is decremented
    a2(1,i-3) = i;  % filling vector from rear
end
a2

%  press the run icon and look at the output
%  in command window

% creating matrix AA and BB
for i = 1:3
    AA(1,i) = a1(1,i);
    AA(2,i) = a2(1,i);
    BB(i,1) = a1(1,i);
    BB(i,2) - a2(1,i);
end

AA       % print the value of AA in the window
BB       % print the value of BB

%  press the run icon and look at the output
%  in command window

% instead of the for statement we could have
AAA(1,:) = a1;  % first row of AAA is the vector a1
AAA(2,:) = a2;  % second row of AAA is vector a2

BBB(:,1) = a1';% first column of BBB is transpose of a1
BBB(:,2) = a2';% second column of BBB is transpose of a2

AAA
BBB

%  press the run icon and look at the output
%  in command window
```

```
who     % list all the variable in the workspace
whos % list all variables with their type and storage

% press the run icon and look at the output
% in command window

% consider writing code for Matrix multiplication
% in most programming languages
% multiply two matrices (only if column of first matrix
% must match row of second matrix)

szAA = size(AA) %    size of AA
szBB = size(BB);%    size of BB
if (szAA(1,2) == szBB(1,1))% ONLY if column and row match
    for i = 1:szAA(1,1)
        for j = 1:szBB(1,2)
            CC(i,j) = 0.0;  % initialize value to zero
            for k = 1:szAA(1,2)
                CC(i,j) = CC(i,j) + AA(i,k)*BB(k,j);
                % add to the variable and replace it
            end     % end of k - loop
        end     % end of j - loop
    end % end of i - loop
end % end if - loop
CC

% press the run icon and look at the output
% in command window

% % Note the power of MATLAB derives from its ability to
% % handle matrices very easily

CCC = AA*BB
% % this completes the script session
```

If you have been running the commands as you have been typing the file is already saved. Otherwise save the above file (script1.m). Add the directory to the MATLAB path as indicated before. Run the script file by typing **script1** at the command prompt. The commands should all execute and you should finally see the MATLAB prompt in the Command window.

Note: You can always revert to the interactive mode by directly entering commands in the MATLAB window after the prompt. In the Command window:

```
>> who
>> clear C      % discards the variable C from the workspace
>>              % use with caution. Values cannot be recovered
```

```
>> help clear
>> exit
```

This session illustrated the following:

- Use of the editor
- Creating scripts
- Running scripts
- Error debugging (recommended activity)
- Programming concepts
- Loop constructs, **if** and **for** loops
- Loop variable and increments
- Array access
- **Clear** statement

1.2.6 Creating a Code Snippet

In this section, we will examine the other type of m-file, which is called the **function *m-file***, which will define a function in Matlab. For those familiar with other programming languages such as C, Java, or FORTRAN, these files represent functions or subroutines. They are primarily used to handle some specific set of calculations. They also provide a way for the modular development of code, as well as code reuse. These code modules are used by being called or referred in other sections of the code, say through a script file we looked at earlier. The code that calls the function m-file is the **calling program/code**. There are three essential parameters in developing the *function m-file*:

1. What input is necessary for the calculations (information passed to the function)
2. What are the specific calculations that must take place (code in the function)
3. What information must be returned to the calling program (information returned after processing in the function)

Matlab requires the structure of the *function m-file* to follow a prescribed format. The variables used in the function files are created when the function is entered and dies when the value is returned from the function. Variables with the same names outside the function are unaffected.

We will use the editor to develop two files: the *script m-file* that will contain the command, which will call the *function m-file*. The *function m-file* will perform a polynomial curvefit. This exercise is called curve fitting. A couple of chapters later this problem will be identified as a problem in **unconstrained optimization**. The *function m-file* will require a set of *x,y data*, together with the *order of the polynomial* to be fit. The technique is based the method of least squares. For now, the calculations necessary to accomplish the exercise are considered known. It involves solving a linear equation with the normal matrix and a right-hand vector obtained using the data points. The **output** from the m-file will be the coefficients representing the polynomial. The responsibility of the *script m-file* is to acquire the information needed, and use the output to compare the curve against the data through a plot, and evaluate the quality of the fit.

Before we start to develop the code, the first line of this file must be formatted as specified by MATLAB. In the first line, the first word starting from the first column is the word *function*. It is followed by the set of values returned by the function [*returnval*]. Next, an *equal to* sign is followed by the **name** [*mypolyfit*] of the function, with the parameters **passed to** the function within parenthesis (*XY, N*). The file is recommended to be saved with the same name as the name of the function [(*mypolyfit.m*)]. The comments between the first line and the first executable statement will appear if you type **help name** (*help mypolyfit*) in the Command window. The reason for the name *mypolyfit.m* is that MATLAB has a built-in function **polyfit**. Open the editor to create the file containing the following code:

```
function returnval = mypolyfit(XY, N)
%
% These comments will appear when the user types
% help myployfit in the Command window
% This space is intended to inform the user how to interact
% with the program, what it does and what are the input
% and output parameters
%%%%%%%%%%%%%%%%%%%%%%%%%%%%%%%%%%%%%%%%%%%%%%%%%%%%%%%%%%%%%
%       mypolyfit performs the Least Square Error
%       fit of polynomial of order N
%
%       x vs y - data found is [:,2] matrix XY
%
%       It returns the vector of N + 1 coefficients
%       starting from the constant term
%%%%%%%%%%%%%%%%%%%%%%%%%%%%%%%%%%%%%%%%%%%%%%%%%%%%%%%%%%%%%
% Applied Optimization with MATLAB Programming
% P. Venkataraman
% Second Edition,  John Wiley and Sons.
%%%%%%%%%%%%%%%%%%%%%%%%%%%%%%%%%%%%%%%%%%%%%%%%%%%%%%%%%%%%%
%

for i = 1:N+1
    a(i) = 0.0; % initialize the coefficient to zero
    % this will be returned if there are
    % insufficient data points
end

sz = size(XY);  % find the number of data points
NDATA = sz(1,1);    % number of data points
if NDATA == 0
```

```
        fprintf('Error: There is no data to fit');
        returnval = a;  % zero value returned
        return;     % return back to calling program
end
% the fprintf prints formatted information to the screen

if NDATA < N
        fprintf('Too few data points for good fit');
        fprintf('\nError: Returned without execution');
        returnval = a; % zero value returned
        return
end


%%%%%%%%%%%%%%%%%%%%%%%%%%%%%%%%%%%%%%%%%%%%%%%%%%%%%%%%%
% The processing starts here.
% The coefficients are obtained as solution to
% the Linear Algebra problem
%              [A][c] = [b]
% Matrix [A] is the Normal Matrix
% Please consult any refernce on Numerical Analysis
%%%%%%%%%%%%%%%%%%%%%%%%%%%%%%%%%%%%%%%%%%%%%%%%%%%%%%%%%


for i = 1:N+1;
        % set up the right hand side
        b(i) = 0.0;
        for m = 1:NDATA; % loop over all data points
            b(i) = b(i) + XY(m,2)*XY(m,1)^(i-1);
        end     % loop m

        % set up the normal matrix
        for j = 1:N+1;
            if j >= i
                power = (i-1) + (j-1);
                A(i,j) = 0.0;   % initialize
                for k = 1:NDATA;   % sum over data points
                    A(i,j) = A(i,j) + XY(k,1)^power;
                end % k loop
            end     % close if statement
            A(j,i) = A(i,j); % exploiting Matrix symmetry
        end     % end j loop
end     % end i loop
%    if the x-points are distinct then inverse is not a
%    problem.  Otherwise debug the matrix [A]

returnval = inv(A)*b';
```

Save the file as *mypolyfit.m*. To use the function we will use the following script file *ScriptPolyfit.m*.

```
%  ScriptPolyfit.m
%
% This is a script file for the running mypolyfit.m
% Chapter 1. Section 1.2.5 Creating a Code Snippet
%
%%%%%%%%%%%%%%%%%%%%%%%%%%%%%%%%%%%%%%%%%%%%%%%%%%%%%%%%
% Applied Optimization with MATLAB Programming
% P. Venkataraman
% Second Edition,  John Wiley and Sons.
%%%%%%%%%%%%%%%%%%%%%%%%%%%%%%%%%%%%%%%%%%%%%%%%%%%%%%%%
% The script will:
%
% create x - y data (a polynomial of order 3)
%
% calls mypolyfit function to obtain coefficients
%
% Compares the original and fitted data on a plot
%
% calculates the error
%
% Also saves the data in an existing directory
%

clear   % clear all values - clean start
clc     % position cursos at the top of the screen
format compact % print single spacing on the screen
               % default is double spacing

% Examine the help available with mypolyfit that you set up
help mypolyfit

% create data points for fitting
for i = 1: 20;
      x = i/20;
      XY(i,1) = x;
      XY(i,2) = 2 + 3.0*x - x*x - 3*x^3;
end

coeff = mypolyfit(XY,3)'  % prints the coefficients on
                          % the screen
% a cubic polynomial was deliberately created to
% check the results. You should get back
```

```
% the coefficients you used to generate the curve
% this is a good test of the program

% MATLAB provides a function called polyval that
% will evaluate the polynomial.  However to use it we must
% reverse the coefficient vector.  We cam use the
% MATLAB function "fliplr" on the coefficient
% vector
cflip = fliplr(coeff)    % flip the coefficient vector
yfit = polyval(cflip,XY(:,1));  % values of y at the
                                % same x values

error = sum((yfit - XY(:,2)).^2);
% sum of the square of the difference between
% fitted value and original value
% NOTE: element by element operator .^

% Plot the original data and the fitted data
plot(XY(:,1),XY(:,2),'ro',XY(:,1),yfit,'b-');
% original data is red circles
% fitted data is the blue line
xlabel('x');  % label x axis
ylabel('y');  % label y axis
legend('Original data','fitted data'); % the legend
text(0.1,1.2,strcat('Error squared =',num2str(error)));
% drop a piece of text at the location x = 0.1, y = 1.2
% the text will combine the string'Error squared" and
% the error value after converting it into a string
title('Using function mypolyfit.m');

save C:\OptBook\Chapter1\XY.dat XY -ascii -double
% (] (line continued from above)
% this will save the values of XY in the file XY.dat
% as ascii text file in double precision values in the
% directory
% C:\OptBook\Chapter1\
% Be sure to create the directory before you run this
% script
```

This concludes the exercise where a function was written to calculate the coefficients of the polynomial used to fit a curve to some *x-y* data. The type of file is the *function m-file*. It needs to be used in a certain way. The code was tested using a cubic polynomial using data generated by a cubic polynomial so the same coefficients are obtained if the code is correct. A plot is used to compare the results. The error is printed on the plot.

1.2.7 Creating a Program

This section introduces some user interface elements that you can use as part of your programs. This section is optional and can be considered as advanced programming. If you were a C++ programmer you would be doing this after many months of experience. It is included here to illustrate that it is not difficult, even for those starting out programming in MATLAB, to think about incorporating advanced resources available in MATLAB. In this section, we will essentially be performing the same calculations as in the last section, including calling **mypolyfit.m**. We will develop a program that will read **x-y data**, curvefit the data using a polynomial, and compare the original and fitted data graphically. There are several ways for you or the users to interact with the code you develop. The basic method is to prompt users for information at the prompt in then Command window. This is the quickest. This is probably what you will use when developing the code. Once the code has been tested, depending on usefulness it might be relevant to consider using more sophisticated custom elements like input boxes and file selection boxes. The book will continue to use these elements throughout as appropriate. While the input elements used in this code are new commands, the program will mostly use commands that have been introduced earlier. In sequential order, the events in this program are as follows:

1. To read the **x-y data** saved earlier using a **file selection** box
2. To read the order of fit using an **input dialog** box
3. To use the **mypolyfit** function developed in the last section to obtain the coefficients
4. To obtain the coordinates of the fitted curve
5. To graphically compare the original and fitted data
6. To report on the fitted accuracy on the figure itself. The new script file will be called **prog_pfit.m**.

Start the text editor to create the file called **prog_pfit.m**. In it, enter the following code:

```
% Program for fitting a polynomial curve to xy data
% Name:  prog_fit.m  ( a script file)
%%%%%%%%%%%%%%%%%%%%%%%%%%%%%%%%%%%%%%%%%%%%%%%%%%%%%%%%%%%%%
% Applied Optimization with MATLAB Programming
% P. Venkataraman
% Second Edition,  John Wiley and Sons.
%%%%%%%%%%%%%%%%%%%%%%%%%%%%%%%%%%%%%%%%%%%%%%%%%%%%%%%%%
%
% Chapter 1, Section 1.2.6
% The program looks for a file with two column ascii data
```

```
% with extension  .dat.  The order of the curve is obtained
% from user. The original and fitted data are compared with
% relevant information displayed on the same figure.
% The program demonstrates the use of the file selection
% box, an input dialog box, creating special text strings
% and displaying them

clear  % clear all values - clean start
clc    % position cursos at the top of the screen
format compact % print single spacing on the screen
               % default is double spacing

[file,path] =uigetfile('*.dat','All Files');
% uigetfile opens a file selection box
% checkout  help uigetfile
% the string variable file will hold the filename
% the string variable path will have the path information
% open the file saved in the last session

if isstr(file) % if a file is selected
    loadpathfile = ['load ',path file];
    % creates a string that will be evaluated using eval
    % loadpathfile is a string variable concatenated with
    % three strings "load ", path and file
    % note the space after load is important

    eval(loadpathfile)
    % evaluates the string enclosed -
    % It executes Matlab load command. This will import the
    % xy-data
    % the data will be available in the workspace as a
    % variable with the same name as the filename without
    % the extension (this assumes you selected the xy-data
    % using the file selection box)
    % XY matrix is avaiable in the workspace

    NDATA = length(XY(:,1));   % number of data points
    clear path loadpathfile
    % we will not need these variables
    % get rid of these variables to free memory
end  % if statement

% Use of an input dialog box to get the order
% of polynomial to be fitted
PROMPT = {'Enter the Order of the Curve'};
```

```
% PROMPT is a string Array with one element
% note the curly brackets
TITLE = 'Order of the Polynomial to be Fitted';
% a string variable
LINENO = 1;   % a data variable

% lets get the value of order of polynomial to
% be fitted from the user
NS = inputdlg(PROMPT, TITLE, LINENO);
% the input dialog captures the user input in NS
% NS is a string Array
% check help input dialog for more information

N = str2num(NS{1,1});
% the string is converted to a number- the order

clear PROMPT TITLE LINENO   % deleting variables

% call function mypolyfit and obtain the coefficients
coeff = mypolyfit(XY,N);

% generate the fitted curve and obtain the squared error
% Here is another (inefficient) way to evaluate
% the y values of fit and the error
err2 = 0.0;
for i = 1:NDATA   % for each data point
    for j = 1:N + 1
        a(1,j) = XY(i,1)^(j-1);
    end
    y(i) = a*coeff;   % the data for the fitted curve
    err2 = err2 + (XY(i,2) -y(i))*(XY(i,2)-y(i));
    % the square error
end

% plotting
plot (XY(:,1),XY(:,2),'ro',XY(:,1),y,'b-');
% original data are red o's
% fitted data is blue solid line
xlabel('x');
ylabel('y');

% create strings using available information
% for the title
strorder = setstr(num2str(N));
% convert the order of curve to a string
```

```
% setstr assigns the string to strorder
titlestr = ['Polynomial curvefit of order ',strorder, ...
    ' of file ', file];
% the three dots at the end are continuation marks
% the title will have the order and the file name
title(titlestr)
legend('original data', 'fitted data');
% you should see the same plot as in the previous exercise

errstr1 = num2str(err2);
errstr2 = ['squared error = ', errstr1];
gtext(errstr2);
% this places the string errstr2 which is obtained
% by combining the string 'squared error' with
% the string representing the value of the error,
% wherever the mouse is clicked on the plot.
% moving the mouse over the figure you should
% see location cross-hairs
clear strorder titlestr errstr1 errstr2 a y x i j
clear NDATA NS coeff XY file err2

% This finishes the exercise
```

Run the program by first running MATLAB in the directory where these files are, or adding the path to locate the files. At the command prompt, type *prog_pfit*. The program should execute, require interactions, and display a figure similar to Figure 1.10. The appearance may be slightly different, depending on the platform MATLAB is being run.

This finishes the MATLAB section of the chapter. The section has introduced MATLAB in a robust manner. A broad range of programming experience has been initiated in this chapter. All new commands have been identified with a brief explanation in the comments. It is important that you use the opportunity to type in the code yourself. That is the only way the use of MATLAB will become familiar. The practice also will lead to fewer syntax errors. The writing of code will significantly improve the ability to debug and troubleshoot. Although this section has maintained a separate section on the use of MATLAB in different ways, subsequent chapters will see an integration of the use of various resources that MATLAB contains.

1.2.8 Application Bibliography

The following listing is a small sample of titles that explore in greater detail, the spread of the application of optimization to new areas. All of books appeared in the period between the two editions of this book.

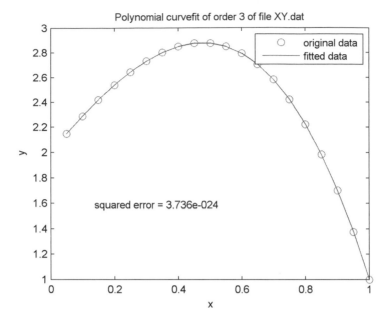

Figure 1.10 Original and fitted data.

Allstot, David James, Kiyong Choi, Lenn Schramm, and Jinho Park, *Parasitic-Aware Optimization of CMOS RF Circuits*. New York: Springer, 2003.

Baldick, Ross, *Applied Optimization: Formulation and Algorithms for Engineering Systems*. London: Cambridge University Press, 2006, 786 pages.

Bellingham, Richard and Russell J. Campanello, "HR Optimization: From Personnel Administration to Human and Organizational Capital Development," *HRD Products*, 2004.

Biegler, Lorentz T., *Large-Scale Pde-Constrained Optimization*. New York: Springer, 2003.

Craven, Bruce Desmond, M. Sardar, and N. Islam, *Optimization in Economics and Finance: Some Advances in Nonlinear, Dynamic, Multicriteria and Stochastic Methods*. New York: Springer, 2005.

Dorigo, Marco and Thomas Stutzle, *Ant Colony Optimization*. Boston: MIT Press, 2004.

Du, Ding-Zhu, *Combinatorial Optimization in Communication Networks*. New York: Springer, 2006.

Fullér, Róbert, and Christer Carlsson, *Fuzzy Reasoning in Decision Making and Optimization*. New York: Springer, 2002.

Hartmann, Alexander K. and Rieger Heiko, *New Optimization Algorithms in Physics*. Hoboken, NJ: Wiley-VCH, 2004.

Laporte, Emmanuel and Patrick Le Tallec, *Numerical Methods in Sensitivity Analysis and Shape Optimization*. New York: Springer, 2003, 216 pages.

Leondes, Cornelius T., "Methods in Diagnosis Optimization," *World Scientific*. 2005.

Lu, Bing, Sachin S. Sapatnekar, and Dingzhu Du, *Layout Optimization in VLSI Design*. New York: Springer, 2001.

Momoh, James A., *Electric Power System Applications of Optimization*. Marcel Dekker, 2001.

Narayan, S. Rau, *Optimization Principles: Practical Applications to the Operation and Markets of the Electric. Power Industry*. Hoboken, NJ: Wiley-IEEE, 2003, 360 pages

Pardalos, Panos M. and Christodoulos A. Floudas, *Optimization in Computational Chemistry and Molecular Biology: Local and Global Approaches*. New York: Springer, 2000.

Putman, Richard E., *Industrial Energy Systems: Analysis, Optimization, and Control*, ASME Press, 2004.

Ralf Korn, Elke, *Option Pricing and Portfolio Optimization: Modern Methods of Financial Mathematics*. American Mathematical Society, 2001.

Rhyder, Robert F., *Manufacturing Process Design and Optimization*. Marcel Dekker, 1997.

Torres, Nãestor V. and Eberhard O. Voit, *Pathway Analysis and Optimization in Metabolic Engineering*, London: Cambridge University Press, 2002.

Vofs, Stefan, Stefan Voss, and David L. Woodruff, *Introduction to Computational Optimization Models for a Production Planning in a Supply Chain*. New York: Springer, 2006.

PROBLEMS

Many problems here were defined by students in my course on optimization as part of their projects. **(For problems 1.1 to 1.5 please use your imagination for domain knowledge.)**

1.1 Identify several possible optimization problems related to an automobile. For each problem, identify all the disciplines that will help establish the mathematical model.

1.2 Identify several possible optimization problems related to an aircraft. For each problem, identify all the disciplines that will help establish the mathematical model.

1.3 Identify several possible optimization problems related to a ship. For each problem, identify all the disciplines that will help establish the mathematical model.

1.4 Identify several possible optimization problems related to a microsystem used for control. For each problem, identify all the disciplines that will help establish the mathematical model.

1.5 Define a problem with respect to your investment in the stock market. Describe the nature of the mathematical model.

1.6 Define the problem and establish the mathematical model for the I-beam, holding up an independent single family home. Assume a single beam in the middle of the basement parallel to the long side of the house.

1.7 Define the mathematical model for an optimum overhanging traffic light. Decide what you want to optimize. Define your parameters. (See Figure Problem 1.7.)

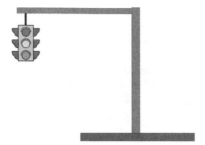

Figure Problem 1.7

1.8 Define the problem and identify a mathematical model for scheduling and optimization of your daily routine activity. Plan around six routine activities. Associate each with cost. Identify a couple of constraints. Keep the model linear.

1.9 Define the problem for a laminar flow in a pipe for maximum heat transfer driven given a specific pump.

1.10 Define a chemical engineering problem to mix various mixtures based on ingredients of limited availability to meet specified demands.

1.11 Find the rectangle of the largest area that can fit within an ellipse of semi-major axis a and semi-minor axis b. Set up the optimization problem, Make any assumptions you need. Refer to Figure Problem 1.11.

1.12 Find the circle of the largest area in a semi-ellipse (right of the y-axis) with semi-major axis a and semi-minor axis b. Set up the optimization problem, Make any assumptions you need (Figure Problem 1.12.)

1.13 Find the quadratic polynomial between the limits $x = 0$ and $x = 3$, enclosing minimum area with x-axis, with a slope of 3 at $x = 0$ and a slope of -1 at $x = 3$. Set up the optimization problem.

1.14 Consider a beam with given material properties (E and I), known cross-section area (A), and of length L. Design the triangular load distribution spread over the entire length of the beam, with the total

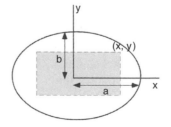

Figure Problem 1.11

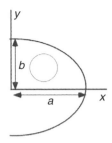

Figure Problem 1.12

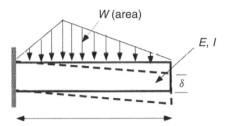

Figure Problem 1.14

load of W, which will cause the least deflection of the end. (See Figure Problem 1.14.)

1.15 The energy crisis of 2008 is forcing the gas company to consider ways to provide short time relief to consumers by trying to use the same infrastructure, with minor modification like a new pump, to increase the flow of gas from the wells. The gas flows from the wells to the utilities through large pipes with pumping stations placed at equidistant locations (L). The pressure difference between two pumping stations can be considered to be $p_1 - p_2$. Figure (Problem 1.15) illustrates the layout and the cross-section description. The volume flow rate Q (cubic feet/hr) can be calculated as:

$$Q = 1350D^{2.5} \left(\frac{p_1 - p_2}{LG} \right)^{0.5}$$

Where D is the diameter of the pipe; G is the specific gravity of the gas (dimensionless); L is the pipe length (yards); p_1 is the inlet pressure (in of H_2O); p_2 is the outlet pressure (in of H_2O); This simple equation is accredited to Dr. Pole in 1851 (Ref: www.psig.org/papers/2000/0012.pdf). However, for any cross section of the flow/pipe (see figure), if the outside pressure is ignored, the tangential and radial stresses are calculated

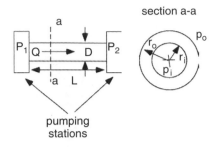

section a-a

Figure Problem 1.15

(Shigley, Mischke, Budynas, *Mechanical Engineering Design*) as:

$$\sigma_t = \frac{p_i r_i^2}{r_o^2 - r_i^2}\left(1 + \frac{r_o^2}{r^2}\right)$$

$$\sigma_r = \frac{p_i r_i^2}{r_o^2 - r_i^2}\left(1 - \frac{r_o^2}{r^2}\right)$$

Note that $\sigma_t > \sigma_r$ and σ_t is maximum at r_i, set up an optimization problem for increasing the flow rate. Identify parameters for the problem based on existing infrastructure (use the library or internet) and make suitable assumptions.

1.16 A thin-walled spherical gas container (radius: r, thickness: t) is required to be designed for an internal pressure of 5 MPa. Find the smallest mass of the container if the factor of safety is 3 and the material used is structural steel. The stresses in the thin-walled structure can be evaluated as follows:

$$\sigma(normal) = \frac{pr}{2t}; \quad \tau(shear) = \frac{pr}{4t}$$

Set up the optimization problem in standard format.

1.17 A closed-end, thin-walled cylindrical pressure vessel is to be designed for minimum mass. It should contain at least 25 m^3 of gas at a pressure of 3.5 MPa. The circumferential/hoop stress is not to exceed 210 MPa, while the circumferential strain is not to exceed 0.001. The material is structural steel. The design variables are inside radius r and thickness t. Draw the figure expressing the problem. Take the maximum hoop stress as follows:

$$\sigma_t|_{max} = \frac{p(2r + t)}{2t}$$

Set up the optimization problem in standard format.

1.18   A chrome vanadium spring ($G = 79.3$ GPa; $\rho = 7,860$ kg/m^3), of minimum mass is needed to carry a uniform pressure load $p = 250$ kPa acting over a diaphragm of diameter of 0.1 m as shown in Figure Problem 1.18 giving a compressive load of F. The maximum shearing stress is 1.5 GPa. The maximum compression of the spring is 0.1 m. The surge frequency should be 60 Hz or higher to avoid resonance. The solid length of the spring (compressed fully) must be larger than $L = 40$ mm because of the block. The spring is fully compressed when carrying the load. The design variables are n the number of coils; d the spring wire diameter, D the mean diameter of the spring. The outer limits on the spring are 0.1m. We define three relations before assembling the constraints.

$$C = \frac{D}{d}; \quad \text{spring index}$$

$$k = \frac{Gd^4}{8D^3(n-2)}; \quad \text{spring constant}$$

$$K = \frac{(4C+2)}{(4C-3)}; \quad \text{Bergsträsser factor}$$

The constraint equations are

$$nd \geq 0.04; \quad d + D \leq 0.1$$

$$\frac{KF^2D}{\pi kd^3} \leq 1.5 \text{ GPa (shear stress)}$$

$$\frac{F^2D^3(n-2)}{kd^4G} \leq 0.1\,m \text{ (maximum compression)}$$

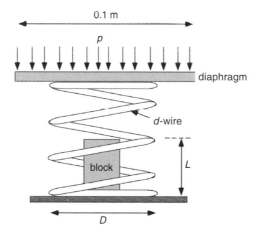

Figure Problem 1.18

$$\frac{d}{2\pi D^2 (n - 2)} \sqrt{\frac{G}{2\rho}} \geq 60 \text{ Hz}; \quad \text{(surge frequency)}$$

And the side constraints:

$$4 \leq n \leq 12; \quad 0.06 \leq D \leq 0.09[m]; \quad 0.001 \leq d \leq 0.012[m]$$

Set up the optimization problem in the standard format.

1.19 Bellville springs, or conical disc springs, can handle large loads in tight spaces. Linear, regressive, and progressive load-deflection characteristics can be designed through stacking (*SAE Spring Design Manual*, Transactions SAE, 1990). They have nonlinear load-deflection and load-stress characteristics. A single spring is used in the following problem and the geometry is defined in the Figure (Problem 1.19). There are four design variables: D_i, D_o, h, t. The assumptions in the following development are: the angle α is small; cross-section do not warp throughout the range of deflection; fully elastic behavior during deflection; maximum stress at the inside joint as shown in the Figure Problem 1.19 (Almen and Lazlo, *The Uniform—Section Disc*, Transactions of ASME, 1936). The functions involved in the problem are:

$$W = \rho \pi t \left(\frac{D_o + D_i}{2} \right) \sqrt{h^2 + \left(\frac{D_o - D_i}{2} \right)^2} \quad \text{(Weight)}$$

Almen and Lazlo define the load-deflection and stress-deflection formulas for the spring as

$$P = \frac{Ey}{(1 - v^2) M (D_o/2)^2} [(h - y/2)(h - y)t + t^3]$$

$$S = \frac{Eh}{(1 - v^2) M (D_o/2)^2} [C_1(h/2) + C_2 t]$$

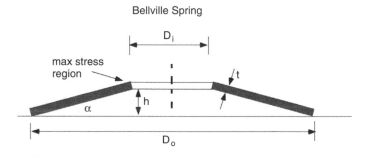

Bellville Spring

Figure Problem 1.19 Bellville spring optimization.

P is the axial load (lb); y is the deflection (inches); E is the modulus of elasticity (psi); S is the maximum tensile stress (psi); ρ is the specif weinght (lb/in^3); v is the Poisson's ratio. The values for M, C_1, C_2 can be calculated by

$$M = \frac{6.0}{\pi \ln \dfrac{D_o}{D_i}} \left[\frac{D_o/D_i - 1}{D_o/D_i} \right]$$

$$C_1 = \frac{6.0}{\pi \ln \dfrac{D_o}{D_i}} \left[\frac{D_o/D_i - 1}{\ln(D_o/D_i)} - 1 \right]$$

$$C_2 = \frac{6.0}{\pi \ln \dfrac{D_o}{D_i}} \left[\frac{D_o/D_i - 1}{2} \right]$$

Set up an optimization problem for minimum weight, with a capacity of generating an axial load of at least 200 lb for a deflection of 0.08 in. The stress must be less than the tensile stress of the material. Take the factor of safety as 2.2. The range of D_o is between 3 in and 10 in. Use appropriate side constraints for the other variables. Choose a material too.

1.20 This is a variation on Example 1.2. Rectangular planks of any cross-section dimensions a and b can be nailed to construct an I-beam like that in Figure Problem 1.20. The nail spacing is 4 in. The wood is Douglas fir ($E = 1.6 \ 10^6$ psi, $G = 0.6 \ 10^6$ psi, $v = 0.33$, $\gamma = 0.016$ ib/in^3). Find the allowable tensile and shear stress for design. Consider the load of 500 lb being carried at the end of 48 in. It is cantilevered at the other end. Design the beam of minimum weight. The maximum load is at the fixed end. The shear in the nails is limited to 100 lb. Develop the mathematical model. You can follow Example 1.2, except that the shear stress constraint is replaced by the shear in the nail. The shear force in the nail is the shear flow at the junction, multiplied by the nail spacing.

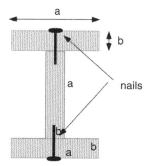

Figure Problem 1.20

1.21 What is the smallest gear ratio (r_C/r_B) for the transmission shafts in Figure Problem 1.21 if shear stress in AB and AC are less than 1,200 psi, and the twist of the end D must be less than 0.16 rad. The length of the shaft AB is 12 inches and its diameter is 0.25 in. The length of the shaft CD is 24 inches and its diameter is 0.12 in. Use your own additional information if required.

1.22 In July, Venkat's Fruit Orchard usually sees a scheduling problem. Both cherries and blueberries ripen around the same time. A cartload of cherries sells for $2,750 while the cartload of blue berries fetches $1,750. The farm must deliver at least a half cartload of cherries and one cartload of blueberries to the local supermarket. Three persons can pick two cartloads of cherries in a day while one person will pick a cartload of blueberries in the same time. Only 46 people show up for work during the day. One cartload of cherries is contained in 4 pallets, while the same amount of blueberries occupies 6 pallets. There are 251 pallets. Each picker is paid $45 per day. The packaging and storage can handle, at most, 30 cartloads per day. Set up the LP problem for the Orchard to maximize profits.

1.23 The energy crisis of 2008 is keeping the refineries very active trying to adjust the product mix due to demand and supply. Largely, the decision is about buying two varieties of crude oil and selling two kinds of product: gasoline and diesel. They can buy light crude at $135 per barrel, while heavy crude is discounted at $95 per barrel. Consider one barrel as 40 gallons. They can sell gasoline at $4.25/gallon and diesel at $4.85 a gallon. A barrel of light crude will yield 0.5 barrels of gasoline and 0.4 barrels of diesel. A barrel of heavy crude will yield 0.4 barrels of gasoline and 0.2 barrels of diesel. Totals supply of light crude is limited to 500,000 barrels, while heavy crude is unlimited in supply. In a day the refinery can only handle 2 million gallons of gasoline and 3 million gallons of diesel. Set up the LP programming problem to maximize daily profits.

1.24 Consider the standard bridge circuit shown in Figure Problem 1.24. The power delivered by the batteries is P_b and the power consumed by the resistors is P_r. Matching the two sets up the constraint

$$P_b - P_r = 0$$

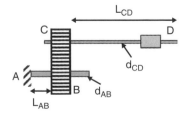

Figure Problem 1.21

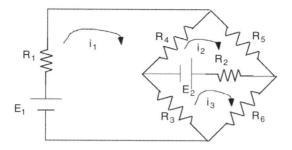

Figure Problem 1.24   A bridge circuit.

Show that the solution to the optimization problem of Minimizing P_r with respect to the currents (i_1, i_2, i_3) and subject to the constraint will yield the Kirchhoff's loop equation (Donald A. Pierre, *Optimization Theory with Applications*). The expression for the power is the following:

$$P_b = E_1 i_1 + E_2(i_3 - i_2)$$

$$P_r = i_1^2 R_1 + (i_2 - i_3)^2 R_2 + (i_1 - i_3)^2 R_3 + (i_1 - i_2)^2 R_4 + i_2^2 R_5 + i_3^2 R_6$$

1.25   A gating circuit using a switch and two diodes is shown in Figure Problem 1.25. When the switch (S_1) is open the load current i_L flows through the load R_L. When the switched is closed the power (P_R) is dissipated through the resistor R. It is necessary to limit this power dissipation to P_m. The current and the dissipated power (Donald A. Pierre, *Optimization Theory with Applications*) are:

$$i_L = \frac{V}{R + R_L}; \quad P_R = \frac{V^2}{R}$$

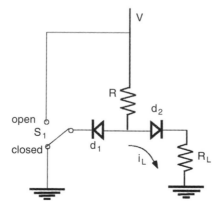

Figure Problem 1.25   A gate circuit.

Set up an optimization problem to maximize the load current with respect to R and V subject to the constraint on the dissipated power.

1.26 A clipping circuit with the given input-output characteristics is shown in the Figure Problem 1.26. For specified levels of V_{o1}, V_{o2}, *and* R_L the resistances R_1, R_2, and the voltage E are to be selected so that the desired input-output characteristics must be obtained. Furthermore, the power drain (P) from the voltage supply (E) must be as low as possible (Donald A. Pierre, *Optimization Theory with Applications*). The functional relations are:

$$P = \frac{(E - V_{o1})^2}{R_2} + \frac{V_{o1}^2}{R_1} + \frac{V_{o1}^2}{R_L}$$

$$V_{o1} = R_1 \left[\frac{E - V_{o1}}{R_2} - \frac{V_{o1}}{R_L} \right]; \quad V_{o2} = \frac{(E - V_{o2})R_L}{R_2}$$

Obtain the relations above. Set up the optimization problem in standard format.

1.27 The ratio of the $|V_2(s)/V_1(s)|$ needs to be maximized for the circuit shown in the Figure Problem 1.27. Note that $s = j\omega$. However, the ratio of the voltages must match at two given frequencies. Hence the constraint:

$$\left| \frac{V_2(j\omega_1)}{V_1(j\omega_1)} \right| = \left| \frac{V_2(j\omega_2)}{V_1(j\omega_2)} \right|$$

must be satisfied. The resistances are fixed while the capacitors and inductors are variables. Set up the constrained optimization problem and express the voltage ratio as a real valued function.

1.28 Consider a plane in level flight where lift (L) is equal to weight (W) and thrust (T) is equal drag (D). The forces are shown in Figure Problem 1.28. The lift and drag can also be calculated from the nondimensional lift and

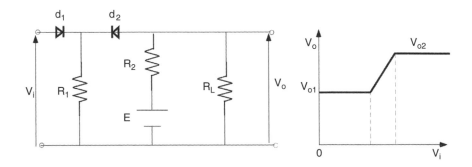

Figure Problem 1.26   A clipping circuit.

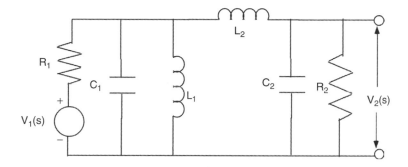

Figure Problem 1.27 Matching transfer function.

Figure Problem 1.28 Aircraft in level flight.

drag coefficients (C_L **and** C_D) through

$$L = \frac{1}{2}\rho V^2 SC_L; \quad D = \frac{1}{2}\rho V^2 SC_D$$

In standard design, the relation between the coefficients can be expressed through the parabolic drag polar:

$$C_D = C_{D0} + K C_L^2$$

where ρ is the atmospheric density at the flight altitude; V is the speed of flight; S is the reference area of the wing; C_{D0} is the zero lift drag; and K is induced drag factor, all of which are constants for the flight. Find the speed for minimum drag. See Figure Problem 1.28.

1.29 Since the shown aircraft is powered by a piston propeller engine, maybe it is better to fly at the speed for minimum power, where the power is calculated as

$$P = DV$$

Find the speed for minimum power.

1.30 Another way of looking at Problem 1.28 is to work with the nondimensional relations. A measure of aircraft performance is aerodynamic efficiency E, which is the ratio of C_L over C_D. Find the maximum aerodynamic efficiency and the speed corresponding to this efficiency.

Problems 1.31 to 1.33 can be handled by assuming a series solution as in Example 1.4, or using Bezier functions (Chapter 9, Chapter 11), or any other type of parametric representation for the variables. These problems are more naturally defined through the calculus of variation. These problems can be postponed until those chapters if only Bezier functions are going to be used.

1.31 The problem involves designing a two-dimensional wing for minimum drag (Miele, *Theory of Optimum Aerodynamic Shapes*). The linearized theory for a 2D symmetrical wing, at zero angle of attack, at a given free stream Mach number (M) leads to the pressure coefficient described by

$$C_p = \frac{2\dfrac{dy}{dx}}{\sqrt{M^2 - 1}}$$

The function $y(x)$ is the description of the airfoil upper surface, as shown in Figure Problem 1.31. The aerodynamic drag, for a given free stream dynamic pressure (q) can then be expressed as follows:

$$D = 2q \int_{x_i}^{x_f} C_p \frac{dy}{dx} dx = \frac{4q}{\sqrt{M^2 - 1}} \int_{x_i}^{x_f} \left(\frac{dy}{dx}\right)^2 dx$$

From the figure

$$x_i = 0; \quad x_f = L; \quad y_i = 0; \quad y_f = 0$$

Consider that the airfoil volume (A) is given, leading to the constraint

$$\frac{A}{2} = \int_{x_i}^{x_f} y(x) \, dx$$

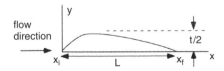

Figure Problem 1.31   Minimum pressure drag.

If the torsional stiffness (I_c) of the thin-skin structure is also given, then

$$\frac{I_c}{2} = \int_{x_i}^{x_f} y^2(x)\, dx$$

If the bending stiffness is prescribed (I_b) is prescribed then

$$\frac{3I_b}{2} = \int_{x_i}^{x_f} y^3(x)\, dx$$

Express the surface as a Bezier function and identify the optimization problem for minimum drag subject to

a. given volume
b. given torsional stiffness
c. given bending stiffness
d. given volume and torsional stiffness
e. given volume and bending stiffness

Note that this problem suggests that using a parametric definition for the function/trajectory we can convert problem from the calculus of variations to design optimization problems.

1.32 Here, we set up a problem in designing a body of revolution for minimum drag under the Newtonian pressure law as illustrated in the Figure Problem 1.32. The Newtonian pressure law in hypersonic aerodynamics is (Miele, *Theory of Optimum Aerodynamic Shapes*)

$$C_p = 2 \sin^2\theta$$

where θ is the local inclination of the body shape. Nevertheless, the functional to be minimized is

$$I = \int_{x_i}^{x_f} \frac{y(y')^3}{1+(y')^2}\, dx + \frac{y_i^2}{2}; \quad x_i = 0; \quad x_f = L; \quad y_f = d/2$$

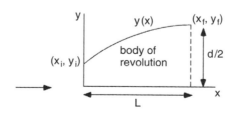

Figure Problem 1.32   Minimum pressure drag for body of revolution.

Express the surface as a Bezier function and identify the optimization problem for minimum drag.

1.33   Consider the RL circuit shown in Figure Problem 1.33. The initial time is 0. The final time T is specified. It is required to maximize the output voltage at the final time. There are two constraints. The first is the differential constraint due to the circuit equation. The second is that the energy delivered over the time interval T is specified as E_s. The following functional equations can be easily derived:

$$v_o(T) = \int_0^T R\frac{di}{dt}dt; \quad \frac{di}{dt} = \frac{v_i}{L} - \frac{R}{L}i; \quad E_s = \int_0^T v_i i \, dt$$

For a fixed R and L, find the functions $v_i(t)$ and $i(t)$ that will solve the optimization problem. Represent $v_i(t)$ and $i(t)$ as Bezier functions.

1.34   The $\Phi X \Psi$ fraternity is going to capitalize on the mouth-watering pizza sauce, accidentally created by one of its members, by organizing a weekly fundraiser for its charities. It plans two varieties of pizza based on two ingredients, sausage and cheese. Of course, each pizza will have the pie, base sauce, sausage, and cheese. The ingredients are in cups for each type of pizza. A cup of sausage costs $0.75, a cup of cheese $0.5, while a cup of sauce is $1.00. The pie costs $0.50 each.

Pizza	Sell Price	Sausage	Cheese	Sauce
Supergooey	$ 7.50	1.5	2.75	2.0
Justgooey	$ 7.50	2.5	1.5	1.5

The fraternity can obtain only 120 cups of sausage, 100 cups of cheese, and 120 cups of sauce, and any amount of pies. Find how many pizzas of the two kinds it must sell for maximum profit.

1.35   A two-dimensional aluminum truss is shown in Figure Problem 1.35. All of the members have the same annular cross-section. For the loading shown

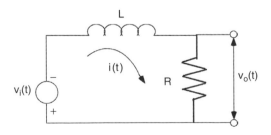

Figure Problem 1.33   A circuit problem.

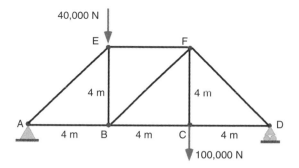

40,000 N

E   F

4 m   4 m

A   D

4 m   B   4 m   C   4 m

100,000 N

Figure Problem 1.35   Truss Problem.

determine the truss of minimum mass. Ensure that the normal stresses are within the elastic limit, while the truss does not fail in buckling.

1.36   A shell and tube heat exchanger is a popular device to transfer energy. The cross-section of one is shown in Figure Problem 1.36. The diameter of the shell (D) is specified as 0.75 m. The length of the exchanger (L) is 2.5 m. The shell is made from a plate that costs $9.50 per unit area. The tubes are cut from stock and cost $1.25 per meter. The energy exchanged by the device can be calculated as $0.5*sqrt(n)*L^{0.75}/d^{0.2}$ kW. Here n is the number of tubes and d is the diameter of the tube. To make sure the tubes fit in the shell we impose an area constraint that the total area of the tubes must be less than half the cross-sectional area of the shell. Just to be sure, we constrain the diameter by requiring $sqrt(n)*d$ must be less than the diameter of the shell. Set up an optimization problem for minimum cost with the understanding that the heat exchanger must transfer at least 10 kW of energy.

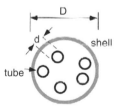

D

d

shell

tube

Figure Problem 1.36   Heat exchange design.

1.37   Fred wants to invest his $100,000 in four mutual funds whose holdings and return is shown in the Table Problem 1.37. He wants his choice to reflect a growth strategy that requires that he has at least 60 percent invested in domestic stocks, at most 20 percent in international stock, and wants his exposure to real estate to be at most 10 percent. How should he invest his money for maximum return?

Table Problem 1.37

Mutual Fund	Domestic Stocks	International Stocks	Bonds	Real Estate	Expected Return
F1	0.2	0.6	0.2	0.0	6 %
F2	0.6	0.1	0.3	0.1	3 %
F3	0.4	0.5	0.0	0.1	4 %
F4	0.5	0.1	0.1	0.3	1 %

1.38   The four-bar link is shown in Figure Problem 1.38. Find the link length L_1 and L_2 to maximize the velocity at D when θ is 90° and ω is 50 rad/s. Use reasonable side constraints.

1.39   Solve Problem 1.38 with a constraint on acceleration while the angular velocity ω is increasing at a rate of 2 rad/sec^2.

1.40   The local Pizza Shoppe wants to create a template it can use to cut three different sizes of the pie base each time it rolls out the rectangular area of the dough as shown in Figure Problem 1.40. The Shoppe is experimenting with metric pizzas to combat obesity. The three sizes are grande ($R_1 = 0.2$ m), medium ($R_2 = 0.15$ m), and personal ($R_3 = 0.1$ m). Set up an optimization problem for locating the three bases (one each) such that there is the least wastage.

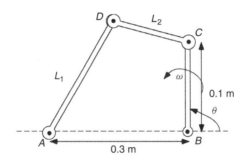

Figure Problem 1.38   Four bar linkage.

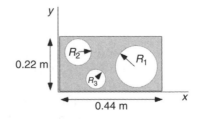

Figure Problem 1.40   Pizza base problem.

2

GRAPHICAL OPTIMIZATION

The examples in this book are mostly structured around two design variables. First, it allows us to represent the information graphically so that the solution can be obtained by inspection. Second, it allows us to develop the numerical techniques using planar vectors, which can be instinctively extended to n-dimensional vectors without much formal exposure. Third, it allows us to explore the mathematics involved in optimization through geometry and graphics. The last reason is most useful for understanding the algorithm and its progress toward the solution. In practical situations, which usually involve more than two variables, there is no opportunity for neat graphical correlation because the limitations of representation. Three variables can be illustrated through three-dimensional graphics by slices, sequence of two-dimensional plots, or movies using MATLAB, but is not as effective in describing optimization ideas as the two-variable problems. This edition includes an example in three variables in this chapter primarily as an opportunity to explore more graphics programming. For more than three variables, it becomes cumbersome to explore the solution through graphics. On those occasions, it is necessary to review the numerical information or numbers to decide on convergence or its lack thereof. It is generally assumed that what happens with many variables is an extension of the geometric features that are observed in problems involving two variables. It will require effort to use your imagination for several variables but it becomes easier with practice.

MATLAB can create publication quality graphics. It can also display the same numerical information in a variety of plots. Contour plots will provide the best graphical representation of the optimization problem in two variables. The points on any contour (or curve) have the same value of the function. Many software packages can create and display these plots. In this book, we will use MATLAB for graphical optimization.

This chapter, in the new edition, has shed the subsection on creating GUI in MATLAB. The GUI feature has evolved into a powerful programming component since the last edition and the new GUI creation tool is quite sophisticated and far more complex and capable than can be addressed in a section in this book. MATLAB provides a separate documentation and has many examples for users for creating a GUI. It still requires dealing with *Handle Graphics*, which is touched on in this chapter.

2.1 PROBLEM DEFINITION

The standard format for optimization problems was established in Chapter 1. It is reintroduced here for reference:

$$\text{Minimize} \quad f(x_1, x_2, \ldots x_n) \tag{2.1}$$

$$\text{Subject to:} \quad h_1(x_1, x_2, \ldots x_n) = 0$$

$$h_2(x_1, x_2, \ldots x_n) = 0 \tag{2.2}$$

$$h_l(x_1, x_2, \ldots x_n) = 0$$

$$g_1(x_1, x_2, \ldots x_n) \le 0$$

$$g_2(x_1, x_2, \ldots x_n) \le 0 \tag{2.3}$$

$$g_m(x_1, x_2, \ldots x_n) \le 0$$

$$x_i^l \le x_i \le x_i^u \quad i = 1, 2, \ldots, n \tag{2.4}$$

In this chapter, for convenience and comprehension, the right-hand side is allowed to have a numeric constant different from zero if it is part of problem formulation. The first example chosen for illustration is a simple one using elementary functions whose graphical nature is well known. This simple example will permit examination of the MATLAB code that will generate the curves and the solution. It will also establish the format for the display of solution to the two-variable problems.

2.1.1 Example 2.1

The first example, Example 2.1, will have two equality constraints and two inequality constraints.

$$\text{Minimize} \quad f(x_1, x_2) = (x_1 - 3)^2 + (x_2 - 2)^2 \tag{2.5}$$

$$\text{Subject to} \quad h_1(x_1, x_2) : 2x_1 + x_2 = 8 \tag{2.6a}$$

$$h_2(x_1, x_2): (x_1 - 1)^2 + (x_2 - 4)^2 = 4 \tag{2.6b}$$

$$g_1(x_1, x_2): x_1 + x_2 \leq 7 \tag{2.7a}$$

$$g_2(x_1, x_2): x_1 - 0.25x_2^2 \leq 0 \tag{2.7b}$$

$$0 \leq x_1 \leq 10; \quad 0 \leq x_2 \leq 10; \tag{2.8}$$

In this definition, we have two straight lines (h_1, g_1), two circles (f, h_2), and a parabola (g_2). Note that two equality constraints and two variables imply that there is no scope for optimization, since the equality constraints will determine the solution. This will be true if the constraints are linearly independent, which is true in this example. Linear independence means that the second constraint is not the same as the first constraint–disguised through elementary arithmetic operations like addition and multiplication. This example is created to demonstrate the code for creation of the graphics in MATLAB. You can copy the segments of the code created for this example as a template for all the other problems used to explore graphical solutions in the book.

Figure 2.1 illustrates the graphical solution to this problem. The figure also displays the inequality constraints using hash marks. The hash marks are drawn using a custom function called *drawHashMarks*. Its use is explained in the code. Please make sure the function is available in the path if you are planning to use it. This is a work in progress and does not always work. In that case, an alternate procedure for identifying inequality constrained is suggested and implemented in a different code.

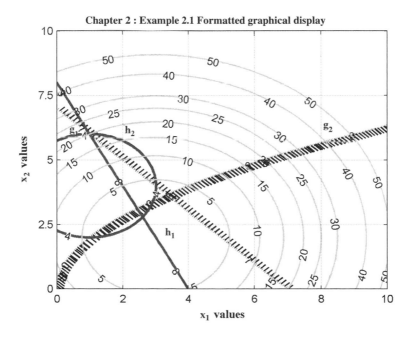

Figure 2.1 Graphical solution for Example 2.1.

2.1.2 Format for the Graphical Display

The graphical solution to Example 2.1, as seen in Figure 2.1, is generated using MATLAB [1, 2]. Later on we will walk through the code for generating the figure. When inequality constraints are distinguished by hash marks the hashed side indicates the *infeasible* region. Please note that the graphical solution is incomplete if the inequality constraints are not distinguished in some manner, or if the feasible region is not visible, since constraint satisfaction has a higher priority than optimization.

In the graphical display of solutions, the objective function is drawn for several contours so that the direction for the minimum can be identified. If you are running the code in MATLAB, you can take advantage of the color for the clarity of solution. The print version requires figures in black and white, and here we will use *LineWidth* and *LineStyle* properties to distinguish the various functions. The color reference is for the graphics you see on screen. The objective functions will be a thin solid in green color. Each equality constraint is drawn as a single curve at the numerical value of the right-hand side. They are solid blue line of thicker line width. They are sometimes identified interactively by having the user place an identifier through moving the cross hairs using the mouse to the appropriate location and clicking the left mouse button. Each inequality constraint is similarly drawn for a value on the right side, as a thick red dotted line. Each inequality constraint may be hashed on one side determined by the user by calling the *drawHashMarks* function, if the function is successful in drawing the hash marks. Another way to denote an equality constraint is to draw a thick blue solid line parallel to the constraint, but in the *infeasible* region. They will also be identified interactively. The design region is the domain in the figure. The feasible region should be visible and the solution can be identified through the figure. The relevant region can be zoomed for better accuracy. The data for the plots are obtained using MATLAB's matrix operations after establishing the plotting mesh. This exploits MATLAB's natural speed for matrix operations.

2.2 GRAPHICAL SOLUTION

MATLAB possesses a powerful visualization engine that permits the solution of the two-variable optimization problem by inspection. There are three ways to take advantage of the graphic features of MATLAB. The first is the use of MATLAB's high-level graphing routines for data visualization. This will be the primary way for solving graphical optimization problems in this book. This will also be the way to incorporate graphical exploration of numerical techniques in the book. For more precise control over the display of data, MATLAB allows user interaction through an object-oriented system identified in MATLAB as Handle Graphics. The third use of the MATLAB graphics engine and custom programming is to use the Handle Graphic system to develop a Graphical User Interface (GUI) for the program or m-file. MATLAB's online documentation can assist you get started with developing the GUI. This is omitted from this version of the book. All of the plotting needs can be met by the high-level graphics functions available in MATLAB.

2.2.1 MATLAB High-Level Graphics Functions

There are three useful windows during a typical MATLAB session. The first is the MATLAB Command window through which MATLAB receives instruction and displays information. The second window is the Editor window where m-files are coded. The third is the Figure Window where the graphic elements are displayed. There can be more than one Figure window on the screen. The Figure window is the target of the high-level graphics functions. The MATLAB user interface in recent versions is very rich and sophisticated and displays a vast amount of information from the workspace. It also allows easy manipulation and code retrieval. The users are encouraged to become more familiar with it. This book will mainly stay focused on optimization and the required code, leaving other explorations to the user.

The graphic functions in MATLAB allow you plot in 2D or 3D. They allow contour plots in 2D and 3D, mesh and surface plots—bar, area, pie charts, histograms, animation and gradient plots. Subplots can also be displayed using these functions. These functions also permit operation with images and 3D modeling. They allow basic control of the appearance of the plot through properties of color, line style, line width, markers, axis ranges, and aspect ratio of the graph. For three-dimensional plots, they allow shading, lighting, transparency, and view control among the additional features. They permit annotation of the graph in several ways. Some of these functions will be used in the next section when we develop the *m-file* for the first example. One way to create the plots for graphical optimization is to create a basic plot through code and then use MATLAB *plot editor* to make it look pretty. A more efficient way is to include all of the graphics command in the code so that it happens without intervention. The latter one is adopted in this book. Note, once you have standardized the appearance of your plots, it is a matter of using the same piece of code except for the lines of code that defines the problem. Hence, there is a serious payoff for understanding the code in the next session.

The two main graphical elements that are typically controlled using the high-level graphics functions are the *figure* and the *axes*. Using Handle Graphics you can control most of the other graphical elements, which also include elements used in the GUI. The *figure* function or command creates a Figure window with a number starting at one, or will create a new Figure window incrementing the window count by one. Normally, all graphics functions are targeted to the current Figure window, which is selected by clicking it with the mouse or executing the command *figure (number)*, where number is the number of the Figure window that will have the focus. Graphic functions or commands will automatically create a window if none exists.

Remember, you can find more information about usage of commands through typing *help command name* at the MATLAB prompt, or through the Help Browser window. We have used the words *Handle Graphics* liberally so far. The word *handle* appears on many platforms and in many applications, particularly those that deal with graphics. They are widely prevalent in object-oriented programming practice. MATLAB's visualization system is object-oriented. Most graphical elements are considered as objects. The "handle" in MATLAB is a system/software created identification number associated normally with a variable, which can identify the specific graphic object. If this handle is available, then the properties of the object (e.g. line size, marker type, color, etc.) can be viewed, set, or reset if necessary. In MATLAB Handle Graphics, this is the way to customize the graphical elements. In high-level graphics functions, this is used in a limited way. In this chapter, we will use it mainly to change the characteristics of some of the graphical elements on the figure.

To understand the concept of *handles* we will run the following piece of code interactively. It deals with creating plots, which you are familiar from Chapter 1. The comments in the code segment indicate the new features that are being emphasized.

Start MATLAB and *interactively* perform the following:

```
>> x = 0:pi/40:2*pi;  % create x vector
>> y = x.*sin(x);     % create y vector
>> plot(x,y,'b-');    % plot y vs x in solid blue line
>> grid;       % draw a grid
>> h = plot(x,y,'b-') % h is the handle to the plot

>> % a new plot is created in the same Figure Window
>> % a numerical value is assigned to h which should
>> % be written to the screen (no semi-colon)
>> % We can use the variable name h or its value
>> % to refer to the plot again
>> set(h,'LineWidth',2);     % this should make your plot
>>     % thicker blue
>> set(h,'LineWidth',3,'LineStyle',':','Color','r')

>> % The handle is used to refer to the object
>> % whose property is being changed
>> % Usually Property information occurs in pairs of
>> % property-name/property-value
>> %

>> get(gca)    % this will list the property of the axes
>> % of the current plot. Note there are a significant
>> % amount of properties you can change to customize the
>> % appearance of the plot
>> % gca is the handle for the current axes

>> set(gca,'ytick',[-5,-2.5,0,2.5,5])
```

```
>> % you have reset the ytick marks on the graph
>> set(gca,'FontName','Arial','FontWeight',..
>>       'bold','FontSize',14)
>>
>> % Changes the font used for marking the axes
>> set(gca,'Xcolor','blue')
>> % changes the x-axis to blue
>> % concludes the demonstration of handles
```

From this session it is evident that to fine-tune the plot you create, you need to first identify the object handle. You use the handle to access the object property. You change the property by setting its value. We will see many more graphics function as we obtain the graphical solution to Example 2.1 in the following section.

2.2.2 MATLAB Plot Editor

We changed properties through calling the higher graphics functions. You can do the same through the MATLAB Plot Editor. Close the previous figure, clear all of the values, and position the cursor at the top of the screen.

```
>>  close
>>  clear
>>  clc
```

Create the same plot
```
>> x = 0:pi/40:2*pi;    % create x vector
>> y = x.*sin(x);       % create y vector
>> plot(x,y,'b-');      % plot y vs x in blue color
>> grid;
>> plot(x,y,'b-')       % h is the handle to the plot
```

In the Figure window select from the menu

```
Edit -> Axes Properties
```

Now you can change the axis information such as labels, limits, ticks, colors, title, grid, and so on, including additional properties. Try it.

Double click on the graph

You will see in Figure 2.2 that you can change the appearance of the plot itself in several ways.

You can invoke the plot editor by clicking on the white arrow next to the printer icon, OR

```
>>  plotedit   % in the Command window
```

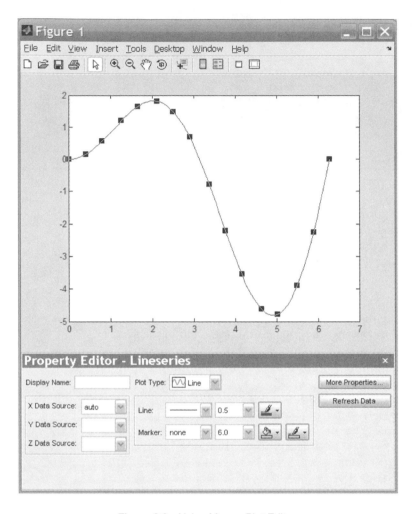

Figure 2.2 Using MATLAB Plot Editor.

2.2.3 Example 2.1 – Graphical Solution

Figure 2.1 is the graphical representation of Example 2.1. The range for the plots matches the side constraints for the problem. The intersection of the two equality constraints identifies 2 possible solution to the problem, approximately (2.6, 2,8) and (1, 6). The inequality constraint g_2 makes the point (2.6, 2.8) unacceptable. Point (1, 6) is acceptable with respect to both constraints. The solution is therefore at (1, 6). While the objective function was not used to determine the solution, contours of the objective function are drawn indicating the direction for the minimum of the objective function. The solution can be identified by inspection of the assembled plots without reference to the terminology or the techniques of optimization. This is straight forward with one or two design variables in the problem. The effort increases exponentially with additional variables.

The code for Example 2.1 is available in the file **Example2_1_.m**[*]. You are strongly recommended to type the code so that you can run it incrementally to understand the graphical features being addressed. Use the editor to create the file. *In the following code some of the commands are on two lines because of Font size and page size limitations. Please keep them one line.* If you want to break them up, please remember to use a continuation symbol (...).

Example2_1.m

```
% Chapter 2: Graphical Optimization
%_____
%%%%%%%%%%%%%%%%%%%%%%%%%%%%%%%%%%%%%%%%%%%%%%%%%%%%%%
% Applied Optimization with Matlab Programming
% Dr. P.Venkataraman
% second Edition,  John Wiley
%%%%%%%%%%%%%%%%%%%%%%%%%%%%%%%%%%%%%%%%%%%%%%%%%%%%%%
%_____
%  Example 2.1 (modified graphics)(Sec 2.1- 2.2)
%
%   graphical solution using MATLAB (two design variables)
%
%   Minimize    f(x1,x2) = (x1-3)**2 + (x2-2)**2
%
%               h1(x1,x2) = 2x1 + x2 = 8
%               h2(x1,x2) = (x1-1)^2 + (x2-4)^2 = 4
%               g1(x1,x2): x1 + x2 <= 7
%               g2(x1,x2): x1-0.25x2^2 <= 0.0
%
%          0 <= x1 <= 10; 0 <= x2 <= 10
%
% The hash marks are drawn by a function called
%_____
% function [XL YL] = drawHashMarks(c1,choice)
%_____
%  inputs are: c1 and choice
%    c1 (the x,y information for the curve)
%    choice input parameter must be
%       'b' for hash on the BOTTOM or OUTSIDE a closed curve
%        't' for hash on the TOP or INSIDE closed curve
%    returns line matrix that can be use to draw hash marks
%    Comment: not tested extensively -
%                      has problems with closed curves
%        the author welcomes any improvement in the code
```

[*]Files to be downloaded from the web site are indicated by boldface courier type.

```
%
%%%%%%%%%%%%%%%%%%%%%%%%%%%%%%%%%%%%%%%% management functions
clear   % clear all variable/information in the workspace-
%           use CAUTION
clc     % position the cursor at the top of the screen
format compact   % avoid skipping a line when writing to the
%                  command window/screen
warning off   % don't report any warnings like divide by %
   %                zero etc.


%%%%%%%%%%%%%%%%%%%%%%%%%%%%%%%%%%%%%%%%%%%%%%%%%%%%%%%%%%%%%%
% define inline functions (usually defined before the code)
%%%%%%%%%%%%%%%%%%%%%%%%%%%%%%%%%%%%%%%%%%%%%%%%%%%%%%%%%%%%%%
% These functions are evaluated based on input parameters.
% This is useful for functions that have simple
% calculations and are unique and will not see much use
% Alternately you can use external functions for
% calculations
% NOTE: that special MATLAB operations are used for
% multiplication, division, and exponention
% .*      ./     .^
% this is element by element operations and are called
% ARRAY operations
%%%%%%%%%%%%%%%%%%%%%%%%%%%%%%%%%%%%%%%%%%%%%%%%%%%%%%%%%%%%%%

% this will calculate the objective function for X1 and X2
% values
% watch out for code continuation ........................
obj_ex21 = ...
inline('(X1-3).*(X1-3) +(X2-2).*(X2-2)','X1','X2');

% these functions will calculate the equality constraints
eqcon1_ex21 = inline('2.0*X1 + X2','X1','X2');

% watch out for code continuation ........................
eqcon2_ex21 = ...
inline('(X1-1).*(X1-1) + (X2-4).*(X2-4)','X1','X2');

% these constraints will calculate the inequality
% constraints
ineqcon1_ex21 = inline('X1 + X2','X1','X2');
ineqcon2_ex21 = inline('X1-0.25*X2.^2','X1','X2');


%%%%%%%%%%%%%%%%%%%%%%%%%%%%%%%%%%%%%%%%%%%%%%%%%%%%%%%%%%%%%%
```

```
%%% end of inline functions
%%%%%%%%%%%%%%%%%%%%%%%%%%%%%%%%%%%%%%%%%%%%%%%%%%%%%%%%%%%%%%%%%

%%%%%%%%%%%%%%%%%%%%%%%%%%%%%%%%%%%%%%%%%%%%%%%%%%%%%%%%%%%%%%%%%
%%%  establish the domain
% x1 and x2 are vectors filled with numbers starting
% at 0 and ending at 10.0 with values at intervals of 0.1

x1=0:0.1:10;  % the semi-colon at the end prevents the echo
x2=0:0.1:10;    % these are also the side constraints
%%%%%%%%%%%%%%%%%%%%%%%%%%%%%%%%%%%%%%%%%%%%%%%%%%%%%%%%%%%%%%%%%

% generates matrices X1 and X2 (matrices)
% corresponding to x1 and x2 (vectors)

% these are the x1 and x2 values at all the intersections
% of x and y values of the vectors x1 and x2 OR
% this is a mesh of x1 and x2 values

[X1 X2] = meshgrid(x1,x2);

% the data are generated using array operations in MATLAB
% this means that values for the entire domain are
% generated in a single operation
% unlike using for/do loops in other programming languages

%%%%%%%%%%%%%%%%%%%%%%%%%%%%%%%%%%%%%%%%%%%%%%%%%%%%%%%%%%%%%%%%%
%%%  generate the data for plotting

f1 = obj_ex21(X1,X2);% the objecive function is evaluated
%       over the entire mesh
ineq1 - ineqcon1_ex21(X1,X2);% the inequality g1 is
%        evaluated over the mesh
ineq2 = ineqcon2_ex21(X1,X2);% the inequality g2 is
%                         evaluated over the mesh
eq1 = eqcon1_ex21(X1,X2);% the equality 1 is evaluated
%       over the mesh
eq2 = eqcon2_ex21(X1,X2);% the equality 2 is evaluated
%       over the mesh

%%%%%%%%%%%%%%%%%%%%%%%%%%%%%%%%%%%%%%%%%%%%%%%%%%%%%%%%%%%%%%%%%
%%% create the various plots
% NOTE: the equality and inequality constraints are not
% written with 0 on the right hand side.
% If you do write them that way you would have to include
```

```
% [0,0] in the contour commands

%%% draw inequality constraint 1
% ineq1 is plotted [at the contour value of 7]
[c1,h1] = contour(x1,x2,ineq1,[7,7],'r:');
clabel(c1,h1);  % label the contour with value
[XL1 YL1] = drawHashMarks(c1,'t'); % draw hash on top
hl1 = line(XL1,YL1,'Color','k','LineWidth',1);% hash
set(h1,'LineWidth',2);

% interactively place the identifier and make it pretty
% will place the string 'g1' on the crosshair where mouse
% is clicked
k1 = gtext('g1');
set(k1,'FontName','Times','FontWeight','bold', ...
       'FontSize',14,'Color','red')

hold on % allows multiple plots

%%% draw inequality constraint 2
% ineq1 is plotted [at the contour value of 0]
[c2,h2] = contour(x1,x2,ineq2,[0,0],'r:');
clabel(c2,h2);
[XL2 YL2] = drawHashMarks(c2,'b');
hl2 = line(XL2,YL2,'Color','k','LineWidth',1);
% interactively place the identifier and make it pretty
set(h2,'LineWidth',2)
k2 = gtext('g2');
set(k2,'FontName','Times','FontWeight','bold', ...
       'FontSize',14,'Color','red')

%%% draw equality constraint 1
[c3,h3] = contour(x1,x2,eq1,[8,8],'b-');
clabel(c3,h3);
set(h3,'LineWidth',2)
% interactively place the identifier and make it pretty
k3 = gtext('h1');
set(k3,'FontName','Times','FontWeight','bold', ...
       'FontSize',14,'Color','blue')

%%% draw equality constraint h2
[c4,h4] = contour(x1,x2,eq2,[4,4],'b-');
clabel(c4,h4);
set(h4,'LineWidth',2)
% interactively place the identifier and make it pretty
```

```
k4 = gtext('h2');
set(k4,'FontName','Times','FontWeight','bold', ...
        'FontSize',14,'Color','blue')

%%% draw contours of objective function at these values
[c,h] = contour(x1,x2,f1,[5 10 15 20 25 30 40 50],'g');
clabel(c,h);
set(h,'LineWidth',1)

%%% label for x-axes
xlabel(' x_1 values', 'FontName','times', ...
        'FontSize',12,'FontWeight','bold');
%%% label for y-axes
ylabel(' x_2 values', 'FontName','times', ...
        'FontSize',12,'FontWeight','bold');

%%% change the tick marks
set(gca,'xtick',[0 2 4 6 8 10])
set(gca,'ytick',[0 2.5 5.0 7.5 10])

%%% place information in the plot
k5 = gtext({'Chapter 2: Example 2.1', ...
        'Formatted graphical display'})
set(k5,'FontName','Times','FontSize',12,'FontWeight','bold')

grid
hold off
```

You should see the plot of Figure 2.1 except for the location of the identifiers, which will vary depending on where you click the left mouse butto. Going over the code, you will see that much of the code is plot *formatting* commands. Most of the computations occur when you call the inline functions. Once you have a plot that you like, you can standardize it by reusing the same code except for the calculation of the data.

If the plot is acceptable, you can print the information on the figure to a file and later dump it to the printer (on the PC you can use the menu to print it). Typing ***help print*** in the Command window should list a set of commands you can use to save the file. The following saves the figure as an encapsulated postscript file using the command line. This will be the way to save through programming.

```
>> print   -depsc2   plot_ex_2_1.eps
```

Alternatively you can save the figure (***Save*** command on the ***File*** menu) as a MATLAB *figure (.fig)* file that you can open through the Command window and continue to edit. You can also save it many other graphics formats. You can use *Export Setup* ... from the *File* menu on the Figure window to change size, resolution, and other properties.

2.3 ADDITIONAL EXAMPLES

The following additional examples will serve to illustrate both optimization problems, as well as additional features of MATLAB that will be useful in developing graphical solutions. The graphical routines in MATLAB are powerful and easy to use. They can graphically display the problems in several ways with very simple commands. The useful display is, however, determined by the user. The first example in this section, Example 2.2, is a problem in unconstrained optimization. The next example is a structural engineering problem of reasonable complexity. The third example demonstrates optimization in the area of heat transfer design. The fourth example is an optimization problem in three variables.

2.3.1 Example 2.2 — Different Ways of Displaying Information

This example illustrates several different ways of graphically displaying a function of two variables. The example is an illustration of global optimization in Reference 3. The single objective function is:

$$f(x_1, x_2) = a * x_1^2 + b * x_2^2 - c * \cos(p * x_1) - d * \cos(q * x_2) + c + d \quad (2.9)$$

with

$$a = 1, b = 2, c = 0.3, d = 0.4, p = 3\pi, q = 4\pi$$

Figures 2.3 to 2.6 (on pages 85 and 86) are the graphical display of solutions in the section. The m-file associated with this example is **Example2_2.m**.

```
Example2_2.m
% Chapter 2: Graphical Optimization
% Example 2.2 Sec.2.3
%_____
%%%%%%%%%%%%%%%%%%%%%%%%%%%%%%%%%%%%%%%%%%%%%%%%%%%
% Applied Optimization with Matlab Programming
% Dr. P.Venkataraman
% Second Edition,  John Wiley
%%%%%%%%%%%%%%%%%%%%%%%%%%%%%%%%%%%%%%%%%%%%%%%%%%%
%_____
% Example 2.2 will introduce various ways to display
% graphical information-like 3D plots, 3D contours,
% filled 2D contours and gradient information
%
```

```
% Minimize:
% f(x1, x2 ) =   a x1^2 + b x2^2-c cos(p x1)-
% d cos(q x2) +   c + d
% with a = 1,   b = 2,   c = 0.3,   d = 0.4, p = 3pi, q = 4 pi
%
%_____


%%%%%%%%%%%%%%%%%%%%%%%%%%%%%%%%%%%%%%%%%%%%%%%%%%%%%%%%%%%%%%%%
%%% management functions
clear  % clear all variable/information in the workspace -
clc    % position the cursor at the top of the screen
format compact % avoid extra a line in the command window
warning off  % don't report any warnings
%%%%%%%%%%%%%%%%%%%%%%%%%%%%%%%%%%%%%%%%%%%%%%%%%%%%%%%%%%%%%%%%

%%%   establish the domain
% x1 and x2 are vectors filled with numbers starting
% at 0 and ending at 10.0 with values at intervals of 0.1
x1=-1:0.01:1;   % don't echo info to screen
x2= 1:0.01:1;   % these are also the side constraints

%%%%%%%%%%%%%%%%%%%%%%%%%%%%%%%%%%%%%%%%%%%%%%%%%%%%%%%%%%%%%%%%
% generates matrices X1 and X2 (matrices)
% corresponding to x1 and x2 (vectors)
% these are the x1 and x2 values at all the intersections
% of x and y values of the vectors x1 and x2
% this is a mesh of x1 and x2 values
[X1 X2] = meshgrid(x1,x2);
%%%%%%%%%%%%%%%%%%%%%%%%%%%%%%%%%%%%%%%%%%%%%%%%%%%%%%%%%%%%%%%%
% since we are evaluating one function, we can just
% calculate the values directly instead of using a FUNCTION
% or an INLINE function
%    f(x1,x2) = a*x1^2 + b*x2^2 -c*cos(p*x1)- ..
%                d*cos(q*x2) + c + d
% NOTE: that special MATLAB operations are used for
% multiplication, division, and exponention
% .*     ./    .^
% this is element by element operations and are called
% ARRAY operations

a = 1; b = 2; c = 0.3; d = 0.4; p = 3.0*pi; q = 4.0*pi;
f1 = a*X1.*X1 + b*X2.*X2 -c*cos(p*X1)-..
   d*cos(q*X2) + c + d;


%%%%%%%%%%%%%%%%%%%%%%%%%%%%%%%%%%%%%%%%%%%%%%%%%%%%%%%%%%%%%%%%
```

```
% First Plot   (Figure 2.3)

% filled contour with default colormap
% f1 = the function is evaluated over the entire mesh
% NOTE: Multiple minimums (LOCAL MINIMUM) with
% the absolutely best minimum (GLOBAL MINIMUM)
%%% filled contour plot
[c1,h1] = contourf(x1,x2,f1,..
[0 0.1 0.25 0.5 0.75 1.0 1.25 1.5 1.75 2.0 2.25 ...
2.5 2.75 3.0]);
clabel(c1,h1);
%%% place a colorbar and defines the axes tick marks
colorbar  % illustrates the scale
set(gca,'xtick',[-1 -0.5 0.0 0.5 1.0])
set(gca,'ytick',[-1 -0.5 0.0 0.5 1.0])

%%% axis labels and title
xlabel(' x_1 values','FontName','times','FontSize',12);
ylabel(' x_2 values','FontName','times','FontSize',12);
title({'Filled Labeled Contour Plot','default color'}, ..
    'FontName','times','FontSize',10)
grid

%%%%%%%%%%%%%%%%%%%%%%%%%%%%%%%%%%%%%%%%%%%%%%%%%%%%%%%%%%%%%%%%
% Second Plot   (Figure 2.4)
% a new Figure Window is used to draw this plot
% Basic contour plot with gradient information
% Gradient information is generated for a larger mesh to
% avoid clutter.  However this information is drawn over
% the contours in the first plot
% grid is not used for readability

figure    % open a new Figure Window
% create coarse mesh for the gradient information
y1 = -1:0.1:1.0;  % x1 values
y2 = -1:0.1:1;    % x2 values
[Y1,Y2] = meshgrid(y1,y2);
% generate function values for this new mesh
f2 = a*Y1.*Y1 + b*Y2.*Y2 -c*cos(p*Y1)-..
    d*cos(q*Y2) + c + d;
%%% Note the advantage of an inline function or a function
% file
%%% draw the smooth contours from previous plot
%%% These are not filled contours
[c1,h1] = contour(x1,x2,f1, ..
```

```
[0 0.1 0.25 0.5 0.75 1.0 1.25 1.5 1.75 2.0 2.25 ...
2.5 2.75 3.0]);
clabel(c1,h1);  % label contours
hold on  % add additional plot information without overwrite
%%% generate numerical gradients using a step of 0.1
%%% using the coarse function
[GX, GY] = gradient(f2,0.1);
%%% draw arrows for the vectors-the gradients
%%% Note the arrows indicate direction of increasing
% function values
quiver(Y1,Y2,GX,GY);
hold off
%%% set ticks, labels, and title
set(gca,'xtick',[-1 -0.5 0.0 0.5 1.0])
set(gca,'ytick',[-1 -0.5 0.0 0.5 1.0])
xlabel(' x_1 ','FontName','times','FontSize',12, ...
    'FontWeight','b');
ylabel(' x_2 ','FontName','times','FontSize',12, ...
    'FontWeight','b');
title({'2D Contour','with Gradient Vectors'}, ...
    'FontName','times','FontSize',10)
%%%%%%%%%%%%%%%%%%%%%%%%%%%%%%%%%%%%%%%%%%%%%%%%%%%%%%%%%%%%%%%%%%%
% Third Plot  (Figure 2.5)
%%% a mesh plot of the function
figure;

%%% change the colormap for a different range of colors
%%% to display the graphical information
colormap(gray);

mesh(x1,x2,f1);
grid
set(gca,'xtick',[-1 -0.5 0.0 0.5 1.0]);
set(gca,'ytick',[-1 -0.5 0.0 0.5 1.0]);
set(gca,'XGrid','on','YGrid','on','ZGrid','on');
colorbar
xlabel(' x_1 ','FontName','times','FontSize',12, ...
    'FontWeight','b');
ylabel(' x_2 ','FontName','times','FontSize',12, ...
    'FontWeight','b');
zlabel(' f ','FontName','times','FontSize',12, ...
    'FontWeight','b');
title({'Mesh Plot-Note many minimums','colormap-gray'}, ...
    'FontName','times','FontSize',10,'FontWeight','b')
```

```
%%%%%%%%%%%%%%%%%%%%%%%%%%%%%%%%%%%%%%%%%%%%%%%%%%%%%%%%%%%%%%%%%
% Fourth Plot   (Figure 2.6)
%%% a surface plot using the same data
%%% with a default colormap
figure;
colormap(jet);
hs = surf(x1,x2,f1); % handle to the surface object

shading interp;
colorbar
set(gca,'xtick',[-1 -0.5 0.0 0.5 1.0]);
set(gca,'ytick',[-1 -0.5 0.0 0.5 1.0]);
set(gca,'XGrid','on','YGrid','on','ZGrid','on');
title({'Surface Plot','colormap-jet/default'}, ..
    'FontName','times','FontSize',10,'FontWeight','b');
xlabel(' x_1 ','FontName','times','FontSize',12, ...
    'FontWeight','b');
ylabel(' x_2 ','FontName','times','FontSize',12, ...
    'FontWeight','b');
zlabel(' f ','FontName','times','FontSize',12, ...
    'FontWeight','b');
view(-35,45)
```

The brief comments in the code should provide an explanation of what you see on the figure. Figures can be further customized as seen in the previous section. From an optimization perspective, Figure 2.4 provides the best information about the nature of the problem. The 2D contour curves identify the neighborhood of the local minimum. The gradient vectors indicate the direction of the function's steepest rise at the point, so peaks and valleys can be distinguished. The contour themselves can be colored without being filled.

The quiver plot shown in Figure 2.4 also provides a mechanism to indicate the feasible region when dealing with inequality constraints, since they indicate the direction in which the constraint function will increase. If several functions are being drawn, then the clutter produced by the arrows may diffuse the clarity. You are encouraged to use the powerful graphical features of MATLAB to your benefit at all times without losing sight of the objective of yours effort. MATLAB graphics has a lot more features than will be covered in this chapter. The exposure in this chapter should be sufficient for you to confidently explore many other useful graphics commands.

The 3D mesh and surface plots have limited usefulness. These plots can be used to reinforce some of the features found in Figure 2.4. The information in these plots may be improved by choosing a camera angle that emphasizes some aspect of the graphical description. This exploration is left to the reader as an exercise. Using *help view* in the MATLAB command window should get you started in this direction.

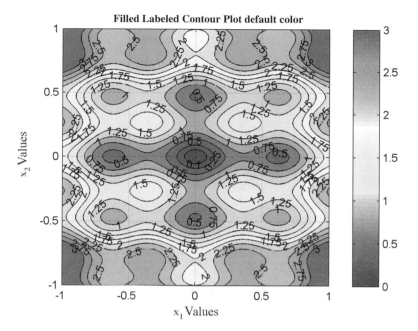

Figure 2.3 Filled Contours with Colorbar: Example 2.2.

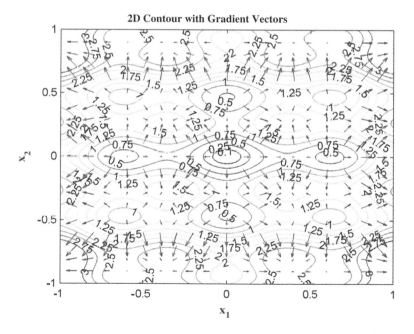

Figure 2.4 Contour with Gradient Vectors: Example 2.2.

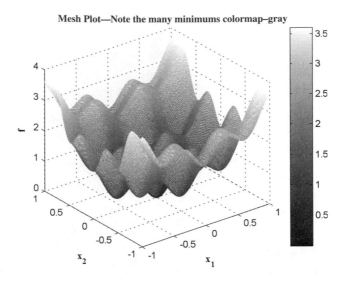

Figure 2.5 Mesh plot with Colorbar: Example 2.2.

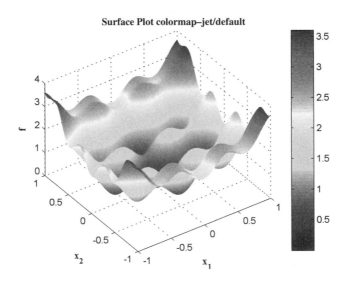

Figure 2.6 3D Surface Plot with Colorbar: Example 2.2.

2.3.2 Example 2.3 — Flagpole Design

The next example is a complex one from structural engineering design that is relevant in civil/mechanical/aerospace engineering applications. It appeared as a problem in another book; it is developed in detail here. The problem is to redesign the basic tall flagpole in view of the increase in wind speeds during extreme weather conditions due to global warming. In recent catastrophic events, the wind speeds in tornadoes have been measured at over 350 miles per hour. High speeds appear to be the norm rather than an unusual event.

Design Problem: Minimize the mass of a standard 10 m tubular flagpole to withstand wind gust of 350 miles per hour. The flagpole will be made of structural steel. Use a factor of safety of 2.5 for the structural design. The deflection of the top of the flagpole should not exceed 5 cm. Additional constraints are developed during the formulation. The problem is described in Figure 2.7.

Mathematical Model: The mathematical model is developed in detail for completeness and to provide a review of useful structural and aerodynamic relations.[5,6] The relations are expressed in original symbols rather than in standard format of optimization problems to provide an insight into problem formulation.

Design Parameters: The structural steel has the following material constants[5]:

E (modulus of elasticity): $200E+09$ Pa

σ_{all} (allowable normal stress): $250E+06$ Pa

ι_{all} (allowable shear stress): $145E+06$ Pa

γ (material density): $7,860 \, kg/m^3$

Additional constants/parameters for calculation:

FS (factor of safety): 2.5

g (gravitational acceleration) $= 9.81 \, m/s^2$

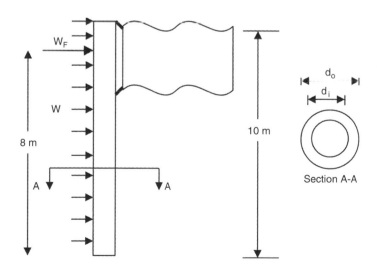

Figure 2.7 Flag Pole Design: Example 2.3.

For the aerodynamic calculations, the following are considered:

ρ (standard air density): $1.225 \, \text{kg/m}^3$
C_d (drag coefficient of cylinder): 1.0
W_F (Flag wind load at 8 m): 5,000 N
V_W (wind speed): 350 mph (156.46 m/s)

The geometric parameters are:

L_p: the location of flag wind load (8 m)
L: length of the pole (10 m)
δ_{all}: permitted deflection (5 cm)

Design Variables: The design variables are shown in Figure 2.7:

d_o: outside diameter (x_1) [Note: x's are not used in the model]
d_i: inside diameter (x_2)

Geometric Relations: The following relations will be useful in later calculations.

A: area of cross-section $= 0.25 * \pi * (d_o^2 - d_i^2)$
I: diametrical moment of inertia $= \pi * (d_o^4 - d_i^4)/64$
Q/t: first moment of area above the neutral axis divided by thickness
$= (d_o^2 + d_o d_i + d_i^2)/6$

Objective Function: The objective function is the weight of the 10 m uniform flagpole:

$$\text{Mass} : f(x_1, x_2) : L * A * \gamma * g \tag{2.10}$$

Constraint Functions: The wind load per unit length (F_D) on the flagpole is calculated as:

$$F_D = 0.5 * \rho * V_W^2 * C_d * d_o$$

The bending moment at the base of the pole due to this uniform wind load on the entire pole is

$$M_W = 0.5 * F_D * L * L$$

The bending moment due to the wind load on the flag is:

$$M_F = W_F * L_p$$

Bending (normal) stress at the base of the pole:

$$\sigma_{\text{bend}} = 0.5 * (M_W + M_F) * d_o/I$$

Normal stress due to the weight is

$$\sigma_{\text{weight}} = \gamma * g * L$$

Total normal stresses to be resisted for design is the sum of the normal stresses computed. Incorporating the factor of safety and the allowable stress from material values, the first inequality constraint can be set up as:

$$g_1(x_1, x_2): \sigma_{\text{bend}} + \sigma_{\text{weight}} \leq \sigma_{\text{all}}/FS \tag{2.11}$$

The maximum shear load in the cross-section is

$$S = W_F + F_D * L$$

The maximum shear stress in the pole is

$$A - S * Q/(I * t)$$

The second inequality constraint based on handling the shear stresses in the flagpole is:

$$g_2(x_1, x_2): \tau \leq \tau_{\text{all}}/FS \tag{2.12}$$

The third practical constraint is based on the deflection of the top of the pole. The deflection of the top of the pole due to a uniform wind load on the pole is

$$\delta_W = F_D * L^4/(8 * E * I)$$

The deflection at the top due to the flag wind load at L_p

$$\delta_F = (2 * W_F * L^3 - W_F * L * L * L_p)/(E * I)$$

The third constraint translates to:

$$g_3(x_1, x_2): \quad \delta_W + \delta_F \leq \delta_{\text{all}} \tag{2.13}$$

To discourage solutions where $d_o < d_i$, we will include a geometric constraint

$$g_4(x_1, x_2): \quad d_o - d_i \geq 0.001 \tag{2.14}$$

Side Constraints: This defines the design region for the search.

$$2 \text{ cm} \leq d_o \leq 100 \text{ cm}; \quad 2 \text{ cm} \leq d_i \leq 100 \text{ cm} \tag{2.15}$$

MATLAB Code: The m-files for this example are given below. An important observation in this problem, and most structural engineering problems in particular, is the order of magnitude of the quantities in the constraining equations. The stress constraints are of the order of 10E+06, while the displacement terms are of the order of 10E-02. Most numerical techniques struggle to handle this range. Typically, such problems need to be normalized before being solved. It is essential in applying numerical techniques used in optimization. The example is plotted using the script file **Example2_3.m**. The design functions are available in independent m files **obj_ex_2_3.m**, **g1_ex2_3.m**, **g2_ex2_3.m**, **g3_ex2_3.m**, and **g4_ex2_3.m**.

Attention is drawn to two important observations in graphically solving this example. First, the code for **drawHashMarks.m**, which is a work in progress, was not very effective for this example because of the steep slopes of the constraints. Instead of hash marks, the code identifies the infeasible region by plotting an additional contour for each constraint, whose value increases constraint limit (right-hand-side value). A thick blue curve indicates this infeasible region.

Second, the design functions, which are calculated using array operations, are further *filtered* using a relational operation to accept values only when the outside diameter is greater than the inside diameter. For example, the objective function over the whole domain is calculated using the following statement:

$$f1 = obj\_ex2\_3(X1,X2).*(X1>X2); \text{ \% objective} \tag{2.16}$$

Here, the operation (X1 > X2) returns true (*value of 1*) if the outside diameter is greater than inside diameter, or false (*value of 0*) if it is not so. The objective function is zero if the condition is not met. This also makes the fourth inequality constraint (g_4) redundant.

```
Example2_3.m  (the main script file)
% Chapter 2: Graphical Optimization
% Example 2.3 Section 2.3.2
%_____

%%%%%%%%%%%%%%%%%%%%%%%%%%%%%%%%%%%%%%%%%%%%%%%%%%%%
% Applied Optimization with Matlab Programming
% Dr. P.Venkataraman
% Second Edition,  John Wiley
%%%%%%%%%%%%%%%%%%%%%%%%%%%%%%%%%%%%%%%%%%%%%%%%%%%%
%_____

% Example 2.3 Design of a flag ploe
%
% Minimize the mass of a standard 10 meter tubular flagpole
% to withstand wind gust of 350 miles per hour.The flagpole
% will be made of structural steel. Use a factor of safety
% of 2.5 for the structural design. The deflection of the
% top of the flagpole should not exceed 5 cm.  Additional
```

```
% constraints are developed during the formulation. The
% problem is described in Figure 2.7.
%
%_____


%%%%%%%%%%%%%%%%%%%%%%%%%%%%%%%%%%%%%%%%%%%%%%%%%%%%%%%%%%%%%%%%%
%%% management functions
clear  % clear all variable/information in the workspace
clc    % position the cursor at the top of the screen
format compact  % no line skip iin the Command Window
warning off  % No warnings like divide by zero etc.
%%%%%%%%%%%%%%%%%%%%%%%%%%%%%%%%%%%%%%%%%%%%%%%%%%%%%%%%%%%%%%%%%


%%%%%%%%%%%%%%%%%%%%%%%%%%%%%%%%%%%%%%%%%%%%%%%%%%%%%%%%%%%%%%%%%
%%%   Problem Data
%_____
% global statementshares same information between
% various m-files, specially parameter values
% This ensures same values in all the files
global ELAS SIGALL TAUALL GAM FS GRAV
global RHO CD FLAGW SPEED LP L DELT
%_____
% Initialize values
ELAS = 200e+09;          % Pa
SIGALL = 250E+06;        % Pa
TAUALL = 145e+06;        %Pa
GAM = 7860;              % kg/m3
FS = 2.5;                % factor of safety
GRAV = 9.81;             % gravitational acceleration
%_____
RHO = 1.225;               % kg/m3
CD = 1.0  ;                % drag coefficient
FLAGW = 5000;              % N
SPEED = 156.46 ;           % m/s
%_____
LP = 8;                    % m
L = 10;                    % m
DELT = 0.05;               % m
%_____
g1val = SIGALL/FS; % right hand value for g1
g2val = TAUALL/FS; % right hand value for g2
g3val = DELT;      % right hand value for g3
g4val = 0.001;     % right hand value for g4
%_____
%%%   establish the domain (SIDE CONSTRAINTS)
```

```
% x1 and x2 are vectors filled with numbers starting
%

x1=0.02:0.005:1;    % don't echo information to the screen
x2=0.02:0.005:1; % a way to avoid x1 = x2

%%%%%%%%%%%%%%%%%%%%%%%%%%%%%%%%%%%%%%%%%%%%%%%%%%%%%%%%%%%%%
% generates matrices X1 and X2 (matrices)
% corresponding to x1 and x2 (vectors)

% these are the x1 and x2 values at all the intersections
% of x and y values of the vectors x1 and x2
% this is a mesh of x1 and x2 values

[X1 X2] = meshgrid(x1,x2);


%%%%%%%%%%%%%%%%%%%%%%%%%%%%%%%%%%%%%%%%%%%%%%%%%%%%%%%%%
%%%  Calculate Functions

f1 = obj_ex2_3(X1,X2).*(X1>X2);         % objective

%%%  NOTE WE ARE ONLY USING FUNCTION VALUES
%%% WHERE X1 > X2
%%% THAT IS: OUTSIDE DIAMETER IS GREATER
%%% THAN INSIDE DIAMETER

ineq1 = g1_ex2_3(X1,X2).*(X1>X2);    % g1

ineq2 = g2_ex2_3(X1,X2).*(X1>X2);    % g2

ineq3 = g3_ex2_3(X1,X2).*(X1>X2);    % g3

%%% because we are only working with values
%%% where x1 > x2 g4 constraint is unnecessary
%%% not plotted
ineq4 = g4_ex2_3(X1,X2);             % g4
%%%%%%%%%%%%%%%%%%%%%%%%%%%%%%%%%%%%%%%%%%%%%%%%%%%%%%%%%
%%%  draw contours of the problem

%%% draw multiple contours of the objective function
[C1,h1] = contour(x1,x2,f1,[10000,50000,100000, ...
          150000, 200000, 250000, 300000],'g-');

clabel(C1,h1);  %  label contours
```

```
set(gca,'xtick',[0 0.2 0.4 0.6 0.8 1.0]) % xtick values
set(gca,'ytick',[0 0.2 0.4 0.6 0.8 1.0]) % ytick values
xlabel('outside diameter','FontName','times','FontSize',12);
ylabel('inside diameter','FontName','times','FontSize',12)
grid
hold on   % there will be addtional plots on this figure


%_____
%%%  draw a single contour of the inequality constrint g1
[C2,h2] = contour(x1,x2,ineq1,[g1val,g1val],'r:');
set(h2,'LineWidth',2);
%clabel(C2,h2);  % avoid label for clarity


% the hash marks program does not work because
% of the steep contours
%[XL2 YL2] = drawHashMarks(C2,'b');
%hl2 = line(XL2,YL2,'Color','k','LineWidth',1);


%
% interactively place q1 on plot
k1 = gtext('g1');
set(k1,'FontName','Times','FontWeight','bold', ...
    'FontSize',14,'Color','red')

% draw the infeasible region for this constraint
%  using thich blue color
[C21,h21]=contour(x1,x2,ineq1,[1.1*g1val,01.1*g1val],'b-');

set(h21,'LineWidth',3);
%_____


%_____
%%%  draw a single contour of the inequality constrint g2

[C3,h3] = contour(x1,x2,ineq2,[g2val,g2val],'r:');
set(h3,'LineWidth',2);
%clabel(C3,h3);  % avoid label for clarity


% the hash marks program does not work because
% %ot enough points
% [XL3 YL3] = drawHashMarks(C3,'b');
% hl3 = line(XL3,YL3,'Color','k','LineWidth',1);


%
% interactively place g2 on plot
```

```
k2 = gtext('g2');
set(k2,'FontName','Times','FontWeight','bold', ...
    'FontSize',14,'Color','red')

% draw the infeasible region for this constraint
%  using thich blue color
[C31,h31] = contour(x1,x2,ineq2,[1.1*g2val,1.1*g2val],'b-');
set(h31,'LineWidth',3);
%_____


%_____
%%%  draw a single contour of the inequality constrint g3

[C4,h4] = contour(x1,x2,ineq3,[g3val,g3val],'r:');
set(h4,'LineWidth',2);
%clabel(C4,h4);  % avoid label for clarity

% the hash marks program does not work because
% of the steep contours
% [XL4 YL4] = drawHashMarks(C4,'b');
% h14 = line(XL4,YL4,'Color','k','LineWidth',1);

% interactively place g3 on plot
k3 = gtext('g3');
set(k3,'FontName','Times','FontWeight','bold', ...
    'FontSize',14,'Color','red')
[C41,h41] = contour(x1,x2,ineq3,[1.1*g3val,1.1*g3val],'b-');
set(h41,'LineWidth',3);
%_____


%_____
%  unnecessary to draw g4 constraint
% [C5,h5] = contour(x1,x2,ineq4,[g4val,g4val],'r-');
% %clabel(C4,h4);  % avoid label for clarity
% set(h5,'LineWidth',3);

% [XL5 YL5] = drawHashMarks(C5,'t');
% h15 = line(XL5,YL5,'Color','k','LineWidth',1);

%
% % interactively place g4 on plot
% k4 = gtext('g4');
% set(k4,'FontName','Times','FontWeight','bold', ...
```

```
%       'FontSize',14,'Color','red')
%
% the equality and inequality constraints are not written
% with 0 on the right hand side. If you do write them
% that way
% you would have to include [0,0] in the contour commands
%
%
hold off
%
% %%%%%%%%%%%%%%%%%%%%%%%%%%%%%%%%%%%%%%%%%%%%%%%%%%%%%%%%%%%%%
% %%%   print values for Problem constants
%
fprintf('\n Graphical Optimization of the Flag Pole')
fprintf('\n_____')
fprintf('\n Elasticity       [Pa] =   %9.3e  ',ELAS)
fprintf('\n Sigma allowable [Pa]     %9.3c  ',SIGALL)
fprintf('\n Gam           [Pkg/m^3]  %9.3f  ',SIGALL)
fprintf('\n Factor of Safety         %9.5f  ',FS)
fprintf('\n Gravitation  [m/s^2]     %9.5f  ',GRAV)
fprintf('\n Flag Drag at      [m]    %9.5f  ',LP)
fprintf('\n Length of Pole   [m]     %9.5f  ',L)
fprintf('\n Air density [kg/m^3]     %9.5f  ',RHO)
fprintf('\n CD of Pole               %9.5f  ',CD)
fprintf('\n Wind Load        [N]     %9.5f  ',FLAGW)
fprintf('\n Wind Speed     [m/s]     %9.5f \n\n ',SPEED)

fprintf('\n Right hand side for g1 [Pa] %9.5f  ',g1val)
fprintf('\n Right hand side for g2 [Pa] %9.5f  ',g2val)
fprintf('\n Right hand side for g3  [m] %9.5f  ',g3val)
fprintf('\n Right hand side for g4  [m] %9.5f\n\n',g4val)
%
```

obj_ex2_3.m (the objective function)

```
function retval = obj_ex2_3(X1,X2)
% global statement is used to share info between
% various m-files
% objective function
global ELAS SIGALL TAUALL GAM FS GRAV
global RHO CD FLAGW SPEED LP L DELT
```

```
AREA = 0.25* pi*(X1.^2-X2.^2);
retval = L * AREA * GAM * GRAV;
```

g1_ex2_3.m   (the first constraint)

```
function retval = ineq1_ex3(X1,X2)
% global statement is used to share same info between
% various m-files
% normal stress constraint
global ELAS SIGALL TAUALL GAM FS GRAV
global RHO CD FLAGW SPEED LP L DELT

AREA = 0.25* pi*(X1.^2-X2.^2); % matrix
INERTIA = pi*(X1.^4-X2.^4)/64; % matrix
FD = 0.5*RHO*SPEED*SPEED*CD*X1;
MW = 0.25*FD*L*L.*X1./INERTIA;
MF = 0.5*FLAGW * LP*X1./INERTIA;

SIGW = GAM*GRAV*L;

retval = MW + MF + SIGW;
% SIGW is added to all matrix elements
```

g2_ex2_3.m   (the second constraint)

```
function retval = ineq2_ex3(X1,X2)
% global statement is used to share info between
% various m-files
% shear stress constraint
global ELAS SIGALL TAUALL GAM FS GRAV
global RHO CD FLAGW SPEED LP L DELT

AREA = 0.25* pi*(X1.^2-X2.^2);
INERTIA = pi*(X1.^4-X2.^4)/64;
FD = 0.5*RHO*SPEED*SPEED*CD*X1;
S = FLAGW + (FD * L);
Q = (X1.*X1 + X1.*X2 + X2.*X2)/6.0;
retval = S.*Q./INERTIA;
```

g3_ex2_3.m   (the third constraint)

```
function retval = ineq3_ex3(X1,X2)
% global statement is used to share info between
```

```
% various m-files
% displacement constraint
global ELAS SIGALL TAUALL GAM FS GRAV
global RHO CD FLAGW SPEED LP L DELT

AREA = 0.25* pi*(X1.^2-X2.^2);
INERTIA = pi*(X1.^4-X2.^4)/64;
FD = 0.5*RHO*SPEED*SPEED*CD*X1;
dw = FD*L^4./(8*ELAS*INERTIA);
df = (2.0*FLAGW*L^3-FLAGW*L*L*LP)./(ELAS*INERTIA);

retval = dw + df;
```

g4_ex2_3.m   (the fourth constraint)

```
function retval = ineq4_ex3(X1,X2)
% design acceptability constraint
retval = X1-X2;
```

Figure 2.8 displays the graphical solution to the problem. The optimal solution is not very clear. Looking at the figure, an intriguing point is in the neighborhood of 0.6. Figure 2.9 is obtained by zooming in near the neighborhood of the design variable at the value of 0.6. You can use the zoom button in the Figure window. See *help zoom* for instruction on its use. It can also be achieved by typing *zoom* at the workspace prompt and using the mouse to drag a rectangle around the region that needs to be enlarged. The zoomed figure, once again, is not very clear about the solution. There appears to be a solution at the intersection of the constraint g_1 and g_3. The graphical way to resolve the solution is to plot the solution in the neighborhood of the minimum. Run the original code with the variation in the design variables between 0.6 and 0.75. It would also be helpful to reduce the **LineWidth** to 1. A visual estimate of the solution in this region is an outside diameter of 0.68 m and an inside diameter of 0.65 m. This may not be the best solution, but is a solution that is feasible with two active constraints.

The graphics in the example were created using the same statements encountered in earlier examples. Color contours were used to establish the feasible region. The zoom feature was employed to obtain a better estimate of the solution.

2.3.3   Example 2.4 — Fin Design for Heat Transfer Application

This example is from the area of heat transfer. The problem is to design a triangular fin of the smallest volume that will at least deliver specified fin efficiencies. The graphical feature of this code is very similar to Example 2.3. In this example, the inequality constraints are computed and returned from a single-function m-file rather than from separate files considered in the previous example. Another new feature in this example is to invoke special mathematical functions, the Bessel

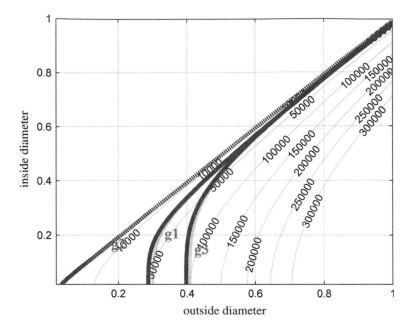

Figure 2.8   Graphical solution: Example 2.3.

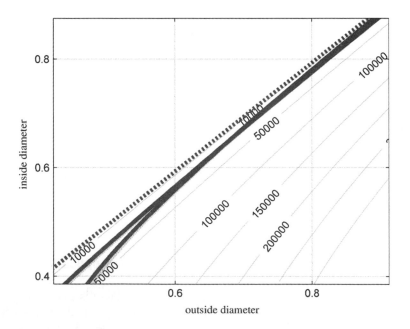

Figure 2.9   Graphical solution (zoomed): Example 2.3.

functions that are available in MATLAB. This example also illustrates a problem where the optimization problem can be adequately defined but the solution is easily determined from the equality and side constraints. This means that the problem can easily accommodate additional demanding constraints. In large, complex mathematical models with many design variables, it is not easy to ensure at least one of the inequality constraints are active.

Design Problem: Minimize the amount of material used in the design of a series of identical triangular fins that cover a given area and operate at or above the specified efficiencies.

Mathematical Model: This type of problem should be able to accommodate several design variables. In this section, the problem is set up to have two design variables. The fin material is aluminum.

Design Parameters: For aluminum

$h - 50$ W/m^2 [convection coefficient]
$k = 177$ W/m $-°$ K [thermal conductivity]
$N = 20$ [Number of fins]
$W = 0.1$ m [width of the fins]

Fin gap is the same as the base length of the triangular fin.

Design Variables:

b: base of the triangular fin
L: height of the triangular fin

Figure 2.10 illustrates the geometry of the fins.

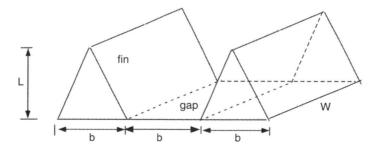

Figure 2.10   Fin Design for Heat Transfer: Example 2.4.

Geometric Relations: The following are some of the area calculations that are used later in the development of the constraints.

$$A_f = (2N - 1)*b*W$$ footprint of the fin and gap
$$A_c = 2*W*[L^2 + (b/2)^2]^{1/2}$$ fin area for heat transfer
$$A_b = (N - 1)*b*W$$ gap area
$$A_t = N*A_c + A_b$$ total area for heat transfer

Objective Function: The total volume of material for the fin is

$$f(b, L) = 0.5*N*W*b*L \tag{2.17}$$

Constraint Functions: The heat transfer equations for the fin are available in[7]. Fins are typically mounted on a specified area. The first constraint is one on area.

$$h(b, L): A_f = 0.015m^2 \tag{2.18}$$

Note: This constraint essentially fixes the value of b. In the graph, this constraint will be a straight line parallel to the L axis. If this were to happen in a problem with several design variables, it would be prudent to eliminate this variable from the mathematical model by identifying it as a design parameter.

The efficiency of a single fin can be established as

$$\eta_f = (1/mL)I_1(2mL)/I_0(2mL)$$

where

$$m = (2h/kL)^{1/2}$$

and I's are Bessel Equations of the *First Kind*. The first inequality constraint is

$$g_1(b, L): \eta_f \geq 0.95 \tag{2.19a}$$

The overall efficiency for the heat transfer is:

$$\eta_o = 1 - N*(A_f/A_t)(1 - \eta_f)$$

The final inequality constraint is:

$$g_2(b, L): \eta_o \geq 0.94 \tag{2.19b}$$

Side Constraints: The side constraints are

$$0.001 \leq b \leq 0.005 \tag{2.20}$$

$$0.01 \leq L \leq 0.03$$

MATLAB Code: Figure 2.11 displays the graphical solution to the problem. The *drawHashMarks* function is used in this example. However, the region for the

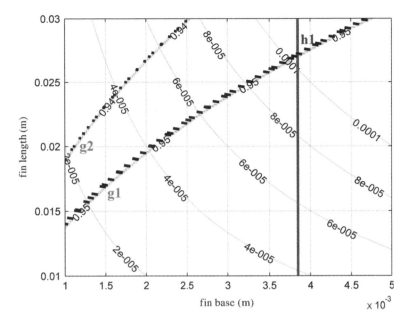

Figure 2.11   Graphical Solution: Example 2.4.

hash marks must be specified as *b* even as the marks are being drawn on the top. This was discovered after initially identifying the infeasible region as in the previous example. The code included below (**Example2_4.m**) also includes drawing the solid blue line in the infeasible region as in the previous example but is commented.

Example2_4.m

```
% Chapter 2: Graphical Optimization
% Example 2.4 Section 2.3.3
%_____
%%%%%%%%%%%%%%%%%%%%%%%%%%%%%%%%%%%%%%%%%%%%%%%%%%%%%%%
% Applied Optimization with Matlab Programming
% Dr. P.Venkataraman
% Second Edition,  John Wiley
%%%%%%%%%%%%%%%%%%%%%%%%%%%%%%%%%%%%%%%%%%%%%%%%%%%%%%%
%_____
% Example 2.4 Design of a triangular fin
%
% Minimize the amount of material used in the design of
% a series of  identical triangular fins that cover
% a given area and operate at or above the specified
% efficiencies.
%_____
```

```
%%%%%%%%%%%%%%%%%%%%%%%%%%%%%%%%%%%%%%%%%%%%%%%%%%%%%%%%%%%%%%%
%%% management functions
clear  % clear all variable/information in the workspace
clc    % position the cursor at the top of the screen
format compact  % no line skip iin the Command Window
warning off  % No warnings like divide by zero etc.
%%%%%%%%%%%%%%%%%%%%%%%%%%%%%%%%%%%%%%%%%%%%%%%%%%%%%%%%%%%%%%%

%%%%%%%%%%%%%%%%%%%%%%%%%%%%%%%%%%%%%%%%%%%%%%%%%%%%%%%%%%%%%%%
%%%  Problem Data
%_____
% global statementshares same information between
% various m files, specially parameter values
% This ensures same values in all the files
global N H K W AREA
%_____

% Initialize values
N = 20;                      % number of fins
W = 0.1;                         % width of fins
H = 50.0;   % convection coefficient W/m*m
K = 177.0;  % thermal conductivity W/m-K
AREA = 0.015; % available fin foot print area
%_____
% right hand limits for the functions
h1val = AREA;  % equality constraint h1
g1val = 0.95;   % g1
g2val = 0.94;   % g2
%_____
%%%  establish the domain (SIDE CONSTRAINTS)
% x1 and x2 are vectors filled with numbers starting
%
x1=0.001:0.0001:0.005;
x2=0.01:0.001:0.03;% x1 and x2 are vectors

[X1 X2] = meshgrid(x1,x2);
% generates matrices X1 and X2 corresponding to
% vectors x1 and x2

%%%%%%%%%%%%%%%%%%%%%%%%%%%%%%%%%%%%%%%%%%%%%%%%%%%%%%%%%%%%%%%
%%%  Calculate Functions
f1 = obj_ex2_4(X1,X2);  % objective function

% Constraints are evaluated
eq1 = h1_ex2_4(X1,X2);% h1
```

```
% returns both g1 and g2
[ineq1, ineq2]  = g_ex2_4(X1,X2); % g1 and g2

%%%%%%%%%%%%%%%%%%%%%%%%%%%%%%%%%%%%%%%%%%%%%%%%%%%%%%%%%%%%%
%%%  draw contours of the problem

%%% draw multiple contours of the objective function

[C1,h1] = contour(x1,x2,f1,[0.00001, 0.00002, 0.00004, ...
     0.00006, 0.00008, 0.0001],'g-');
clabel(C1,h1);
set(gca,'xtick',[0.001 0.0015 0.002 0.0025 0.003 0.0035 ...
     0.004 0.0045 0.005])
set(gca,'ytick',[0.010 0.015 0.02 0.025 0.03 0.125]);
xlabel('fin base (m)','FontName','times','FontSize',12);
ylabel('fin length (m)','FontName','times','FontSize',12)
grid
hold on

%_____
%%%  draw a single contour of the equality constraint h1

[C2,h2] = contour(x1,x2,eq1,[h1val,h1val],'b-');
set(h2,'LineWidth',2);

k1 = gtext('h1');
set(k1,'FontName','Times','FontWeight','bold', ...
    'FontSize',14,'Color','blue')

%_____

%_____
%%%  draw a single contour of the inequality constraint g1

[C3,h3] = contour(x1,x2,ineq1,[g1val,g1val],'r:');
set(h3,'LineWidth',2);
clabel(C3,h3);  % avoid label for clarity

%the hash marks program works only
% if 'b' is used
[XL3 YL3] = drawHashMarks(C3,'b');
hl3 = line(XL3,YL3,'Color','k','LineWidth',2);

%
% interactively place g2 on plot
```

```
k2 = gtext('g1');
set(k2,'FontName','Times','FontWeight','bold', ...
    'FontSize',14,'Color','red')
% draw the infeasible region for this constraint
%  using thich blue color
% [C31,h31] = contour(x1,x2,ineq1, ...
%      [0.999*g1val,0.999*g1val],'b-');
% set(h31,'LineWidth',3);
%_____


%_____
%%%  draw a single contour of the inequality constrint g2

[C4,h4] = contour(x1,x2,ineq2,[g2val,g2val],'r:');
set(h4,'LineWidth',2);
clabel(C4,h4);  % avoid label for clarity

%the hash marks program works only
% if 'b' is used
[XL4 YL4] = drawHashMarks(C4,'b');
h14 = line(XL4,YL4,'Color','k','LineWidth',2);

%
% interactively place g2 on plot
k3 = gtext('g2');
set(k3,'FontName','Times','FontWeight','bold', ...
    'FontSize',14,'Color','red')
% draw the infeasible region for this constraint
%  using thich blue color
% [C41,h41] = contour(x1,x2,ineq2, ...
%      [0.999*g2val,0.999*g2val],'b-');
% set(h41,'LineWidth',3);

%_____
hold off

% %%%%%%%%%%%%%%%%%%%%%%%%%%%%%%%%%%%%%%%%%%%%%%%%%%%%%%%%%%%%%%%%
% %%%  print values for Problem constants
%

fprintf('\n Graphical Optimization of Triangular Fin')
fprintf('\n _____')
fprintf('\n Convection Coefficient   [W/m^2] =   %9.3e ',H)
```

```
fprintf('\n Thermal Conductivity    [W/m-K]    %9.3e  ',K)
fprintf('\n Number of fins           [    ]     %4.2i  ',N)
fprintf('\n Fin Width           [m]    %9.5f ',W)
fprintf('\n Area of base of fin    [m^2] %9.5f ',AREA)

fprintf('\n Right hand side for h1 [m^2] %9.5f ',h1val)
fprintf('\n Right hand side for g1 []    %9.5f ',g1val)
fprintf('\n Right hand side for g2 []    %9.5f\n\n ',g2val)
```

obj_ex2_4.m   The objective function

```
function retval = obj_ex2_4(X1,X2)
% volume of the fin
global N H K W AREA

 retval = 0.5*N*W*X1.*X2
```

h1_ex2_4.m   The equality constraint

```
function retval = h1_ex2_4(X1,X2)
% the equality constraint on area
global N H K W AREA

retval = (2.0*N-1)*W*X1;
```

g_ex2_4.m   The inequality constraints

```
function [ret1, ret2] = g_ex2_4(X1,X2)
% returns both the inequality constraints
global N H K W AREA

%  evaluate 2*m*L
c = 2*sqrt(2.0)*sqrt(H/K)*X2./sqrt(X1);

% value of g1
ret1 = (besseli(1,c)./((0.5*c).*besseli(0,c)));

Ac = 2.0*W*sqrt((X2.*X2 + .25*X1.*X1));
Ab = (N-1)*W*X2;
At = N*Ac + Ab;
Ar = Ac./At;
```

```
% value of g2
ret2 = (1.0-N *Ar.*(1-ret1));
```

These files, with user interaction to identify constraint types, should create Figure 2.11. Can you identify the solution? It is at the intersection of h_1 and g_1.

2.3.4 Example 2.5 — Shipping Container with Three Design Variables

This example will explore some creative ways to employ MATLAB graphics to determine the solution to a constrained three design variable problem. Graphically describing functions of three variables is quite challenging, even in MATLAB. It requires special views of the plots to communicate the information, in the problem and will be different for different problems. Trying to establish the optimum, with constraint information makes it even more challenging. The example has simple functions so it will be easy to understand the plots.

Design Problem: Today's concern over the waste, recycling, and the environment has manufacturers trying to adopt new packaging materials to deliver their products. One such case involves using biodegradable cartons made from recycled materials. The cost of the material is based on the surface area of the rectangular container that will house the product. The cost per unit area is $1.5 per square meter. The different products can be accommodated by a single container. The container must hold a volume of 0.032 m³. The perimeter of the base must be less than or equal to 1.5 m. Its sides are scaled geometrically to hold information labels. The width should not exceed three times the length. Its height must be less than two thirds the width. Its length and width are less than 0.5 m. Find the container of minimum cost.

Mathematical Model: For this model, we have a single nonlinear objective function, a nonlinear equality constraint, three linear inequality constraints, and two side constraints. The design variables are shown in Figure 2.12.

Design Parameters:

Cost = $1.5/m² [Material cost]]
Vol = 0.032 m³ [Volume constraint]
Per = 1.5 m [Perimeter constraint]

Design Variables:

a: length of container
b: width of container
c: height of container

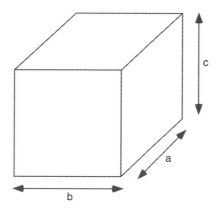

Figure 2.12 Example 2.5 — problem definition.

Objective Function: The surface area of each side can be easily calculated. The objective function is

$$\text{Minimize } f(a, b, c) = 2^{*}\text{Cost}^{*}(a^{*}b + b^{*}c + c^{*}a) \qquad (2.21)$$

Constraint Functions: The volume constraint can be easily written as an equality constraint:

$$h(a, b, c): a^{*}b^{*}c = \text{Vol} \qquad (2.22)$$

The perimeter inequality constraint is expressed as

$$g_1(a, b, c): 2^{*}(a + b) \leq \text{Per} \qquad (2.23a)$$

The geometric constraints on the sides is set up as

$$g_2(a, b, c): b - 3^{*}a \leq 0 \qquad (2.23b)$$

$$g_3(a, b, c): c - (2/3)^{*}b \leq 0 \qquad (2.23b)$$

Side Constraints: The side constraints are

$$0 \leq a \leq 0.5 \qquad (2.24a)$$

$$0 \leq b < 0.5 \qquad (2.24b)$$

MATLAB Code: Figure 2.13 displays the graphical solution to the problem. The code is available in **Example2_5.m**. No function files are used for this example. We will use $x = a$, $y = b$, and $z = c$ so that we can relate to the plots. Instead of contours, we will be drawing *iso-surfaces* — points having the same value of the function in three-dimensional space. The version used at the time of writing this code allowed only one iso-surface in each plotting instance. Additionally, using a single color did not make the surfaces distinct. It was necessary to use an edge color for better graphical definition of the surface. In order to make the equality constraint distinct, the magenta color was chosen for identifying the infeasible region for the inequality constraint. The actual constraint is in red. The figure in the book is annotated to identify the constraints from the infeasible region. A parallel surface is used to define the infeasible region.

Example2_5.m
```
% Chapter 2: Graphical Optimization
% Example 2.5 Section 2.3.4
%_____
%%%%%%%%%%%%%%%%%%%%%%%%%%%%%%%%%%%%%%%%%%%%%%%%%%%%%%%
% Applied Optimization with Matlab Programming
% Dr. P.Venkataraman
% Second Edition,  John Wiley
%%%%%%%%%%%%%%%%%%%%%%%%%%%%%%%%%%%%%%%%%%%%%%%%%%%%%%%
%_____
% Example 2.5 - Shipping Container with
%                  Three Design Variables
%
% Today's concern over the waste, recycling, and
% the environment has manufacturers trying to adopt
% new packaging materials to deliver their products.
% One such case involves using bio-degradable
% cartons made from recycled materials.  The cost
% of the material is based on the surface area of
% the rectangular container that will house the product.
% The cost per unit area is $ 1.5 per square meter.
% The different products can be accommodated by a
% single container. The container must hold a
% volume of 0.032 m3.  The perimeter of the base
% must be less than or equal to 1.5 m.  Its sides
% are scaled geometrically to hold information
% labels. The width should not exceed three
% times the length.  Its height must be less
% than two thirds the width.  Its length and
% width are less than 0.5 m.  Find the container
% of minimum cost.
%_____
%%%%%%%%%%%%%%%%%%%%%%%%%%%%%%%%%%%%%%%%%%%%%%%%%%%%%%%%%%%%%%%
%%% management functions
clear  % clear all variable/information in the workspace
clc    % position the cursor at the top of the screen
format compact  % no line skip iin the Command Window
warning off  % No warnings like divide by zero etc.
%%%%%%%%%%%%%%%%%%%%%%%%%%%%%%%%%%%%%%%%%%%%%%%%%%%%%%%%%%%%%%%
```

```
%%%%%%%%%%%%%%%%%%%%%%%%%%%%%%%%%%%%%%%%%%%%%%%%%%%%%%%%%%%
%%%   Problem Data
%_____
% global statementshares same information between
% various m-files, specially parameter values
% This ensures same values in all the files
global k1 k2 k3 k4 k5
%_____


% Parameter values
k1 = 0.032;     % volume constraint (Vol)
k2 = 1.5;          % perimeter constraint (Per)
k3 = 0.5;       % upper limit on a
k4 = 0.5;       % upper limit on b
k5 = 1.5;       % cost per area (Cost)


% By parameterizing the problem constants
% you can design another container
% just by changing the k values
% power of programming


%_____
%%%   establish the domain (SIDE CONSTRAINTS)
% x, y, z are vectors
%
x = linspace(0,k3,50);
y = linspace(0,k4,50);
z = linspace(0,k3,50);

[X, Y, Z] = meshgrid(x,y,z);
% create a volume mesh in 3D where
% functions will be evaluated
% you can check size of the
% volumes and functions

%%%%%%%%%%%%%%%%%%%%%%%%%%%%%%%%%%%%%%%%%%%%%%%%%%%%%%%%%%%
%%%   Calculate Functions-no function m-files used
%%% functions are set up with zero right hand side

obj = 2*(X.*Y + Y.*Z + Z.*X);% objective function
h = X.*Y.*Z -k1;             % equality constraint
g1 = 2*(X + Y )-k2;   % ineaquality constraint g1
g2 = Y-3*X;           % inequality constraint g2
g3 = 3*Z-2*Y;         % inequlaity constraint g3
```

```
%%%%%%%%%%%%%%%%%%%%%%%%%%%%%%%%%%%%%%%%%%%%%%%%%%%%%%%%%%%%%%
%%% draw Iso-surfaces of the problem
%%% draw multiple contours of the objective function
minobj = min(min(min(obj)));
maxobj = max(max(max(obj)));
isoobj = linspace(minobj,maxobj,10);

for i = 1:length(isoobj)
    isoobjval = isoobj(i);
    % MATLAB allows only one isosurface
    % we will draw 10 between minimum value
    % of objective and maximum value
    p1 = patch(isosurface(x,y,z,obj,isoobjval), ...
        'FaceColor','g','EdgeColor','y');
    % without edge color the surfaces
    % do not stand out
    axis tight  % axis are set to data extant
    set(gcf,'Renderer','zbuffer');
    lighting phong

    if(1 == 1)
        hold on
    end
end
ylabel('y', ..
    'FontWeight','b','Color','b', ...
    'VerticalAlignment','bottom');
xlabel('x', ..
    'FontWeight','b','Color','b', ...
    'VerticalAlignment','bottom');
zlabel('z', ..
    'FontWeight','b','Color','b', ...
    'VerticalAlignment','bottom');

% Please read the title.
% It is used for communicating information
title({'\bf\itIso-surfaces of objective function',
    '\bf\itBetween minimum and maximum value',
    '\bf\itLowest value near (0,0,0)'});

set(gca,'Color','white','XColor','k', ...
        'YColor','k','ZColor','k');
view(37,30)
grid
pause(5) %hold for 5 seconds
```

```
hold on
%_____
%%% draw a single iso-surface of  h
p2 = patch(isosurface(x,y,z,h,0), ...
    'FaceColor','b','EdgeColor','none');
title({'\bf\itIso-surface of Equality constraint',
    '\bf\itSolution must lie on this surface',
    '\bf\itIso-surface value is 0'});
pause(5)
%_____
%_____
%%% draw Iso-surface of g1
% actual constraint
p3 = patch(isosurface(x,y,z,g1,0), ...
    'FaceColor','r','EdgeColor','none');
title({'\bf\itIso-surface of q1',
    '\bf\itIso-surface value is 0'});
pause(5);

% infeasible region
p4 - patch(isosurface(x,y,z,g1,0.1), ...
    'FaceColor','m','EdgeColor','none');
title({'\bf\itInfeasible region of g1',
    '\bf\itIso-surface value is 0.1'});
pause(3)
%_____
%_____
%%% draw Iso-surface of  g2
% actual constraint
p5 = patch(isosurface(x,y,z,g2,0), ...
    'FaceColor','r','EdgeColor','none');
title({'\bf\itIso-surface of g2',
    '\bf\itIso surface value is 0'});
pause(5);

% infeasible region
p6 = patch(isosurface(x,y,z,g2,0.1), ...
    'FaceColor','m','EdgeColor','none');
title({'\bf\itInfeasible region of g2',
    '\bf\itIs-osurface value is 0.1'});

pause(3)
%_____
%_____
%%% draw Iso-surface of  g3
```

```
% actual constraint
p7 = patch(isosurface(x,y,z,g3,0), ...
    'FaceColor','r','EdgeColor','none');
title({'\bf\itIso-surface of g3',
    '\bf\itIso-surface value is 0'});
pause(5)

% infeasible region
p8 = patch(isosurface(x,y,z,g3,0.1), ...
    'FaceColor','m','EdgeColor','none');
title({'\bf\itInfeasible region of g3',
    '\bf\itIso-surface value is 0.1'});
pause(3);


title({'\bf\itGraphical Solution Example 2.5',
    '\bf\itSolution appears to be decided by ',
    '\bf\ith, g3, and f '});


%_____
hold off


% %%%%%%%%%%%%%%%%%%%%%%%%%%%%%%%%%%%%%%%%%%%%%%%%%%%%%%%%%%%%%%%
% %%%   print values for Problem constants

fprintf('\n Example 2.5Graphical Optimization of Container')
fprintf('\n_____')
fprintf('\n Volume constraint   [m^3]  =   %9.3f ',k1)
fprintf('\n Perimeter constraint [m]   =   %9.3f ',k2)
fprintf('\n Cost              [ $/m^2]  =   %9.3f ',k5)
fprintf('\n side constraint (x)[m]     =   %9.5f ',k3)
fprintf('\n side constraint (y)[m]     =   %9.5f\n\n ',k4)
```

Figure 2.13 gives you only an idea for the solution. It is difficult to establish the actual solution at $x = 0.3132$ m, $y = 0.3915$ m, and $z = 0.2610$ m, with an objective function value of \$0.9196. The equality constraint and the inequality constraint g_3 are active. It is possible to rotate the figure and see if a different view yields a better indication of the solution. To rotate, click on the rotate button under the menubar and hold down the left mouse button and drag. It is still difficult to determine the optimum solution. Let us see if two-dimensional representation of the design space helps better understand the solution. We do that in the next script file

MATLAB Code: Another way of graphically solving the problem is to one vary one of the variables (say z) and look at the contours generated in the plane of the other variables (*x-y* plane). This requires user intervention to understand the plots

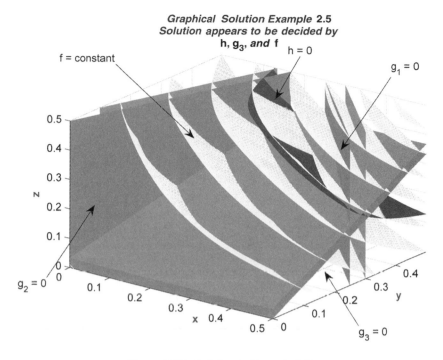

Figure 2.13 Solution to Example 2.5.

and locate the solution. This will also require examination of a large number of two-dimensional plots. The solution most likely will be determined by the pitch of the variation of the z variable. In the following code, the problem is explored by looking at 51 plots with z varying by 0.1. One can go to a finer scale in an adaptive fashion by adjusting the range of z based on the observation of the plots.

Example2_5_2D.m
```
% Chapter 2: Graphical Optimization
% Example 2.5 Section 2.3.4
% using contour plots
%_____
%%%%%%%%%%%%%%%%%%%%%%%%%%%%%%%%%%%%%%%%%%%%%%%%%%%%%%%%
% Applied Optimization with Matlab Programming
% Dr. P.Venkataraman
% Second Edition,  John Wiley
%%%%%%%%%%%%%%%%%%%%%%%%%%%%%%%%%%%%%%%%%%%%%%%%%%%%%%%%
%_____
% Example 2.5 - Shipping Container with
%               Three Design Variables
%
% Today's concern over the waste, recycling, and
```

```
% the environment has manufacturers trying to adopt
% new packaging materials to deliver their products.
% One such case involves using bio-degradable
% cartons made from recycled materials.  The cost
% of the material is based on the surface area of
% the rectangular container that will house the product.
% The cost per unit area is $ 1.5 per square meter.
% The different products can be accommodated by a
% single container. The container must hold a
% volume of 0.032 m3.  The perimeter of the base
% must be less than or equal to 2 m.  Its sides
% are scaled geometrically to hold information
% labels. The width should not exceed three
% times the length.  Its height must be less
% than two thirds the width.  Its length and
% width are less than 0.5 m.  Find the container
% of minimum cost.
%_____


%%%%%%%%%%%%%%%%%%%%%%%%%%%%%%%%%%%%%%%%%%%%%%%%%%%%%%%%%%
%%% management functions
clear  % clear all variable/information in the workspace
clc    % position the cursor at the top of the screen
format compact  % no line skip iin the Command Window
warning off  % No warnings like divide by zero etc.
%%%%%%%%%%%%%%%%%%%%%%%%%%%%%%%%%%%%%%%%%%%%%%%%%%%%%%%%%%
%%%%%%%%%%%%%%%%%%%%%%%%%%%%%%%%%%%%%%%%%%%%%%%%%%%%%%%%%%
%%%  Problem Data

% Parameter values
k1 = 0.032;    % volume constraint (Vol)
k2 = 1.5;      % perimeter constraint (Per)
k3 = 0.5;      % upper limit on a
k4 = 0.5;      % upper limit on b
k5 = 1.5;      % cost per area (Cost)

% By parameterizing the problem constants
% you can design another container
% just by changing the k values
% power of programming
%_____
%%%  establish the domain (SIDE CONSTRAINTS)
% x, y, z are vectors
%
x = linspace(0,k3,50);
```

```
y = linspace(0,k4,50);
z = linspace(0,k3,51);


[X, Y, Z] = meshgrid(x,y,z);
% create a volume mesh in 3D where
% functions will be evaluated
% you can check size of the
% volumes and functions


%%%%%%%%%%%%%%%%%%%%%%%%%%%%%%%%%%%%%%%%%%%%%%%%%%%%%%%%%%%%%%
%%%   Calculate Functions-no function m-files used
%%% functions are set up with zero right hand side

obj = 2*k5*(X.*Y + Y.*Z + Z.*X);% objective function
h = X.*Y.*Z -k1;          % equality constraint
g1 = 2*(X + Y )-k2;     % ineaquality constraint g1
g2 = Y-3*X;               % inequality constraint g2
g3 = 3*Z-2*Y;            % inequlaity constraint g3


%%%%%%%%%%%%%%%%%%%%%%%%%%%%%%%%%%%%%%%%%%%%%%%%%%%%%%%%%%%%%%
%%%   For every value of z
%%%   Draw objective function and constraints
%%%   THE PLOT is PAUSED so YOU can OBSERVE
%%%   the INFORMATION

for i = 1:length(z)
    z1 = z(i);  % value of z

 % get 2D data for each value of z
 % for all of the functions
 %   these values are used for drawing contours

    obj_2D(:,:) = obj(:,:,i);
    h_2D(:,:) = h(:,:,i);
    g1_2D(:,:) = g1(:,:,i);
    g2_2D(:,:) = g2(:,:,i);
    g3_2D(:,:) = g3(:,:,i);

    hf1 = figure('position',[50 50 500 500]);
    % give a handle for the figure so that
    % it can be closed after each iteration

    %_____
    % objective function at the current z
    [C1,h1] = contour(x,y,obj_2D,10,'g-');
```

```
clabel(C1,h1);

grid
ylabel('y ', ..
    'FontWeight','b','Color','b', ...
    'VerticalAlignment','bottom');  % boldface blue
xlabel('x ', ..
    'FontWeight','b','Color','b', ...
    'VerticalAlignment','bottom');  % boldface blue

zstring = num2str(z1);  % convert z value to string
% create a title string using z value
texttitle = ...
    strcat('Functions evaluated at z = ',zstring);
title(texttitle)
axis square
hold on

%  draw the equality constraint
% REMEMBER solution must lie on the constraint
[C2,h2] = contour(x,y,h_2D,[0,0],'b-');
set(h2,'LineWidth',3);

%  draw inequality constraint g1
[C3,h3] = contour(x,y,g1_2D,[0 0],'r:');
set(h3,'LineWidth',3);
clabel(C3,h3);  % label for clarity
% infeasible region for g1
[C31,h31] = contour(x,y,g1_2D,[0.01 0.01],'m-');
set(h31,'LineWidth',2);

% draw inequality constraint g2
[C4,h4] = contour(x,y,g2_2D,[0 0],'r:');
set(h4,'LineWidth',3);
clabel(C4,h4);  % label for clarity
% infeasible region for g2
[C41,h41] = contour(x,y,g2_2D,[0.01 0.01],'m-');
set(h41,'LineWidth',2);

% draw inequality constraint g3
[C5,h5] = contour(x,y,g3_2D,[0 0],'r:');
set(h5,'LineWidth',3);
clabel(C5,h5);  % avoid label for clarity
% infeasible region for g3
[C51,h51] = contour(x,y,g3_2D,[0.01 0.01],'m-');
```

```
set(h51,'LineWidth',2);
hold off


%  the computer pauses for observation
% MUST HIT RETURN/ENTER to continue
pause
close(hf1) % close the figure

end
%%%%%%%%%%%%%%%%%%%%%%%%%%%%%%%%%%%%%%%%%%%%%%%%%%%%%%%%%%%%%%%%%
% % %%%  print values for Problem constants
fprintf('\n Example 2.5Graphical Optimization of Container')
fprintf('\n_____')
fprintf('\n Volume constraint    [m^3]  =   %9.3f  ',k1)
fprintf('\n Perimeter constraint [m]   =   %9.3f  ',k2)
fprintf('\n Cost               [ $/m^2]  =   %9.3f  ',k5)
fprintf('\n side constraint (x)[m]    =   %9.5f  ',k3)
fprintf('\n side constraint (y)[m]    =   %9.5f\n\n  ',k4)
```

As the code is executed, it is only around $z = 0.13$ that you start seeing the equality constraint (solid blue line). Around $z = 0.24$ the possibility of solution takes shape because you can observe the feasible region. Figure 2.14 represents

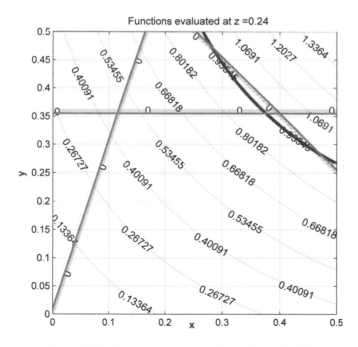

Figure 2.14 Two dimensional solution to Example 2.5.

a possible solution based on the value of the objective function (which cannot be estimated well from the contour values). As you continue to examine additional plots you will notice that at $z = 0.33$, the design region is in violation of constraint g_3. At this level of z the objective function values are quite higher than before for the feasible region. We can rerun the example by adjusting the limits on x, y, and z. to lie around the solution region. For example for the next pass can examine the solution between $z = 0.23$ and $z = 0.30$ with a finer pitch for z. It is possible to therefore identify the solution graphically to the precision you need.

We should be able to conclude that the two-dimensional description of the design space, Figure 2.14, with more effort, is much more effective then the three-dimensional representation of the problem for three variables, Figure 2.13.

2.4 ADDITIONAL MATLAB GRAPHICS

This section provides an opportunity to incorporate more user interaction in the code. One of the conclusions that you may have drawn from the exercises in this chapter is that drawing contours of functions entails the same sequence of actions (and code). The main differences are the different functions for generating the data set, different contour values, and the number of contours. The additional coverage is very modest and largely addresses a little more string handling. It is included here for the sake of completeness, as we have accomplished our goal of graphical optimization in the previous sections. This section can be avoided, postponed, or emphasized for the projects in the course, as it has no additional information for graphical optimization. It does have a lot to offer for understanding MATLAB graphics and programming. You are is encouraged to take some time getting familiar with the version of MATLAB installed on your computer. There are usually significant enhancements in newer versions of MATLAB. You do not have to customize your figure through code alone, as MATLAB allows you to customize your plot and add annotations to it through additional editing features of the Figure window.

In the previous sections, an equally spaced vector was obtained for each variable based on the side constraint. A meshgrid was then generated. All functions were evaluated on this grid. Contour plots were drawn for the functions representing the objective and constraints. Except for the objective function which displayed several contours, the constraints were only drawn for a single value. This procedure suggests the possibility of automation. User input will be necessary for selecting the range for the variables, as well as the functions that will be plotted. These common interactions can be introduced through a program.

2.4.1 Handle Graphics

MATLAB Handle Graphics refer to a collection of low-level graphic routines that actually generate or carry out graphical changes you see in the MATLAB Figure window[8]. For the most part, these routines are transparent, and typical users need not be aware of them. In fact, to understand optimization and to program the techniques in MATLAB, it is not necessary to know about Handle Graphics since using the higher level plot commands such as plot, contour, etc. and so on are sufficient to get the job done. Note in the specific instances of labeling contours, using the *drawHashMarks* function, or making the figure window close every iteration in the last script, that we used handle graphics with knowing about it.

The higher-level graphics commands actually kick in several of the Handle Graphics routines to display the plot on the figure. A noteworthy feature of MATLAB graphics is that it is implemented in an object-oriented manner. What this means is that most of the entities you see on the figure, such as, axes, labels, and text, lines, are all objects. Without going into description or details of object-oriented programming, this implies that most of the graphical items on the figure have properties that can be changed through program code. In the previous sections, we mostly manipulated color, font size, and line width.

In order to change the properties of objects, it is necessary to identify or refer to them. Objects in MATLAB are identified by a unique number referred to as a *handle*. Handles can be assigned to all objects when they are created. Some objects are part of other objects. Objects that contain other objects are referred to as *container* objects. There is a definite hierarchical structure among objects, Figure 2.15. The root of all MATLAB graphic objects is the *Figure* object. The *Figure* object, for example, can hold several GUI controls, also known as the *Uicontrol* object. It also contains the *Axes* object. The *Axes* object in turn has the *Line*, *Text, Image* object, for example. All objects in MATLAB can be associated with a set of properties that is usually based on its function or usefulness. Different object types have different sets of properties. Each of these properties is described though a pair of related information. The first element of this pair is the *Name* for the property and the second is the *Value* corresponding to the property. When the object is created it inherits all of the default properties. For example, the following plot is associated with the variable h_1, which is also the handle to the graphics object created by the *Plot* command—the plot itself. Whether you use the plot command by itself or associate it with h_1, the same

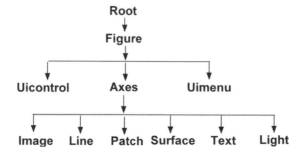

Figure 2.15 Hierarchical representation of graphical objects.

plot appears in the Figure window. But the handle allows you to customize the plot further.

```
h_1 = plot(t, sin(2*t),'-go', ...
          'LineWidth', 2, ...
          'MarkerEdgeColor','k', ...
          'MarkerFaceColor',[0.49 1 0.63], ...
          'MarkerSize',12,)
```

The last four lines above represent 4 properties of the plotted line that is being changed. The property names on the left can be easily understood. The format size units are in points. The color value is a row vector of three color values between zero and one representing the red, green, and blue values respectively. There are several other properties of the plotted line which will be set at their default values. The MATLAB online reference should list all objects and their properties.

In general, object properties can be changed by using the *set* function to change their values. Similarly object properties can be read by the use of the *get* function. To inquire about the property of an object

```
get(handle,'Property Name')
get(h_1,'Color')
```

In the above, h_1 is the handle of the object whose color is desired. Similarly,

```
set(handle, 'Property Name', Property Value')
set(h_1,'Color','r')   or set(h_1,'Color',[1 0 0])
```

in the above the value of *Color* for the object represented by the handle h_1 is set to red in two ways. The second specifies the color red through the vector value. It is expected that the object identified through h_1 should have a *Color* property. MATLAB will inform you of the error in case you are assigning a property/value that is not valid for the object.

2.4.2   Interactive Contour Plots

In this section, we will create contour plots that will prompt the user for the following:

- Range of the plot
- Number of plots
- Contour levels
- Number of contours
- Labeling contours, and also use

- A file selection box/utility that will be used to select the function m-file for the plots

The exercise avoids distinguishing inequality constraints, as it is left as an exercise for the student.

MATLAB Code:Section2_4.m

```
% Chapter 2: Graphical Optimization
% Section 2.4, Interactive Experience
% using contour plots
%_____
%%%%%%%%%%%%%%%%%%%%%%%%%%%%%%%%%%%%%%%%%%%%%%%%%%%%%
% Applied Optimization with Matlab Programming
% Dr. P.Venkataraman
% Second Edition,  John Wiley
%%%%%%%%%%%%%%%%%%%%%%%%%%%%%%%%%%%%%%%%%%%%%%%%%%
%_____
%%%%%%%%%%%%%%%%%%%%%%%%%%%%%%%%%%%%%%%%%%%%%%%%%%%%%%%
%%% management functions
clear  % clear all variable/information in the workspace
clc    % position the cursor at the top of the screen
format compact  % no line skip iin the Command Window
warning off  % No warnings like divide by zero etc.
%%%%%%%%%%%%%%%%%%%%%%%%%%%%%%%%%%%%%%%%%%%%%%%%%%%%%%
% This code will plot two-dimensional contours based
% (i)   range of the variables
% (ii)  number of plots
% (iii) number of contour levels
% (iv)  labeled or unlabeled contours
% (v)   each plot requires an m-file
%       that can conly be dependent on
%       two matrices X1, X2 that are obtained
%       by the use of mesh-grid function
%%%%%%%%%%%%%%%%%%%%%%%%%%%%%%%%%%%%%%%%%%%%%%%%%%%


%%% NOTe:  It is a good idea to set up default values
%%% in case the user just prefers to hit
%%% the return/enter key

%%% input the range of the x1 (x) variable
fprintf('x1 is a vector of 3 values which represents\n ');
fprintf('[ start value, end value, increment]\n')
[x1min x1max x1inc] = ..
    input('Input x1(x)- values-default -[-4 4 0.05]:');
```

```
fprintf('\n');
fprintf('x2 is a vector of 3 values which represents\n ');
fprintf('[ start value, end value, increment]\n')
[x2min x2max x2inc] = ..
    input('Input x2(y)- values-default -[-4 4 0.05]:');
fprintf('\n');
%%%%%%%%%%%%%%%%%%%%%%%%%%%%%%%%%%%%%%%%%%%%%%%%%%%%%%%%
%%% obtain range for x and y
%%%%%%%%%%%%%%%%%%%%%%%%%%%%%%%%%%%%%%%%%%%%%%%%%%%%%%%%

% XInfo, YInfo are introduced for
% introducing default values
Xinfo = [x1min x1max x1inc];
if isempty(Xinfo) % user did not enter information
    % user hits enter without defining x1
    Xinfo = [-4 4 0.05]; % default
end
fprintf('\n')

Yinfo = [x2min x2max x2inc];
if isempty(Yinfo)
    % user hits enter without defining x1
    Yinfo = [-4 4 0.05]; % default
end

%%%%%%%%%%%%%%%%%%%%%%%%%%%%%%%%%%%%%%%%%%%%%%%%%%%%%%%
%%% create the 2D Mesh fo the calculations
%%%%%%%%%%%%%%%%%%%%%%%%%%%%%%%%%%%%%%%%%%%%%%%%%%%%%%%

fprintf('\n')
xvar = Xinfo(1):Xinfo(3):Xinfo(2); % x1 vector
yvar = Yinfo(1):Yinfo(3):Yinfo(2);  % x2 vector
[X1, X2]= meshgrid(xvar,yvar);      % matrix mesh


%%% obtain number of plot information
% set default number of plots to 1
Np = input(' Number of plots-default value = 1:');

if isempty(Np)
    Np= 1;
end
clf;    % clear figure
for Mplot = 1:Np
```

```
% the function will be a function m-file
% of the kind that has been used before
% it will return a matrix based on the input
% of two matrices X1 and X2 based on the
% mesh of the design space
% NOTE: should not have global values

text1 = ['function plotted must be a MATLAB' ...
'\nfunction M-File with 2 matrices for input.'  ...
'\nPlease select function name in the file dialog' ...
    '\nbox and hit return: \n '];
fprintf(text1)

% using the uigetfile dialog box
% Note you can change the first element if you
% know where the function resides
[file, path] = ...
uigetfile('c:\*.m','Files of type MATLAB m-file',300,300);


% check if file is string
% strip the .m extension from the file so it
% can be called by the program
% NOTE: the FUNCTION name mus be same as FILE name
%
if isstr(file)
 functname = strrep(file,'.m','');
else
 fprintf('\n\n');
 text2 = [' You have chosen CANCEL or the file' ...
 '\nwas not acceptable. The program needs a ' ...
 '\n File to Continue Please call Section2_4.m' ...
 '\n again and choose a file  OR press the up-arrow ' ...

 '\n button to scroll through previous commands \n\n' ...

  'Bye !'];
 error(text2); % display message and end program

end

%%% clear variable not useful any more
clear text1 text2; % clears  text1 and text2
```

```
clear Fun maxval minval strcon
clear convalue onevalue labcont labcontU

Fun = feval(functname,X1,X2); % evaluates the function
% at X1, X2
maxval = max(max(Fun));  % maximum value of function
minval = min(min(Fun));  % minimum value of function

fprintf('The contour ranges from\n')
fprintf('MIN: :%9.3f   MAX.: %9.3f  ',minval,maxval);
fprintf('\n');
strcon = ...
    input('Do you want to set contour values ?[ no]:','s');

strconU = upper(strcon);

if strcmp(strconU,'YES') | strcmp(strconU,'Y')
    fprintf(' Input a vector of contour levels ');
    fprintf('\n')
    fprintf('between %9.3f and %9.3f ',minval,maxval);
    fprintf('\n')
    convalue = input(' Input contour level [Vector] :');
    labcont = input(' Labelled contours ? [ no]:','s');
    labcontU = upper(labcont);
    if strcmp(labcontU,'YES')| strcmp(labcontU,'Y')
        [C,h] = contour(xvar,yvar,Fun,convalue,'g-');
        clabel(C,h);
        refresh;
    else
        contour(xvar,yvar,Fun,convalue,'g-');
        refresh;
    end

else
    ncon = input('Input number of contours [20] :');
    if isempty(ncon)
        ncon = 20;
        labcont = input(' Labelled contours ? [ no]:','s');

        labcontU = upper(labcont);
        if strcmp(labcontU,'YES')| strcmp(labcontU,'Y')
            [C,h] = contour(xvar,yvar,Fun,ncon,'g-');
            clabel(C,h);
            refresh;
```

```
        else
            contour(xvar,yvar,Fun,ncon,'g-');
            refresh;
        end


        % handle single contours for constraints
    elseif ncon == 1
    onevalue = input('Input the single contour level: ');

        labcont = input(' Labelled contours ? [ no]:','s');

        labcontU = upper(labcont);
        if strcmp(labcontU,'YES')| strcmp(labcontU,'Y')
            [C,h] = ...
                contour(xvar,yvar,Fun,[onevalue,onevalue],'b-');

            clabel(C,h);
            refresh
        else
        contour(xvar,yvar,Fun,[onevalue,onevalue],'b-');
            refresh;
         end
    else
        labcont = input(' Labelled contours ? [ no]:','s');

        labcontU = upper(labcont);
        if strcmp(labcontU,'YES')| strcmp(labcontU,'Y')
            [C,h] = contour(xvar,yvar,Fun,ncon,'g-');
            clabel(C,h);
            refresh;
        else
            contour(xvar,yvar,Fun,ncon,'g-');
            refresh;
        end


    end
  end
  if Np > 1
    hold on;
    pause(3);
  end
  Hf = gcf;
end
```

```
figure(Hf);
grid
hold off

ylabel('x_2 ', ..
    'FontWeight','b','Color','b', ...
    'VerticalAlignment','bottom');  % boldface blue
xlabel('x_1 ', ..
    'FontWeight','b','Color','b', ...
    'VerticalAlignment','top');  % boldface blue

title('Contours Using Section2-4.m')
axis square
```

The plotting commands have been used before. The new MATLAB commands are *isempty, clf, isstr, strrep, max, min, upper, strcm*. Use the help command to know more about these. Run the script file and understand the sequence of actions, as well as the prompts regarding the contour. In this example, the functions that create the plot do not need global values for calculation transferred from the Command workspace. In other words, the functions that are to be plotted can be calculated independently.

REFERENCES

1. MATLAB. *The Language of Technical Computing: Using MATLAB, Version 5*, The Math Works Inc., 1998.
2. MATLAB. *The Language of Technical Computing: Using MATLAB Graphics, Version 5*. The Math Works Inc., 1996.
3. Bohachevsky, I.O., M.E., Johnson and M.L, Stein, "Generalized Simulated Annealing for Function Optimization", *Technometrics*, (August 1986) Vol. 28, no. 3.
4. Arora, J.S. *Introduction to Optimum Design*, New York: McGraw-Hill, 1989.
5. Beer, F. P., and E.R. Johnston, Jr., *Mechanics of Materials*, 2nd ed. New York: McGraw-Hill, 1992.
6. Fox, R.W., and A.T. McDonald, *Introduction to Fluid mechanics*, 4 ed., John Wiley & Sons, Inc., 1992.
7. Arpaci, V.S., *Conduction Heat Transfer*, Reading, MA: Addison-Wesley, 1966.
8. Hanselman, D., B. Littlefield, *Mastering MATLAB 5, A Comprehensive Tutorial and Reference*, The MATLAB Curriculum Series, Englewood Cliffs, NJ: Prentice Hall, 1996.

PROBLEMS

Many of the graphical enhancements are possible through *plotedit* features in later releases of MATLAB. The following problems are just suggestions.

2.1 Plot the ballistic trajectory from simple two-dimensional mechanics.

2.2 Produce an animated display of the trajectory.

2.3 Create a two-dimensional program that will display a random firing location and a random target location. Allow the user to choose initial velocity magnitude and angle. Plot the trajectory of the projectile and calculate the final error.

2.4 Create a three-dimensional program that will display a random firing location and a random target location. Allow the user to choose initial velocity magnitude and direction angles. Plot the trajectory of the projectile and calculate the final error.

2.5 Draw the boundary layer profile of laminar flow over a flat plate. Draw lines indicating velocity profile at 10 points in the boundary layer.

2.6 Create a program that will plot the velocity profile for user specific inputs. Allow user to express his inputs through a dialog box.

2.7 Graphically solve Example 1.1.

2.8 How will you graphically solve Example 1.2?

2.9 Graphically solve Example 1.3 with three variables.

2.10 Graphically solve Example 2.3 if the pole is 20 m tall. Change the constraints if needed.

2.11 Graphically solve Example 2.4 if the gap is $b/2$. You will need to derive the functions again.

2.12 Explore creating a movie of the two-dimensional solution to Example 2.5. You will have to read the MATLAB documentation.

2.13 Differentiate equality and inequality constraint in the code of Section 2.4. Draw the infeasible region of the inequality constraint.

2.14 A thin-walled spherical gas container (radius: r, thickness: t) is required to be designed for an internal pressure of 5 MPa. Find the smallest mass of the container if the factor of safety is 3 and the material used is structural steel. The stresses in the thin-walled structure can be evaluated as

$$\sigma(normal) = \frac{pr}{2t}; \quad \tau(shear) = \frac{pr}{4t}$$

Set up the optimization problem in standard format and graphically obtain the solution (Problem 1.16).

2.15 In July, Venkat's Fruit Orchard usually sees a scheduling problem. Both cherries and blueberries ripen around the same time. A cartload of cherries sells for $2,750 while the cartload of blue berries fetches $1,750. The farm must deliver at least a half cartload of cherries and one cartload of blue berries to the local supermarket. Three persons can pick two cartloads of cherries in a day, while one person will pick a cartload of blueberries in the same time. Only 46 people show up for work during the day. One cartload of cherry is contained in 4 pallets, while the same amount of blueberries occupies 6 pallets. There are 251 pallets. Each picker is paid

$45 per day. The packaging and storage can handle, at most, 30 cartloads per day. Set up the LP problem for the Orchard to maximize profits. Obtain the solution graphically (Problem 1.22).

2.16 A measure of aircraft performance is aerodynamic efficiency E, which is the ratio of C_L over C_D. Find the maximum aerodynamic efficiency and the speed corresponding to this efficiency assuming that the aircraft performance can be described by a parabolic polar. Solve the problem graphically if $C_{D0} = 0.02$ and $K = 0.02$. You can assume the aircraft is operating at 10,000 ft on a standard day, traveling at a speed of 220 miles/hour, and the reference area of the wing is 174 ft^2. The weight of the aircraft is 3100 lb [Problem 1.30].

2.17 A shell and tube heat exchanger is a popular device to transfer energy. The cross-section of one is shown in Figure Problem 2.17. The diameter of the shell (D) is specified as 0.75 m. The length of the exchanger (L) is 2.5 m. The shell is made from a plate which costs $ 9.50 per unit area. The tubes are cut from stock and cost $ 1.25 per meter. The energy exchanged by the device can be calculated as $0.5*sqrt(n)*L^{0.75}/d^{0.2}$ kW. Here n is the number of tubes and d is the diameter of the tube. To make sure the tubes fit in the shell we impose an area constraint that the total area of the

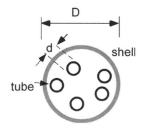

Figure Problem 2.17

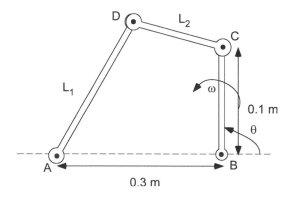

Figure Problem 2.18

tubes must be less than half the cross-sectional area of the shell. Just to be sure, we constrain the diameter by requiring $sqrt(n)*d$ must be less than the diameter of the shell. Set up an optimization problem for minimum cost with the understanding that the heat exchanger must transfer at least 10 kW of energy. Obtain the solution graphically. Consider **n** as a discrete variable.

2.18 The four-bar link is shown in Figure Problem 2.18. Find the link length L_1 and L_2 to maximize the velocity at D when θ is $90°$ and ω is 50 rad/s. Use reasonable side constraints. Obtain the solution graphically.

3

LINEAR PROGRAMMING

Engineering design mostly deals with mathematical models characterized by non-linear equations. The presence of a single nonlinear equation in the model is sufficient to identify the problem from the area of **nonlinear programming (NLP)** or **mathematical programming**. Several of these problems were considered in the previous chapter on graphical optimization. When looking at the graphical solution, it is difficult to escape the feeling that the optimal solution is influenced by the gradient and the curvature of the functions in the mathematical model. This geometrical impact, which did not feature in the discussion, will be an important consideration in the numerical techniques which will come later.

There is an equally important class of problems whose mathematical model is made up exclusively of functions that are only linear. Example 1.3, Equations (1.21) and (1.22) represent just such a model. The same example was modified in Equations (1.29) and (1.30) so that a graphical solution could be discussed through Figure 1.6. It is redrawn as Figure 3.1 in this chapter. The code for generating this figure is introduced later. It is apparent from the graphical description that there is no curvature evident in the figure regarding any of these functions. These problems are termed **linear programming (LP)** problems. These problems are natural in the subject of *operations research*, which covers a vast variety of models used for several kinds of decision-making. Example 1.3 represents a decision-making problem. Typically, *Linear Programming* can be a course by itself (or several courses, for that matter). A linear programming course will largely deal with mathematical concepts of linear algebra, usually in matrix form; deal with a large set of linear equations; and also deal with the computer algorithms for their implementation. In real world applications, the number of variables is huge, even as much as a million or more. A significant portion of such a course would be to develop mathematical models from different areas

130

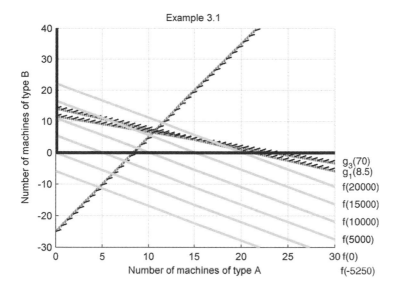

Figure 3.1 Solution to Example 3.1.

of applications, as there is usually one numerical technique that is commonly used to solve LP problems. It is called the **Simplex method** and is based on the algorithm by Dantzig. The method involves mostly elementary row operations typically encountered in **Gauss-elimination methods** that are part of the numerical techniques used in linear algebra. In this chapter, only a limited, but useful discussion is presented. First, it is important to understand LP problems, the modeling issues, and their solution, as they are different from most of the other problems in the book. Second, many of the current numerical techniques for NLP problems obtain their solution by linearizing the solution at the current design point. These linearized equations can be solved by the methods of this chapter. The reader is directed to any of the books on LP for a more detailed description of the models and the techniques, as the presentation in this book is simple and brief.[1-3]

For the new edition, generating the graphical solution and the use of MATLAB to apply the Simplex method is vastly improved and quite impressive for communicating the concepts effectively. The graphics in this chapter are all created using MATLAB and the code is included as part of the download. The code for implementing the Simplex method can generate the same tables that are found in the chapter.

3.1 PROBLEM DEFINITION

Example 1.3 in Chapter 1 will be used to define the terminology associated with the LP problem. The modified version of the problem is used to develop the format of the LP problem so that we can follow through with the graphical solution.

The mathematical model includes two design variables x_1, x_2 which represent the number of component placement machines of type A and B, respectively. The objective is to maximize the number of boards to be manufactured. Constraint g_1 represents the acquisition dollars available. Constraint g_2, represents the floor space constraint. Constraint g_3 represents the number of operators available.

$$\text{Maximize} \quad f(x): 990x_1 + 900x_2 + 5250 \tag{3.1}$$

$$\text{Subject to:} \quad g_1(x): 0.4x_1 + 0.6x_2 \leq 8.5 \tag{3.2a}$$

$$g_2(x): 3x_1 - x_2 \leq 25 \tag{3.2b}$$

$$g_3(x): 3x_1 + 6x_2 \leq 70 \tag{3.2c}$$

$$x_1 \geq 0; \quad x_2 \geq 0$$

The problem defined in Equations (3.1) to (3.2) is a natural representation of the mathematical model in engineering design or any other discipline. This representation can be considered the inequality form of the mathematical model, although it is not expressed in the *standard format* for the LP problem. The solution can be recognized in the graphical description of the functions in Figure 3.1.

3.1.1 Standard Format

The standard format of the LP problem includes only *equality* constraints. It is set up as a minimization problem. In addition, all the variables in the model are expected to be semi-positive (≥ 0) or non-negative.[4] Finally, the constraint limits, the quantity on the right-hand side of the constraint, are required to be positive (>0). In the problem defined in Equations (3.1) and (3.2), several changes must be carried out. The objective function must be converted to the opposite type, and the constraints need to be transformed to the equality type. The variables x_1 and x_2, representing the number of component placing machines of model A and B, respectively, can be expected to be semi-positive (≥ 0); that is, either some machines will be ordered or no machines will be ordered. Note that the discrete nature of the problem is being ignored after all 3.75 machines of type A is not something that would be ordered. The solution will be rounded to a suitable integer. There are extensive discussions from discrete and integer programming to suggest that this may not establish the optimum value. At this time, no justification is advanced for rounding to the nearest integer, except that it is intuitive and convenient to do so at this juncture.

In the objective function $f(x)$, the coefficients 990, 900, are called the **cost coefficients**. Each coefficient associated with a design variable represents the increase (if positive) or decrease (if negative) in the cost per unit change in the related variable. The simplest way to transform the objective function to the required format is to multiply all the terms by -1. Therefore the new objective function looks like this:

$$\text{Minimize} \quad f(x): -990x_1 - 900x_2 - 5{,}250 \tag{3.3}$$

In Equation (3.3), the constant term on the right ($-5{,}250$) can be absorbed into the left-hand side by defining a new objective function $f\_(x)[f(x) + 5{,}250]$, without affecting the optimal values for the design variables. It can be observed in Equation (3.3) that increasing the values of the variables will make the objective more negative, which is good because we are trying to make $f(x)$ as low as possible. If a particular cost coefficient had a positive sign, then increasing the amount of the corresponding variable would lead to the increase of the objective function, which is not desirable if the function needs to be minimized. This idea is exploited in the Simplex method presented a little later.

The inequality constraints have to be transformed to an equality constraint. The simplest way to achieve this change is to introduce an additional semi-positive variable for each inequality constraint. That will create the corresponding equality constraint in a straightforward way. Consider the first constraint:

$$g_1(x): \quad 0.4x_1 + 0.6x_2 \le 8.5 \tag{3.2a}$$

This can be transformed to

$$g_1(x): \quad 0.4x_1 + 0.6x_2 + x_3 = 8.5$$

The new variable x_3 is referred to as the **slack variable**. The definition is quite appropriate, because it takes up the slack of the original constraint terms when it is less than 8.5. The slack x_3 will be positive if the first two terms will be less than 8.5 (for constraint satisfaction) from the original definition in Equation (3.2a). In other words, it is the difference between the constraint limit (8.5) and the first two terms of the constraint. If the values of x_1 and x_2 cause the constraint to be at its limit value then x_3 will be zero. This is the reason it is defined to be semi-positive. By definition, therefore, x_3 is similar to the original variables in the LP problem. Transforming the remaining constraints in a similar manner, the complete equations in standard format is

$$\text{Minimize} \quad f(x): -990x_1 - 900x_2 - 5250 \tag{3.3}$$

$$\text{Subject to:} \quad g_1(x): 0.4x_1 + 0.6x_2 + x_3 = 8.5 \tag{3.4a}$$

$$g_2(x): 3x_1 - x_2 + x_4 = 25 \tag{3.4b}$$

$$g_3(x): 3x_1 + 6x_2 + x_5 = 70 \tag{3.4c}$$

$$x_1 \ge 0, \quad x_2 \ge 0, \quad x_3 \ge 0, \quad x_4 \ge 0, \quad x_5 \ge 0 \tag{3.5}$$

Equations (3.3) to (3.6) express the LP problem in *standard format*. An inconsistency can be observed with respect to the format of the problems defined in the earlier chapters. Previously, $g(x)$ was reserved to define inequality constraints while $h(x)$ defined equality constraints. Here, we are retaining the symbol $g(x)$ for the equality constraints. This shift in definition will be accepted for this chapter only for convenience. In subsequent chapters, the prior meaning for the symbols is restored. Actually, symbols are not required, as the constraint under discussion can also be referred to by equation number. The formulation in Equations (3.3) to (3.5) can be expressed succinctly by resorting to a matrix notation. This conversion is quite straightforward. Before embarking on this transformation, the representation of the objective function needs to be changed again. The constant term is brought to the left-hand side so that the objective function can be expressed with only terms involving the design variables on the right. This is shown as $f_-(x)$. Once again, for convenience the same symbol $f(x)$ is retained in place of this new function $f_-(x)$ in subsequent discussion. These extra symbols have been thrown in to keep the derivation details distinct. This also allows us to develop the mathematical model in a natural way and then reconstruct it to conform to the standard representation in most references. Therefore,

$$\text{Minimize} \quad f(x) \equiv f_-(x): \; f(x) + 5{,}250 = -990x_1 - 900x_2 \qquad (3.6)$$

The following definitions are used for rewriting the problem using matrices. Let the cost coefficients be represented by $c_1, c_2, \ldots c_n$. The coefficient in the constraints is represented by the symbol a_{ij} (using double subscripts). The first subscript identifies the constraint number while the second constraint is the same as the variable it is multiplying. It is also useful to identify the first subscript with the row number and the second subscript with the column number to correspond with operations that use linear algebra. For example, the constraint $g_1(x)$ will be represented as follows:

$$g_1(x): \quad a_{11}x_1 + a_{12}x_2 + a_{13}x_3 = 8.5$$

where $a_{11} = 0.4, a_{12} = 0.6, a_{13} = 1$.

Note since there are five design variables in the problem definition, additional terms with the following coefficients, $a_{14} (=0)$, and $a_{15}(=0)$, are included in the expression for $g_1(x)$ without changing its meaning:

$$g_1(x): \quad a_{11}x_1 + a_{12}x_2 + a_{13}x_3 + a_{14}x_4 + a_{15}x_5 = 8.5$$

The constraint $g_2(x)$, with $a_{21} = 3, a_{22} = -1, a_{23} = 0, a_{24} = 1, a_{25} = 0$, from Equation (3.4b) can be similarly expressed as

$$g_2(x): \quad a_{21}x_1 + a_{22}x_2 + a_{23}x_3 + a_{24}x_4 + a_{25}x_5 = 25$$

Letting the right-hand side values of the constraints be represented by the letter $b_1, b_2, \ldots, b_m$, the standard format in Equations (3.3) to (3.6) can be translated to

$$\text{Minimize} \quad f(x)\colon \boldsymbol{c}^T \boldsymbol{x} \tag{3.7}$$

$$\text{Subject to:} \quad g(x)\colon \boldsymbol{A}\boldsymbol{x} = \boldsymbol{b} \tag{3.8}$$

$$\text{Side constraints:} \quad \boldsymbol{x} \geq 0 \tag{3.9}$$

$\boldsymbol{x}$ represents the column vector of design variables, including the slack variables, $[x_1, x_2, \dots, x_n]^T$ ($\boldsymbol{T}$ represents the transposition symbol). $\boldsymbol{c}$ represents the column vector of cost coefficients, $[c_1, c_2, \dots, c_n]^T$. $\boldsymbol{b}$ represents the column vector of the constraint limits, $[b_1, b_2, \dots, b_m]^T$. $\boldsymbol{A}$ represents the $m \times n$ matrix of constraint coefficients. In this book, all vectors are column vectors unless otherwise identified.

$$A = \begin{bmatrix} a_{11} & a_{12} & \cdot & a_{1n} \\ a_{21} & a_{22} & \cdot & a_{2n} \\ \cdot & \cdot & \cdot & \\ a_{m1} & a_{m2} & \cdot & a_{mn} \end{bmatrix} \quad b = \begin{bmatrix} b_1 \\ b_2 \\ \cdot \\ b_m \end{bmatrix} \quad c = \begin{bmatrix} c_1 \\ c_2 \\ \cdot \\ c_n \end{bmatrix} \quad x = \begin{bmatrix} x_1 \\ x_2 \\ \cdot \\ x_n \end{bmatrix}$$

Comparing equations (3.4) to (3.6) with Equations (3.7) to (3.9) the following relations can be established:

$$x = [x_1, x_2, x_3, x_4, x_5]^T$$

$$c = [-990 \quad -900 \quad 0 \quad 0 \quad 0]^T \tag{3.10a}$$

$$b = [8.5 \quad 25 \quad 70]^T$$

$$A = \begin{bmatrix} 0.4 & 0.6 & 1 & 0 & 0 \\ 3 & -1 & 0 & 1 & 0 \\ 3 & 6 & 0 & 0 & 1 \end{bmatrix} \tag{3.10b}$$

The following additional definitions are used in the standard format. n is the number of design variables. m is the number of constraints. In the example, $n = 5$, and $m = 3$. Note that the slack variables are included in the count for n.

Negative Values of Design Variables: The LP standard mathematical model allows only for non-negative design variables. This does not limit the application of LP in any way. There is a mechanism to handle negative values for variables. If a design variable, say x_1, is expected to be unrestricted in sign, then it is defined as a difference of two non-negative variables:

$$x_1 = x_1^h - x_1^l \tag{3.11}$$

where x_1^h and x_1^l are acceptable LP variables. If the latter is higher than the former x_1 can have negative values. Equation 3.11 is used to replace x_1 in the mathematical model. Note that this procedure also increases the number of design variable in the model.

Type of Constraints: The less than equal to constraint ($\leq$) was handled naturally in developing the LP standard mathematical model. The standard numerical technique for obtaining the solution requires that the slack variables be positive to start the iterative solution. In the standard model, b must also be positive throughout. Now, if there is a constraint that is required to be maintained above a specified value, for example

$$g_1(x): \quad a_{11}x_1 + a_{12}x_2 + a_{13}x_3 \geq b_1 \tag{3.12}$$

then natural approach is to introduce a negative slack variable $-x_4$ to change (3.12) to an equality constraint as (3.13):

$$g_1(x): \quad a_{11}x_1 + a_{12}x_2 + a_{13}x_3 - x_4 = b_1 \tag{3.13}$$

For the numerical technique, the sign associated with x_4 is of the opposite sense. It is possible to change the sign on x_4 by multiplying the equation by -1. This will cause the right-hand constraint limit to be negative, which is not permitted in the standard model. In such cases, another variable, x_5 is included in the expression. In this case, two additional variables are now part of the model (3.14).

$$g_1(x): \quad a_{11}x_1 + a_{12}x_2 + a_{13}x_3 - x_4 + x_5 = b_1 \tag{3.14}$$

The variable x_5 is called an **artificial variable**. Artificial variables are also introduced for each equality constraint that is naturally present in the model. These variables will be illustrated through an example later on in the chapter.

3.1.2   Modeling Issues

The LP program is characterized by the mathematical model defined in Equations (3.7) to (3.9). Due to the linear nature of all the functions involved, the discussion of the solution is possible by the consideration of the coefficients of the functions alone—the variables themselves can be ignored. These coefficients, which are present as vectors or a matrix, can be manipulated to yield solutions by using concepts from linear algebra.

A look at the graphical solution to the LP problem in Figure 3.1 suggests that the solution to the problem, if it exists must lie on the constraint boundary because the curvature of the objective function is zero everywhere (it is linear). Furthermore, the solution to the LP problem is primarily influenced by Equation (3.8). Being a linear equation, topics from linear algebra are necessary to understand some of the implications of the mathematical model.

If $n = m$, that is, the number of unknowns are the same as the number of variables, there can be no scope for optimization, as Equation (3.8) will determine the solution, provided one exists. In the following, many of the concepts are explained through simple graphical illustration, or by consideration of simple examples. They mostly deal with examples of two variables (x_1, x_2). For example, consider two constraints:

$$g_1: \quad x_1 + x_2 = 2$$

$$g_2: \quad -x_1 + x_2 = 1$$

Figure 3.2 illustrates the graphical solution to the problem. The solution is at $x_1 = 0.5$, and $x_2 = 1.5$. This is independent of the objective function (whatever it may be). In matrix form, A and b are

$$A = \begin{bmatrix} 1 & 1 \\ -1 & 1 \end{bmatrix} \quad b = \begin{bmatrix} 2 \\ 1 \end{bmatrix}$$

The existence of solution depends on the rows of A (or columns of A—we will deal with rows only for convenience). If the rows of A *are linearly independent*, then there is a *unique solution* to the system of equations. In this example, you cannot get the second row of A by adding, subtracting, multiplying, or dividing (*elementary row operations*) any row vector to the first row of A. The figures in this section can be drawn using MATLAB by running the MATLAB code identified through the figure number. The lines are drawn using the **NewdrawLine.m*** function m-file. Files must be downloaded from the website. This is true of all files whose code is not printed in this book.

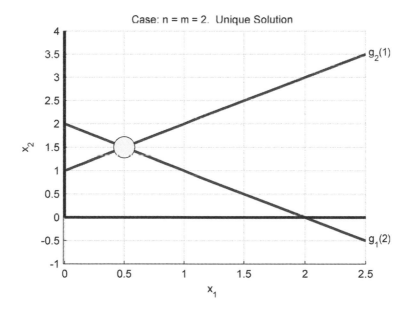

Figure 3.2 Unique solution — linear independence.

*Files to be downloaded from the web site are indicated by boldface courier type.

Now define another problem as follows (only changing g_2 from the previous example). The new problem is

$$g_1: \quad x_1 + x_2 = 2$$
$$g_2: \quad 2x_1 + 2x_2 = 4$$

The new function g_2 represents twice the value of g_1. This means g_2 can be obtained from g_1. Therefore, g_2 is said to depend on g_1. Technically, the two equations are *linearly dependent*—the actual dependence of the functions is not a concern as much as the fact that they *are not linearly independent*. Both g_1 and g_2 establish the same line graphically in Figure 3.3. *What is the solution to the problem?* There are **infinite solutions**, as long as the pair of (x_1, x_2) values satisfies g_1. In Figure 3.3, any point on the line is a solution. Not only (0.5, 1.5), but (2,0, 0,2) are also solutions. From an *optimization perspective*, this is quite good, for unlike the previous case of unique solution, the objective function can be used to find the best solution. It would be better, however, if the choices were finite rather than infinite. Here, even though originally $n = 2$ and $m = 2$, the linear dependence finally reduces the case to $n = 2$, and $m = 1$, there is only one effective constraint.

Although several examples can be constructed to illustrate linear dependence, it appears reasonable to conclude that *for a useful exercise in optimization (linear programming), the number of constraints cannot equal the number of unknowns* $(\boldsymbol{m \neq n})$. It is useful to recognize that this corresponds to the case $\boldsymbol{n > m}$.

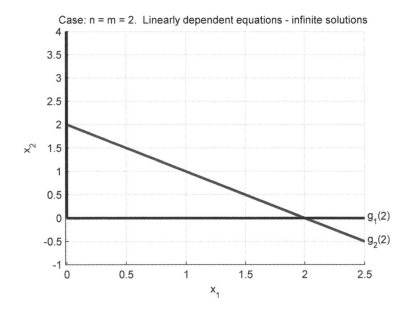

Figure 3.3   Infinite solution — linearly dependent.

If $m > n$, there are more equations than the number of variables. This implies that the system of equations represented by Equation (3.8) is an *inconsistent* set or has a *redundant* set of equations. Consider the following illustration for $m = 3$ and $n = 2$, which uses the same g_1 and g_2 used in the first illustration, while adding a new g_3:

$$g_1: \quad x_1 + x_2 = 2$$

$$g_2: \quad -x_1 + x_2 = 1$$

$$g_3: \quad x_1 + 2x_2 = 1$$

Figure 3.4 illustrates that the set of equations is *inconsistent* since a solution does not exist. If one were to exist, then the three lines must pass through the solution. Since they are all straight lines, there can be only one unique intersecting point. In Figure 3.4, only two of the three lines intersect at different points.

Redefine g_3 as

$$g_3: \quad x_1 + 2x_2 = 3.5$$

This new g_3 is shown in Figure 3.5. Now a unique solution to the problem at (0.5, 1.5) is established. This is the same solution established by consideration of the g_1 and g_2 alone. This implies that g_3 is **redundant**. Actually, g_3 can be obtained by multiplying g_1 by 1.5 and adding it to g_2 multiplied by 0.5. That is

$$g_3 = 1.5g_1 + 0.5g_2$$

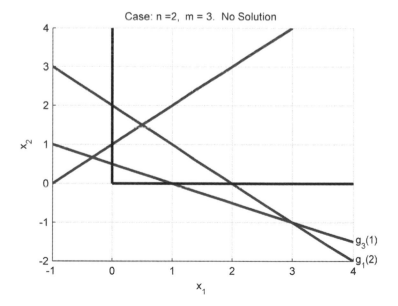

Figure 3.4 No solution — inconsistent set of equations.

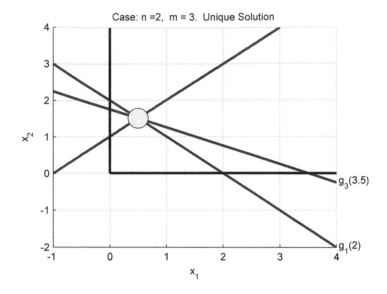

Figure 3.5   Unique solution — linear dependence.

The new line g_3 is obtained by *linearly* combining g_1 and g_2—that is, adding constant multiples of the functions. This is another example of *linear dependence*. Note in this case it is difficult to conclude linear dependency by looking at g_3 because g_3 looks like a different line. If only the solution is considered important, then the redundancy that *is* illustrated by the set

$$g_1: \quad x_1 + x_2 = 2$$

$$g_2: \quad -x_1 + x_2 = 1$$

$$g_3: \quad 2x_1 + 2x_2 = 4 \text{ (same as } g_1\text{)}$$

produces the same result as in the case of linear dependence. This suggests that the concept of *redundancy* can be associated with *linear dependence*.

The discussion of linear dependence and independence was established using simple equations and graphics. Consider the practical problem where you have a lot of variables. It is difficult to verify linear dependence or redundancy through graphics. The same discussion can take place by reasoning on the coefficients themselves. In this case, the coefficient matrix A should lead us to the same conclusion regarding the linear independence of a set of linear equations. The concept of a *determinant* is necessary to proceed in this direction.

Determinant: The *determinant* is associated with a square matrix. For a general 2×2 matrix A, where

$$A = \begin{bmatrix} a_{11} & a_{12} \\ a_{21} & a_{22} \end{bmatrix}$$

The *determinant* is expressed and evaluated as

$$\begin{vmatrix} a_{11} & a_{12} \\ a_{21} & a_{22} \end{vmatrix} = \det(A) = a_{11} \times a_{22} - a_{12} \times a_{21} \tag{3.15}$$

For $n = m$, and with g_1: $x_1 + x_2 = 2$, and g_2: $-x_1 + x_2 = 1$, the matrix A is

$$A = \begin{bmatrix} 1 & 1 \\ -1 & 1 \end{bmatrix}$$

and its *determinant* is

$$\det(A) = |A| = (1) * (1) - (1) * (-1) = 2$$

From theorems in linear algebra,[5] if $\det(A)$ is *not zero*, in which case it is termed **non-singular**, a unique solution exists to the set of equations. This was true for the example just illustrated, as the only solution was located at (0.5, 1.5), shown graphically in Figure 3.1.

Consider the example with g_1: $x_1 + x_2 = 2$, and g_2: $2x_1 + 2x_2 = 4$. In this case

$$\det(A) = |A| = (1) * (2) - (1) * (2) = 0$$

If $\det(A)$ is *zero*, that is **singular**, there are either *no solution* or *infinite solutions*. For this example, there were infinite solutions. Determinant of higher-order square matrices are evaluated by setting up lower-order determinants until they are reduced to a 2×2 determinant, which is evaluated as shown. Any textbook on engineering mathematics or linear algebra should illustrate this technique. It is not reproduced here.

If $n \neq m$, discussion of the existence of solutions requires additional concepts like **rank** of a matrix and **augmented matrix**. Since the case $n > m$ is of interest in optimization only that case is discussed in subsequent illustration. Typically, m is the number of rows of matrix A, while n is the number of columns in A. A useful discussion will need at least three variables and two equations. Three variables will deny the use of graphics to develop the following concepts. Using the set

$$g_1: \quad x_1 + x_2 + x_3 = 3$$

$$g_2: \quad -x_1 + x_2 + 0.5x_3 = 1.5$$

The matrices are

$$A = \begin{bmatrix} 1 & 1 & 1 \\ -1 & 1 & 0.5 \end{bmatrix} \quad b = \begin{bmatrix} 3 \\ 1.5 \end{bmatrix} \quad A^* = \begin{bmatrix} 1 & 1 & 1 & 3 \\ -1 & 1 & 0.5 & 1.5 \end{bmatrix} \tag{3.16}$$

The new matrix A^* is called the *augmented* matrix—the columns of b are added to A. According to theorems of linear algebra (presented here without proof):

- If the *augmented* matrix (A^*) and the matrix of coefficients (A) have the same *rank* $r < n$, then there are *many solutions*.
- If the *augmented* matrix (A^*) and the matrix of coefficients (A) do not have the same *rank a solution does not exist*.
- If the *augmented* matrix (A^*) and the matrix of coefficients (A) have the same *rank* $r = n$, then there is a *unique solution*.

Rank of a Matrix: While there are formal definitions for defining the rank of the matrix, a useful way to determine the rank is to look at the determinant. The rank of a matrix A is the order of the largest non-singular square submatrix of $A^{\#}$—that is, *the largest submatrix with a determinant other than zero*.

In the previous example, the largest *square submatrix* is a 2×2 matrix (since $m = 2$ and $m < n$). Taking the submatrix which includes the first two columns of A, the determinant was previous established to have a value of 2, therefore non-singular. Thus the rank of A is 2 ($r = 2$). The same columns appear in A^* making its rank 2 too. Therefore, infinitely many solutions exist. One way to determine the solutions is to assign ($n - r$) variables arbitrary values and using them determine values for the remaining r variables. The value ($n - r$) is also often identified as the *degree of freedom* (DOF) for the system of equations. In this particular example, the DOF can be identified with a value of 1 (i.e., $3 - 2$). For instance, x_3 can be assigned a value of 1, in which case $x_1 = 0.5$, and $x_2 = 1.5$. On the other hand, if $x_3 = 2$, then $x_1 = 0.25$ and $x_2 = 0.75$. The above cases illustrates that for DOF of 1, once x_3 is selected, x_1 and x_2 can be obtained through some further processing

Another example for illustrating the rank is the inconsistent system of equations introduced earlier where $n = 2$ and $m = 3$, reproduced for convenience as

$$g_1: \quad x_1 + x_2 = 2$$

$$g_2: \quad -x_1 + x_2 = 1$$

$$g_3: \quad x_1 + 2x_2 = 1$$

The matrices for this system are

$$A = \begin{bmatrix} 1 & 1 \\ -1 & 1 \\ 1 & 2 \end{bmatrix}; \quad b = \begin{bmatrix} 2 \\ 1 \\ 1 \end{bmatrix}; \quad A^* = \begin{bmatrix} 1 & 1 & 1 \\ -1 & 1 & 1 \\ 1 & 2 & 1 \end{bmatrix}$$

The rank of A cannot be greater than 2, since $n = 2$ and is less than $m = 3$. From prior calculations, the determinant of the first two rows of A is 2. This is greater than zero, and therefore the rank of A is 2. For A^* the determinant of the 3×3 matrix has to be examined. The determinant of a general $n \times n$ square matrix A can be calculated in terms of the *cofactors* of the matrix. The *cofactor* involves the determinant of a *minor* matrix—which is a $(n-1) \times (n-1)$ matrix obtained by deleting an appropriate single row and a single column from the original $n \times n$ matrix A, multiplied by 1 or -1, depending the total of the row and column values.[5,6] Here, the determinant for the 3×3 matrix is defined as it is quite unlikely a higher order will be used for illustration in the book. The $(+1)$ and (-1) below are obtained by using $(-1)^{i+j}$, where i and j represent the row and column values of the coefficient multiplying the value

$$
A = \begin{vmatrix} a_{11} & a_{12} & a_{13} \\ a_{21} & a_{22} & a_{23} \\ a_{31} & a_{32} & a_{33} \end{vmatrix}
$$

$$
= (+1)a_{11}\begin{vmatrix} a_{22} & a_{23} \\ a_{32} & a_{33} \end{vmatrix} + (-1)a_{12}\begin{vmatrix} a_{21} & a_{23} \\ a_{31} & a_{33} \end{vmatrix} + (+1)a_{13}\begin{vmatrix} a_{21} & a_{22} \\ a_{31} & a_{32} \end{vmatrix} \tag{3.17}
$$

Therefore

$$
|A^*| = \begin{vmatrix} 1 & 1 & 1 \\ -1 & 1 & 1 \\ 1 & 2 & 1 \end{vmatrix} = 1\begin{vmatrix} 1 & 1 \\ 1 & 1 \end{vmatrix} - 1\begin{vmatrix} -1 & 1 \\ 1 & 1 \end{vmatrix} + 1\begin{vmatrix} -1 & 1 \\ 1 & 2 \end{vmatrix}
$$

$$
= (1)(0) + (-1)(-2) + (1)(-3) = -1
$$

A^* is non-singular and its rank is 3, while rank of A is 2. The system of equations will have no solutions from this theorem and can be seen graphically in Figure 3.4. For the sake of completeness, using the previously defined dependent set of equations:

$$
g_1: \quad x_1 + x_2 = 2
$$

$$
g_2: \quad -x_1 + x_2 = 1
$$

$$
g_3: \quad 2x_1 + 2x_2 = 4
$$

which yields the following matrices:

$$
A = \begin{bmatrix} 1 & 1 \\ -1 & 1 \\ 2 & 2 \end{bmatrix}; \quad b = \begin{bmatrix} 2 \\ 1 \\ 4 \end{bmatrix}; \quad A^* = \begin{bmatrix} 1 & 1 & 2 \\ -1 & 1 & 1 \\ 2 & 2 & 4 \end{bmatrix}
$$

The rank of A is 2. Note that using the first and last row of A for the calculation of the determinant yields a *singular* 2×2 submatrix. The determinant of A^* is

$$
|A^*| = \begin{vmatrix} 1 & 1 & 2 \\ -1 & 1 & 1 \\ 2 & 2 & 4 \end{vmatrix} = (1)\begin{vmatrix} 1 & 1 \\ 2 & 4 \end{vmatrix} + (-1)\begin{vmatrix} -1 & 1 \\ 2 & 4 \end{vmatrix} + (2)\begin{vmatrix} -1 & 1 \\ 2 & 2 \end{vmatrix}
$$

$$
= 2 + 6 - 8 = 0
$$

which makes A^* singular. However, the rank of A^* is at least 2 due to the presence of the same submatrix, which determined the rank of A. From this theorem, rank of A is equal to the rank of A^*, which is 2 and this is the same as $n(=2)$. It can be concluded that there is a *unique* solution for the system of equations. A useful observation of A^* matrix in the preceding example is that two of the rows are a constant multiple of each other. These matrices are singular. The same holds true for two columns that are multiples of each other.

Unit Vectors: Real-world optimization problems frequently involve a large number of design variables (*n variables*). In real LP problems, the number of variables can be staggering. For a successful solution to the optimization problem, it is implicitly assumed that the set of design variables is linearly independent. That is, there is no direct dependence among any two variables of the set. In this case, the *abstract space* spawned by the design variables is called the *n-dimensional Euclidean space*. The word *abstract* is used here to signify that it a construct in the imagination, as only three variables can be accommodated in the familiar 3D physical space. The design vector is a point in this space. The Euclidean space is a direct extrapolation of the physical, geometrical three-dimensional space that surrounds us. This space is also termed as the *Cartesian* space. For an illustration, consider the point (2, 3, 2) in Cartesian space or in the *rectangular* coordinate system. In Figure 3.6, the point is marked as P. The triad of numbers within parenthesis denote the length (*values*) along the x_1, x_2, x_3 axes, respectively.

The Cartesian space is identified by a set of mutually perpendicular lines or *axes* marked as x_1, x_2, and x_3 in this book. They are popularly written as *x, y, z*. The point where they intersect is the origin (O) of the coordinate system. The point P can also be associated with the *vector* OP, drawn from the origin O to the point P. Very often, a **point** and a **vector**—two very different entities—are used synonymously. From elementary vector addition principles, the vector OP can be constructed by sequentially joining three vectors along the three coordinate

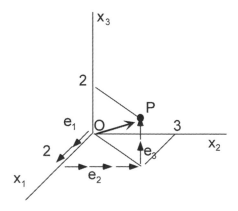

Figure 3.6   Rectangular coordinate system.

axes. In Figure 3.6, this is shown by laying out a vector along the x_1 axis, adding a vector along the x_2 axis to it, and finally adding the third vector along the x_3 axis. The small arrows in the figure are the *unit* vectors. They have a magnitude of **one** and a direction along one of the axis. These unit vectors are identified as e_1, e_2, *and* e_3 in the figure. Mathematically, the point P and the vector addition can be can be represented as (column vectors)

$$\begin{bmatrix} 2 \\ 3 \\ 2 \end{bmatrix} = 2 \begin{bmatrix} 1 \\ 0 \\ 0 \end{bmatrix} + 3 \begin{bmatrix} 0 \\ 1 \\ 0 \end{bmatrix} + 2 \begin{bmatrix} 0 \\ 0 \\ 1 \end{bmatrix} = 2e_1 + 3e_2 + 2e_3 \qquad (3.18)$$

The advantage of unit vectors is that it can establish any point in the Cartesian space through an appropriate triad of numbers signifying coordinate values or location. Actually, any point in the Cartesian space can be determined by *any set of three linearly independent vectors*. These vectors are termed as the **basis vectors**. In this case, the simple connection between the point and the elementary addition, illustrated in Equation (3.20), will not be available. Furthermore, through the methods of linear algebra, these three vectors can be reduced to the unit vectors through elementary row/column operations—also called **Gauss-Jordan elimination** or **reduction**. In *n-dimensional* Euclidean space spanned by the *n*-design variables, the point $P_n(p_1, p_2, \ldots p_n)$, will be represented by a collection of *n* values enclosed in parenthesis. The corresponding *n* unit vectors are defined as:

$$e_1 = \begin{bmatrix} 1 \\ 0 \\ \cdot \\ \cdot \\ 0 \end{bmatrix}; \quad e_2 = \begin{bmatrix} 0 \\ 1 \\ \cdot \\ \cdot \\ 0 \end{bmatrix}; \quad e_j = \begin{bmatrix} 0 \\ \cdot \\ 1 \\ \cdot \\ 0 \end{bmatrix}; \quad e_n = \begin{bmatrix} 0 \\ 0 \\ \cdot \\ \cdot \\ 1 \end{bmatrix} \qquad (3.19)$$

The vector from the origin (O) to P_n is represented as:

$$OP = p_1 e_1 + p_2 e_2 + \ldots + p_n e_n$$

where P_n is the point $(p_1, p_2, \ldots, p_n)$ or expressed as a column vector $[p_1 \ p_2 \ \cdots \ p_n]^T$. These new definitions will be sufficient to understand the techniques used to solve LP problems.

3.2 GRAPHICAL SOLUTION

The graphical discussion is limited to two design variables. The graphical discussion is used here to develop certain geometric ideas related to LP. In LP, the introduction of slack variables to a two-variable problem for the *equality constrained* representation will easily exceed this limit of two variables for graphical illustration, so the graphical solution is illustrated with respect to the original variables in the design problem using the *inequality constrained* representation. MATLAB is once again used for drawing the graphical solution. The problem is not reduced to the standard format for the graphical solution. It is important to do so for a numerical solution. Since the primary graphical element is the *straight line*, a function for drawing lines is developed through the function m-file **NewdrawLine.m** below. This function can be found in the Chapter 3 directory in the download from the publisher's site.

NewdrawLine.m: This function m-file for drawing straight lines is significantly improved from the earlier version of **drawLine.m**. The line to be drawn is represented as

$$ax + by = c$$

Help *NewdrawLine* will output the following information to the Command window

```
Applied Optimization with MATLAB Programming - 2nd Edition
P. Venkataraman
John Wiley - 2008


Chapter 3:  Linear Programming


Draws a line
Draws hash marks if it is an inequality constraint
identifies the constraint at (x2, y2)
---------------------------------------------------------
Lines are represented as: ax + by = c ( c >= 0)
Not more than one coefficient can be zero
 x2 > x1 and they are the plotting limits
in the case y1 and y2 are required but not defined
through the equations
their limits are the same as x1 and x2
---------------------------------------------------------
x1, x2 indicate the range of x for the line
a, b, c are the coefficients of x, y, and intercept
typ indicates type of line being drawn
with respect to constraints if any
        le ( <= )-inequality [less than equal to]
        ge,( >= )-inequality [greater than equal to]
        ee - equality
        nn - objective function
number of points for plot  = 50 [nplot]
  Colors:  objective function [green]
            equality constraint [blue]
     inequality constraint dashed [red], black hash marks
```

```
strval is a constraint label
        like 'h_1', 'g_1', 'g_2', 'f'
        The lines are identified on the right along with
        the value of c

If a == 0 then   a upper limit on y is drawn. The hash marks
are drawn in black and the line is black
%%
If b == 0 then   a upper limit on x is drawn. The hash marks
are drawn in black  and the line is black.  The x2 input
is the y limit (if any)
    The handles to the figures are returned in case you need
    to delete them for animation
```

The usage and restrictions can be seen in the information detailed in the comments. The significant change is that hash marks will be drawn for *inequality* constraints. You can number the constraints through **strval**. You will need to identify if the right-hand side value should be less than or greater than zero. It will also print the value of the constant c, along with the type and number, on the right side of the graph if the line ends in the visible region of the plot. Figures 3.2 to 3.5 have been produced using this new function m-file. It does not return a value. Please make sure to open the file in the editor and go over the code. It is commented to help you understand the code.

3.2.1   Example 3.1

Example 3.1 is reproduced once more for convenience.

$$\text{Maximize} \quad f(X)\text{: } 990x_1 + 900x_2 + 5{,}250 \tag{3.1}$$

$$\text{Subject to:} \quad g_1(X)\text{: } 0.4x_1 + 0.6x_2 \le 8.5 \tag{3.2a}$$

$$g_2(X)\text{: } 3x_1 - x_2 \le 25 \tag{3.2b}$$

$$g_3(X)\text{: } 3x_1 + 6x_2 \le 70 \tag{3.2c}$$

$$x_1 \ge 0; x_2 \ge 0$$

The graphical solution was shown in Figure 3.1. The solution, read from the figure (you can zoom the area of solution) is $x_1^* = 11$, and $x_2^* = 7$. It is the intersection of the active constraints (3.2b) and (3.2c). The actual values are $x_1^* = 10.48$ and $x_2^* = 6.42$. Since the number of machines has to be an integer, the solution is adjusted to a neighboring integer value. The solution was obtained through the code **Example3_1.m**, which is included to show the use of the **NewdrawLine.m** function.

Example3_1.m
```
% Chapter 3: Linear Programming
%--------------------------------------------------
%%%%%%%%%%%%%%%%%%%%%%%%%%%%%%%%%%%%%%%%%%%%%%%%%%%%%
% Applied Optimization with Matlab Programming
% Dr. P.Venkataraman
% second Edition,  John Wiley
%%%%%%%%%%%%%%%%%%%%%%%%%%%%%%%%%%%%%%%%%%%%%%%%%%%%%
%--------------------------------------------------
%  Example 3.1 (Sec 3.2.1)
%
%   graphical solution using matlab (two design variables)
%   the following script should allow the graphical solution
%
%   Maximize    f(x1,x2) = 990x1 + 900x2 +5250
%
%               g1(x1,x2) = 0.4x1 + 0.6x2 <= 8.5
%               g2(x1,x2) = 3x1 - x2 <= 25
%               g3(x1,x2) : 3x1 + 6x2 <= 70
%               x1 >= 0, x2 >= 0
%
%
% The function are drawn by a function called
%---------------------------------------------
%   function [h1, h11] = NewdrawLine(x1,x2,a,b,c,typ,strval)
%---------------------------------------------
%  inputs are:
%    x1 : lower limit of x
%    x2 : upper limit of x
%    a, b, c are the coefficient of the line which is
%            expressed as ax + by = c
%    typ is a string variable
%    'nn' : objective (green)
%    'ee' : equality constraint (blue)
%    'gt' : greater than equal to zero (red dashed with hash)
%    'lt' : less than equal to zero (red dashed with hash)
%
%    strval is a constraint label
%            like 'h_1', 'g_1', 'g_2', 'f'
%            The lines are identified on the right along with
%             the value of c
```

```
%    The handles to the figures are returned in case
%    you need to delete them
%    for animation
%-------------------------------------------------------------


%%%%%%%%%%%%%%%%%%%%%%%%%%%%%%%%%%%%%%%%%%%%%%%%%%%%%%%%%%%%%%%%%
%%% management functions
clear  % clear all variable/information in the workspace
       % - use CAUTION
clc    % position the cursor at the top of the screen
format compact  % avoid skipping a line when writing to
       % the command window
warning off  % don't report any warnings like divide by
       % zero etc.


%%%%%%%%%%%%%%%%%%%%%%%%%%%%%%%%%%%%%%%%%%%%%%%%%%%%%%%%%%%%%%%%%

%%% Draw the constraints
[h1, h11]= NewdrawLine(0,30,0.4,0.6,8.5,'lt','g_1');
[h1, h11]= NewdrawLine(0,30,3,-1,25,'lt','g_2');
[h1, h11]- NewdrawLine(0,30,3,6,70,'lt','g_3');

%%% Draw several values of the objective function
[h1, h11]= NewdrawLine(0,30,990,900,-5250,'nn','f');
[h1, h11]= NewdrawLine(0,30,990,900,0,'nn','f');
[h1, h11]= NewdrawLine(0,30,990,900,5000,'nn','f');
[h1, h11]= NewdrawLine(0,30,990,900,10000,'nn','f');
[h1, h11]=  NewdrawLine(0,30,990,900,15000,'nn','f');
[h1, h11]=  NewdrawLine(0,30,990,900,20000,'nn','f');
xlabel('Number of machines of type A');
ylabel('Number of machines of type B');
title('Example 3.1 ')
grid
axis([0 30 -30 40])

hx = line([0 30], [0 0]);
set(hx,'LineWidth',2.5,'Color','k')
hy = line([0.1 0.1],[0 40]);
set(hy,'LineWidth',2.5,'Color','k')
```

In the code, there are numerous calls to the *NewdrawLine* function to help define the constraints, and several lines for increasing objective function values. The plot is labeled and the x and y axis are draw thicker. The feasible region is clearly identified through the hashed lines. The solution can be located at the intersection of g_2 and g_3.

3.2.2 Characteristics of the Solution

The geometry evident in the graphical solution of Example 3.1 is used to define and explain some of the concepts associated with LP and the corresponding numerical technique. Figure 3.7 is the graphical representation of the constraints involved in Example 3.1 (same as Figure 3.1 without the objective function). In order to relate the geometry to LP concepts, the standard format of LP is necessary. The standard format established before is:

$$\text{Minimize} \quad f(X): \ -990x_1 - 900x_2 - 5250 \tag{3.3}$$

$$\text{Subject to:} \quad g_1(X): 0.4x_1 + 0.6x_2 + x_3 = 8.5 \tag{3.4a}$$

$$g_2(X): 3x_1 - x_2 + x_4 = 25 \tag{3.4b}$$

$$g_3(X): 3x_1 + 6x_2 + x_5 = 70 \tag{3.4c}$$

$$x_1 \geq 0, \quad x_2 \geq 0, \quad x_3 \geq 0, \quad x_4 \geq 0, \quad x_5 \geq 0 \tag{3.5}$$

In Figure 3.7, the constraints $x_1 \geq 0$ and $x_2 \geq 0$ are added as dashed lines to the three functional constraint lines (3.4). The circles are the points of intersection

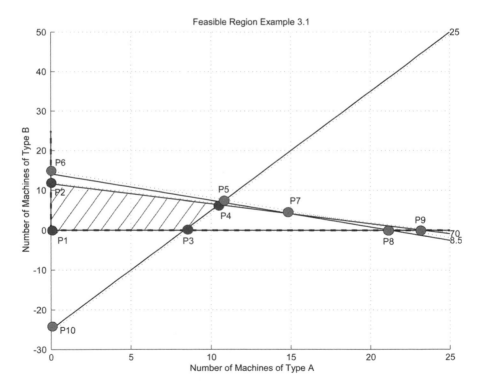

Figure 3.7 Feasible region, Example 3.1.

of five constraints, taken two at a time. This gives 10 points, which are identified as P1, P2, ... P10. The hashed area is the feasible region—that is, the design space in which all the constraints are satisfied. This is the rectangular region with the corners defined by the points P1, P2, P4, and P3. The remaining points are not feasible. All of these points of intersection can be associated with values for the variables. Note that there are five design variables ($n = 5$) and three functional constraints ($m = 3$). For example:

P1: ($x_1 = 0$, $x_2 = 0$, $x_3 = 8.5$, $x_4 = 25$, $x_5 = 70$)

P2: ($x_1 = 0$, $x_2 = 11.67$, $x_3 = 1.5$, $x_4 = 36.67$, $x_5 = 0$)

P3: ($x_1 = 8.333$, $x_2 = 0$, $x_3 = 5.167$, $x_4 = 0$, $x_5 = 45$)

P4: ($x_1 = 10.476$, $x_2 = 6.429$, $x_3 = 0.4524$, $x_4 = 0$, $x_5 = 0$)

P5: ($x_1 = 10.7$, $x_2 = 7.05$ $x_3 = 0$, $x_4 = 0$, $x_5 = -4.4$)

P6: ($x_1 = 0$, $x_2 = 14.17$, $x_3 = 0$, $x_4 = 39.17$, $x_5 = -15.42$)

The values of the variables are obtained as the intersection of the constraints taken two at a time. In this list, for each point, exactly two of the variables are zero. The number two corresponds to the value of ($n - m$); n being the number of variables, m being the number *functional* constraints. Points P5 and P6 are infeasible because one of the variables has a negative value, since in LP problems all the variables must be non negative.

Basic Solution: A **basic solution** is obtained by setting exactly $n - m$ variables to zero. In Figure 3.7, all the points identified by the circles represent basic solutions. The points chosen above are all basic variables. In general, for n design variables and m constraints, the number of basic solutions is given by the combination

$$\left(\frac{n}{m}\right) = \frac{n!}{m!(n - m)!}$$

Basic Variables: The set of variables in the *basic solution* that have non-zero values are called **basic variables**. Correspondingly, the set of variables in the *basic solution* that have the value of zero are called **non-basic variables**. For the point P1, x_1, and x_2 are *non-basic* variables, while x_3, x_4, and x_5 are *basic* variables.

Basic Feasible Solution: This is a *basic solution* that is also feasible. These are the points P1, P2, P3, and P4 in Figure 3.7. Points P5 and P6 represent basic solutions that are not feasible. In LP, the solution to the problem, if it is unique, must be a basic feasible solution. The basic solution can also be considered geometrically as a *corner* point or an *extreme* point of the feasible region. The geometry in Figure 3.7 confirms this.

Convex Polyhedron: A bounded region of the feasible design region—the region defined by the quadrilateral comprising of the points P1, P2, P3, and P4 in Figure 3.7.

The term *convex set* represents a collection of points or vectors having the following property: For any two points in the set (or within a region), if all the points on the line connecting the two points also lie in the same region, the region is called a **convex set**. Imagine any line drawn in the region defined by the quadrilateral whose corner points are P1, P2, P3, and P4. From Figure 3.7, it is clear this line will still be within the region established by the quadrilateral—making it a convex set.

Consider Figure 3.8. Points Q1 and Q2 belong to a bounded region. The line joining Q1 and Q2 includes the line segment Q3Q4 that does not belong to the region. Hence, the polyhedron in Figure 3.8 is *non-convex*.

Optimum Solution: A feasible solution that minimizes the objective function is the optimum solution. This will be point P4 inferred from Figure 3.1, since the objective function is not shown in Figure 3.7. In LP, the optimum solution must be a *basic feasible* solution.

Basis: The basis represents the columns of the coefficient matrix **A** that correspond to the basic variables. They form the basis of the m-dimensional space. They are termed as the basis vectors.

Canonical Form: The basis vectors are reduced to unit vectors through row/column operations (or Gauss–Jordan elimination). The basic feature of the

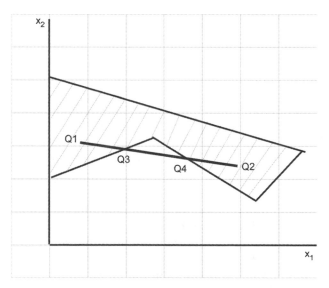

Figure 3.8 Non-convex polyhedron.

numerical technique for LP (Simplex method) is a repetitive procedure starting from an initial basic feasible solution, and determining the best neighboring basic feasible solution that improves the objective. The procedure is carried on until the optimum solution is reached, or if it is determined no solution is possible. The canonical form is used for rapidly identifying the solution. Each iteration in the procedure can be described mathematically as follows:

The starting constraints are organized as this matrix equation

$$[A]_{mxn} [X]_{nx1} = [b]_m \qquad (3.20a)$$

After applying the Gauss–Jordan elimination, the equation (3.20a) is assembled as

$$[I]_{mxm} [X]_m + [R]_{mx(n-m)} [X]_{(n-m)} = [b] \qquad (3.20b)$$

The m-design variables in the first term in equation (3.20b) are the **basic variables**. The identity matrix I can be considered as the complete collection of unit vectors. The set of $(n - m)$ design variables in the second term are the non-basic values. In the **Simplex method**, the non-basic variables are summarily set to zero.

3.2.3 Different Solution Types

There are, at most, four different results that can be expected for the solution of the LP problem: (1) unique solution; (2) infinitely many solutions; (3) unbounded solution, and of course, the possibility that there is (4) no solution.

Unique Solution: The example used for discussion has a unique solution (Figure 3.1). The condition necessary for this to occur is that the objective function and the constraints have dissimilar slopes, and the feasible region is bounded/closed. Geometrically, this can be visually explained as the movement of the line representing the objective function, parallel to itself, in a direction that minimizes the function value, until it remains just in contact with one of the feasible corners (basic feasible solution) of the feasible region. The feasible region is identified distinctly in Figure 3.7. A simple exercise of imagining different objective function with different slopes should convince the reader that it will be possible to establish unique solutions at P1, P2, P3, or P4. The 2D geometry and construction can be used to understand the extension to n-dimensions. Here, the lines will be represented by **hyper-planes**.

Infinite Solution: In order for this to occur, the objective function must be parallel to one of the constraints that determine the feasible region. For example, in Figure 3.7, let the original problem be redefined so that the objective function is parallel to the constraint g_3. Any point on the constraint, and lying between the line segment defined by the points P2 and P4, is an optimal solution to the problem. They will yield the same value of the objective.

Unbounded Solution: In this case, the feasible region is not bounded. In Figure 3.7, if the constraints g_1 and g_2 were not part of the problem formulation, then the feasible region is not bound on the top. Referring to Figures 3.1 and 3.7, the objective function can be shifted to unlimited higher values. In practice, there will be an upper bound on the range of the design variables (not part of standard format) that will be used to close this region, in which case solutions of type (1) or (2) can be recovered. The presence of an unbounded solution also suggests that the formulation of the problem may be lacking. Additional meaningful constraint(s) can be accommodated to define the solution.

No Solution: Figure 3.7 is used to explain this possibility. Consider the direction of inequality in g_1 is changed to the opposite type ($\geq$). The feasible region with respect to the constraints g_1 and g_3 are to the right of $x_1 = 15.0$. This point is not feasible with respect to the constraint g_2. Therefore, there is no point that is feasible. There is no solution to the problem.

In this discussion, the two variable situations provided an obvious classification of the solutions. In practice, LP models are large, with over hundreds of variables. Modeling and transcription errors may unintentionally give rise to the different solutions. The generation of solution is based on numerical techniques of linear algebra, which is often sensitive to the quality of the matrix of coefficients. Filtering out errant data is usually significant exercise in the search of optimal solutions.

3.3   NUMERICAL SOLUTION – THE SIMPLEX METHOD

The standard numerical method for LP problems is based on the algorithm due to Dantzig. It is also referred as the *Simplex method*. The procedure is related to the solution of a system of linear equations. The actual application of the procedure can be associated with the Gauss–Jordan method from linear algebra, where the coefficient rows are transformed through elementary multiplication and addition. Most computer installations carry software that will help solve LP problems. MATLAB also provides procedures to solve LP problems in its *Optimization Toolbox*. In this section, the Simplex method is applied to simple problems primarily to understand the programming and geometric features. LP programming methods are used in the direct techniques for nonlinear problems. The Simplex method is developed in detail using Example 3.1 illustrating several concepts. The calculations are implemented using MATLAB in this chapter but spreadsheet programs like Microsoft *Excel* can also be used to implement the Simplex method used for the simple problems.

3.3.1   Features of the Simplex Method

In this section, the machine selection example, Example 3.1 is used for illustrating and discussing the Simplex method. The method is iterative. Given a starting point, it will march forward through improving designs until it has found the solution or cannot proceed further. For the sake of completeness, the original problem is rewritten here:

$$\text{Maximize} \quad f(X): 990x_1 + 900x_2 + 5250 \tag{3.1}$$

$$\text{Subject to:} \quad g_1(X): 0.4x_1 + 0.6x_2 \leq 8.5 \tag{3.2a}$$

$$g_2(X): 3x_1 - x_2 \leq 25 \tag{3.2b}$$

$$g_3(X): 3x_1 + 6x_2 \leq 70 \tag{3.2c}$$

$$x_1 \geq 0; x_2 \geq 0$$

The problem was transformed to the *standard format* earlier. We will use this format to develop the numerical procedure.

Example 3.1:

$$\text{Minimize} \quad f(X): -990x_1 - 900x_2 - 5,250 \tag{3.3}$$

$$\text{Subject to:} \quad g_1(X): 0.4x_1 + 0.6x_2 + x_3 = 8.5 \tag{3.4a}$$

$$g_2(X): 3x_1 - x_2 + x_4 = 25 \tag{3.4b}$$

$$g_3(X): 3x_1 + 6x_2 + x_5 = 70 \tag{3.4c}$$

$$x_1 \geq 0, x_2 \geq 0, x_3 \geq 0, \quad x_4 \geq 0, \quad x_5 \geq 0 \tag{3.5}$$

x_3, x_4, and x_5 are the slack variables.

The Simplex method is used on the problem being expressed in the standard format. The following information is useful to assist in organizing the calculation, as well as recognizing the motivation for subsequent iterations. These apply for a problem with a unique solution:

- The number of variables in the problem is n. This includes the slack/surplus variables.
- The number of constraints is $m(m < n)$.
- The optimization problem is always to minimize the objective function f.
- The number of basic variables is m (same as the number of constraints).
- The number of non-basic variable is $n - m$.
- The set of unit vectors expected is $e_1, e_2, \ldots, e_m$.
- The column of the right-hand side (b) has positive values greater than or equal to zero.
- The calculations are organized in a table, which is called a *tableau*, or *Simplex tableau*. It will be referred to as the table in this book.

- Only the values of the coefficients are necessary for the calculations. The table therefore contains only coefficient values. The matrix [**A**] in previous discussions is the coefficients in the constraint equations.
- The immediate objective function is the **last row** in the table. The constraint coefficients are written above.
- Each column represents the coefficient corresponding to the same variable, except for the last column which is the value for the right hand side (*b*).
- Row operations consist of adding (subtracting) a definite multiple of the **pivot row** to other rows of the table and replacing the original rows. The pivot row identifies the row in which the unit vector (in a previously identified column) will have the unity value.
- Row operations are needed to obtain the canonical form. Practically, this implies that you should be able to spot the **unit vectors** in the table. If this happens then the current iteration is complete and considerations of the next one must begin

With this in mind, the Simplex method is applied to the machine selection problem. The example will use and clarify the items in this list.

3.3.2   Application of Simplex Method

Simplex Table 1:   In the Table 3.1 (Simplex Table 1), the first row indicates the variable names. The last row is the objective function. The last column is the right-hand side values. Spreadsheets are an efficient way to process the information. If it is being employed, then the symbol f can be removed from the last column and last row to allow normal calculations using numeric values only. The symbol f is avoided in the tables in the book, with the understanding that the interpretation will be made for the solution. The last column is the right-hand side values. The rest of the entries are the coefficients of the constraint equations.

The current iteration is over if the table displays the *canonical* form. In practice the canonical form comprises of spotting the *m-unit vectors* in the table, as well as making sure the entries under the *b-column*, except for the rows representing the objective function(s), are non negative (≥ 0). A glance at Simplex Table 1 of Example 3.1 indicates that the **canonical form** is present. There are three unit vectors (e_1, e_2, e_3) under the variable columns, including the last row.

Table 3.1   Simplex Table 1

x_1	x_2	x_3	x_4	x_5	b
0.4	0.6	1	0	0	8.5
3	−1	0	1	0	25
3	6	0	0	1	70
−990	−900	0	0	0	$f + 5{,}250$

The unit vectors in the table also identify those m-variables that will belong to the *basis*. Those variables will have a non-zero value for this iteration. The remaining, $n - m$, *non-basic* variables are set to zero. This solution is directly interpreted from Table 3.1. The solution, therefore, is:

$$x_1 = 0.0, x_2 = 0.0, x_3 = 8.5, x_4 = 25, x_5 = 70, \text{ and } f = -5,250$$

Has the Optimal Solution Been Obtained? The optimal solution requires that f be reduced as much as possible. For the current solution, the value of x_1 is 0. From Equation (3.3) if x_1 were increased from zero, then the objective function f will decrease further in value. A similar reasoning can be used for x_2. If x_1 and x_2 were not zero, the *solution can be further improved*. Therefore the solution has not converged. The second/next iteration will require assembling another similar table, Table 3.2.

New Variables in the Basis: The only way for x_1 and/or x_2 to have a positive value is if they became a basic variable. Since the number of basic variables is prescribed for the problem, some of the current basic variables must become *non-basic* to accommodate x_1 and x_2. For each iteration, the Simplex method allows *only one pair* of variables to conduct this *basis–non-basis* exchange. Each iteration takes place through a new Table.

To affect the exchang, it is necessary to determine which is the variable that will enter the basis (x_1, or x_2)—**entering basic variable (EBV)**. Also to be established, which of the current basic variables (x_3, x_4, *or* x_5) will leave the basis—**leaving basic variable (LBV)**? In terms of the numerical calculations, in the next Simplex table, the unit vector under the LBV column will be transferred to the EBV column through elementary row/column operations.

Entering Basic Variable: The EBV is chosen by its ability to provide for the largest decrease in the objective function in the current iteration/table. This is determined by the largest negative coefficient in the row associated with the objective function. From Table 3.1, this value is -990 and corresponds to the variable x_1. **EBV is x_1.** A tie can be broken arbitrarily.

Table 3.2   Example 3.1: Simplex Table 2

x_1	x_2	x_3	x_4	x_5	b
0	0.7333	1	−0.1333	0	5.1667
1	−0.3333	0	0.3333	0	8.3333
0	7	0	−1	1	45
0	−1230	0	330	0	13530

Leaving Basic Variable: The LBV is also chosen by its ability to improve the objective. The LBV is only determined *after the EBV has been chosen*. This is important because only the column under the EBV is examined to establish the LBV. The minimum ratio of the values of right hand side (*b-column values*) to the corresponding coefficients in the EBV column, for those coefficient values that are positive, decides the LBV. This minimum value identifies the row which is also the **pivot row**. In this pivot row, the column, which contains the unity value of the unit vector (or a current basic variable), is the column of the leaving basic variable. This is the LBV.

In Table 3.1, the EBV has already been identifies as x_1. The ratios under this column are 21.25, 8.3333, and 23.3333. The least value is 8.3333. This corresponds to the second row of Table 3.1. The unit vector under the x_4 column is in this row (row 2). Therefore, x_4 is the LBV and the second row is the pivot row for Table 3.2. This implies that the second row will be utilized to change the current x_1 column into a unit vector that is currently under x_4. This mechanical procedure is based on the following reasoning. The value of EBV must be such that it will not cause constraint violation with the current values of the other basic feasible variables. The application of the Gauss–Jordan process should shift the column under the LBV to the EBV, leaving the column of the other unit vectors (that are part of the basis) unaltered. Table 3.2 is then constructed, starting with the row 2 as the pivot row.

Construction of Simplex Table 2 (Table 3.2):   The original row 2 (Table 3.1) is

3   −1   0   1   0   25

The value of the first element in the new row will be 1. The row is modified by dividing through by 3. The new second row, also the pivot row, in Table 3.2 will be

1   −0.3333   0   0.33330   8.3333

In Table 3.2, the first element of the first row (Table 3.1) must be reduced to 0. This is always achieved by *adding/subtracting from the row being modified an appropriate multiple of the pivot row*. The modification to the first row is

row1 − {0.4 * (pivot row2)} to yield (action)
0   0.73331   −0.1333   0   5.1667   (result)

The third row will be obtained by

row3 − {3 * (pivot row3)}

The fourth row will be obtained by

row4 + {(990 * (pivot row2)}

Table 3.2 is obtained after these calculations.

The negative value of the coefficient in Table 3.2 in row 4 indicates we need at least one more iteration/table. The EBV for Simplex Table 3 is x_2. From examining the positive ratios of the values in the b column to the coefficient in the x_2 column, the LBV can be identified as x_5. The pivot row is row 3. Table 3.3 will be constructed by transforming the rows of Table 3.2.

Construction Simplex Table 3 (Table 3.3):

row3 = (row3) / 7 (pivot row)
new row 1 = row1 − {0.73333 * (pivot row3)}
new row 2 = row2 + {0.3333 * (pivot row3)}
new row 4 = row4 + {1230 * (pivot row3)}

Table 3.3 is obtained based on the previous operations.

Table 3.3 does not have any negative value in the objective function row. A solution has been achieved. The basic variables have the unit vector in their column. The non-basic variables are set to zero. From Table 3.3 the optimum values can be observed as follows:

$$x_1{}^* = 10.4762, \; x_2{}^* = 6.4285, \; x_3{}^* = 0.4524, \; f^* = -21407.14$$

Note, the value in the last row and last column is interpreted as

$$f^* + 21407.14 = 0$$

3.3.3 Solution Using MATLAB Code

In this edition, we improve on the use of MATLAB for executing the Simplex method. The code is available in **MATLABExample3_1.m**. It is not explained here, but the results of running the well-commented code is produced here. The code should do the following:

- Identify and print the Simplex tables.
- Determine if further iterations are necessary.
- Identify the entering basic variable.
- Identify the pivot row.

Table 3.3   Example 3.1: Simplex Table 3

x_1	x_2	x_3	x_4	x_5	b
0	0	1	−0.0285	−0.1047	0.4524
1	0	0	0.2857	0.0476	10.4762
0	1	0	−0.1428	0.1428	6.4285
0	0	0	154.28	175.71	21407.14

- Identify the leaving basic variable.
- Use matrix manipulations to generate the new matrix using row operations.
- Recognize and print the solution.

The reader is strongly recommended to look over the code, as it contains several new elements, like handling the following:

- Necessary character strings.
- Logical operations to identify particular rows and columns and variables.
- Creating unit vectors.
- Comparing vectors.
- Recognizing basic variables.

The following statements will appear in the Command window as the result of running the script file **MATLABExample3_1.m**.

```
Example 3.1 - The Simplex Method
Calculations using MATLAB

Data Entry - Set up Augmented matrix Ab = [A b]
The values can be checked in Simplex Table 1

Number of variables [n]:    5
Number of constraints [m]:    3


------------------
Simplex Table-1
------------------
          x1         x2         x3         x4         x5          b
       0.400      0.600      1.000      0.000      0.000      8.500
       3.000     -1.000      0.000      1.000      0.000     25.000
       3.000      6.000      0.000      0.000      1.000     70.000
    -990.000   -900.000      0.000      0.000      0.000   5250.000


-----------------------------------
Another Iteration is Necessary !!!!
-----------------------------------


Entering basic Variable (EBV) = x1
Pivot Row:    2
Leaving Basic Variable (LBV) = x4
------------------
Simplex Table-2
------------------
```

x1	x2	x3	x4	x5	b
0.000	0.733	1.000	-0.133	0.000	5.167
1.000	-0.333	0.000	0.333	0.000	8.333
0.000	7.000	0.000	-1.000	1.000	45.000
0.000	-1230.000	0.000	330.000	0.000	13500.000

```
------------------------------------
Another Iteration is Necessary !!!!
------------------------------------


Entering basic Variable (EBV) = x2
Pivot Row:    3
Leaving Basic Variable (LBV) = x5
------------------
Simplex Table-3
------------------
```

x1	x2	x3	x4	x5	b
0.000	0.000	1.000	-0.029	-0.105	0.452
1.000	0.000	0.000	0.286	0.048	10.476
0.000	1.000	0.000	-0.143	0.143	6.429
0.000	0.000	0.000	154.286	175.714	21407.143

```
------------------------------------
Further Iteration is Not Possible
Check for Solution
------------------------------------
 x( 3) =    0.452
 x( 1) =   10.476
 x( 2) =    6.429

Optimal value : -21407.143
```

3.3.4 Solution Using MATLAB's Optimization Toolbox

This section is useful for those that have access to the *Optimization Toolbox* from MATLAB *(ver 7.3)*.

Start MATLAB. In the Command window, type *help linprog*. This will provide information on the use of the linear programming routine to solve the problem. Here, the standard use is employed. It involves specifying values for the cost coefficients (*f-vector*), the coefficient/constraint matrix (*A-matrix*), and the right-hand side vector (*b-vector*). The following sequence of steps illustrates the use of the program. Using the linear programming function

```
>>    f = [-990;-900]
>>    A = [0.4 0.6 ; 3 -1; 3 6]
```

```
>>    b = [8.5 25 70]'
>>    [x, fval] = linprog(f,A,b)
```

The solution as posted in the Command window **Optimization terminated**.

```
x =
    220/21
     45/7
fval =
  -113100/7
```

To this solution must be added the constant $-5{,}250$, which was omitted in the problem definition for MATLAB. The constant just moves the line parallel to itself and is not used in the calculation. Using **format short** will report the solution as decimals instead of rational numbers.

3.4 ADDITIONAL EXAMPLES

In this section, additional examples are presented. These examples illustrate the extension/modification of the Simplex method to handle greater-than equal constraints, negative values for the design variables, equality constraints, and so on. In all the cases, the problem is transformed appropriately and the same Simplex method is then applied.

3.4.1 Example 3.2 — Transportation Problem

The Fresh Milk cooperative supplies milk in gallon jugs from its two warehouses located in Buffalo, New York and Williamsport, Pennsylvania. It has a capacity of 2,000 gallons at Buffalo and 1,600 gallons at Williamsport. It delivers 800 gallons/day to Rochester, New York. Syracuse, New York requires 1,440 gallons/day, and the remaining 1360 gallons are trucked to New York City. The cost to ship the milk to each of the destinations is different and is given in Table 3.4. Establish the shipping strategy for minimum cost.

Problem Formulation: Let x_1 be the number of gallons shipped from Buffalo to Rochester. Let x_2 be the number of gallons shipped from Buffalo to Syracuse.

Table 3.4 Shipping Cost (per gallon)

	Rochester	Syracuse	New York City
Buffalo	4.2	4.5	6.0
Williamsport	4.7	4.6	5.1

The warehousing constraint at Buffalo is 2,000 gallons/day. Therefore

$$x_1 + x_2 \le 2000 \tag{3.21}$$

Amount shipped from Williamsport to Rochester $= 800 - x_1$
Amount shipped from Williamsport to Syracuse $= 1,440 - x_2$
Amount shipped from Williamsport to New York City is

$$1,600 - (800 - x_1) - (1,440 - x_2) \ge 0$$

$$x_1 + x_2 \ge 640 \tag{3.22}$$

The side constraints on x and y are the respective warehouse limits. The shipping cost is

$$\text{Cost} = 4.2 * x_1 + 4.5 * x_2 + 6.0 * (2,000 - x_1 - x_2) + 4.7 * (800 - x_1)$$
$$+ 4.6 * (1,440 - x_2) + 5.1 * (x_1 + x_2 - 640) \tag{3.23}$$
$$\text{Cost} = -1.4 * x_1 - x_2 + 19,120$$

The problem can now be constructed in standard format.

Example 3.2:

$$\begin{aligned} \text{Minimize} \quad & f(x_1, x_2): \ -1.4x_1 - x_2 + 19,120 & (3.24) \\ \text{Subject to:} \quad & g_1(x_1, x_2): x_1 + x_2 \le 2,000 & (3.25a) \\ & g_2(x_1, x_2): x_1 + x_2 \ge 640 & (3.25b) \\ & 0 \le x_1 \le 800; \ 0 \le x_2 \le 1,440 & (3.26) \end{aligned}$$

Figure 3.9 illustrates the graphical solution to Example 3.2. The values for f on the figure include the value of 19,120. Remember, the lines must be expressed as $ax + by = c$ for the plots. The solution is at the intersection of constraint g_1 and the upper limit on the value of x as the objective is to decrease f as much as possible. From Figure 3.9, the solution is:

$$x_1 = 00; \quad x_2 = 1,200, \quad f = 16,800, \text{ or } \$168.00 \tag{3.27}$$

Figure 3.9 can be obtained by running **Example3_2.m**. Note that the limits on x_1 and x_2 are also drawn by the code.

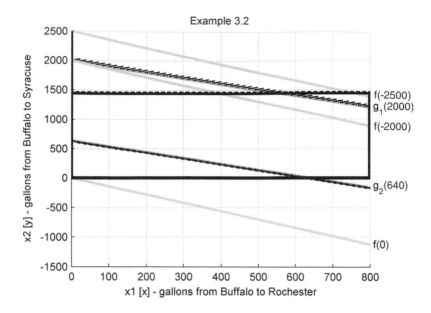

Figure 3.9 Graphical Solution — Example 3.2.

Two-Phase Simplex Method: The main difference between Example 3.1 and Example 3.2 is the $\geq$ constraint in the latter. Since the standard LP problem only requires the design variables be semi-positive (≥ 0), the right-hand constraints on x_1 and x_2 have to be accommodated through additional inequality constraints. In developing the details of the method, only the original variables are identified as x. The additional variables (slack, surplus, artificial) are identified separately to clarify the procedure. The surplus variable creates a problem in directly recognizing the canonical form in the Simplex Table 1. It is necessary to approach the solution through additional calculations developed through *two phases*. Example 3.2 expressed in the standard format of linear programming is:

$$\text{Minimize} \quad f(x_1, x_2): \ -1.4x_1 - x_2 + 19,120 \tag{3.24}$$

$$\text{Subject to:} \quad g_1(x_1, x_2): \ x_1 + x_2 + s_1 = 2,000 \tag{3.28a}$$

$$g_2(x_1, x_2): \ x_1 + x_2 - s_2 = 640 \tag{3.28b}$$

$$g_3(x_1, x_2): \ x_1 + s_3 = 800 \tag{3.28c}$$

$$g_4(x_1, x_2): \ x_2 + s_4 = 1,440 \tag{3.28d}$$

$$x_1 \geq 0, x_2 \geq 0, s_1, s_2, s_3, s_4 \geq 0 \tag{3.29}$$

Here, s_1, s_3, s_4, are the **slack** variables and s_2 is the **surplus** (similar to a slack but used for $\geq$ constraints) variable. In Example 3.1, when the first Simplex table was set up (Simplex Table 3.1), the slack variable were helpful in identifying the canonical form. Here in Example 3.2, it does not work out that way because of the coefficient of s_2 is -1. Multiplying equation (3.28b) by -1 migrates the negative sign to the right-hand side, which is also disallowed as far as the recognition of the canonical form is concerned. This means additional problem-specific preprocessing must take place to identify the initial canonical form. Since the Simplex method is used to handle large problems with several variables, it is more convenient to apply the Simplex procedure in a consistent way. This is accomplished by applying the same Simplex technique in a two-part sequence.

The first part is recognized as *Phase I*. Here, a new kind of variable, *an artificial variable*, is defined for each *surplus* variable in the same equation. Also, a new objective function called an **artificial cost function** or an **artificial objective function** is introduced. The artificial objective is defined as a sum of all the artificial variables in the problem. In *Phase I*, the artificial objective is reduced to zero using the standard Simplex procedure. When this is accomplished, *Phase I* is complete. If the artificial objective depends only on the artificial variables, and if its value is zero, this implies that the artificial variables are **non-basic variables**. This also suggests that these variables were basic variables at the start of the procedure. When *Phase I* is completed, then both the artificial objective function, and the artificial variables are discarded from the table and *Phase II* begins.

Phase II is the standard Simplex technique applied to the table from the end of *Phase I*, neglecting all the artificial elements. The table should be in canonical form. Additional tables are obtained as necessary until the solution is reached.

In Example 3.2 set up earlier, there will be one artificial variable a_1, and an artificial cost function A_f. In the example, only equation (3.28b) will be affected:

$$g_2(x, y): x_1 + x_2 - s_2 + a_1 = 640 \tag{3.30}$$

The artificial cost function will be

$$A_f = a_1 \tag{3.31}$$

Table 3.5 represents the first table in Phase I. The first four rows represent the constraints. The fifth row represents the original objective function. The last row is the *artificial objective function*. This last row used to drive the iterations in Phase I. Table 3.5, is not in canonical form, but replacing the sixth row by the result of subtracting the second row from the sixth row will provide a canonical form. Table 3.6 illustrates the result of such a row manipulation.

Simplex Method, Phase I: In Table 3.6, there are two choices (x_1, x_2) available for the EBV (entering basic variable), as both of them have a coefficient of -1 in the last row. Although the choice can be arbitrary, x_1 is a good choice because it has a larger negative coefficient in the original objective function. The LBV (leaving basic variable) is identified through the minimum positive value of

Table 3.5   Example 3.2: Initial Table, Phase I

x_1	x_2	s_1	s_2	a_1	s_3	s_4	b
1	1	1	0	0	0	0	2000
1	1	0	−1	1	0	0	640
1	0	0	0	0	1	0	800
0	1	0	0	0	0	1	1440
−1.4	−1	0	0	0	0	0	$f - 19120$
0	0	0	0	1	0	0	A_f

Table 3.6   Example 3.2: Simplex Table 1, Phase I (Canonical form)

x_1	x_2	s_1	s_2	a_1	s_3	s_4	b
1	1	1	0	0	0	0	2000
1	1	0	−1	1	0	0	640
1	0	0	0	0	1	0	800
0	1	0	0	0	0	1	1440
−1.4	−1	0	0	0	0	0	$f - 19120$
−1	−1	0	1	0	0	0	$A_f - 640$

the ratio of the values in the b column to the values under the x_1 column. These ratios are 2,000/1, 640/1, and 800/1. The selection identifies the second row as the *pivot* row and a_1 as the LBV. Using the second row as the pivot row, the unit vector $[0\ 1\ 0\ 0\ 0\ 0]^T$ has to be transferred from a_1 column to the x_1 column. Construction of Table 3.7 starts with the second row in Table 3.6, which is also the second row of Table 3.7 because of the coefficient of 1 under the second row under the x column (no scaling is necessary). This is also the pivot row. To

Table 3.7   Example 3.2: Simplex Table 2, Phase I (Canonical form)

x_1	x_2	s_1	s_2	a_1	s_3	s_4	b
0	0	1	1	−1	0	0	1,360
1	1	0	−1	1	0	0	640
0	−1	0	1	−1	1	0	160
0	1	0	0	0	0	1	1,440
0	0.4	0	−1.4	1.4	0	0	$f - 18,224$
0	0	0	1	0	0	0	A_f

obtain a 0 in the first row in the x column, the pivot row is subtracted from the first row in Table 3.6. The new first row is

$$0 \quad 0 \quad 1 \quad 1 \quad -1 \quad 0 \quad 0 \quad 1{,}360$$

To obtain 0 in the x_1 column in the third row, the pivot row is subtracted from the third row of Table 3.6. The fourth row has a 0 in place and hence is copied from Table 3.6. The fifth row is obtained by adding a 1.4 multiple of the pivot row to the fifth row in Table 3.6. The last row is the addition of the pivot row to the last row from Table 3.6. Table 3.7 is compiled as follows:

From Table 3.7, the value of A_f is 0, and a_1 is not a basic variable. This identifies the **end of Phase I**. The a_1 column and the last row are discarded and *Phase II* is started with Table 3.8.

Simplex Method. Phase II: The column with the greatest negative coefficient (-1.4) is s_2. This is the EBV. LBV is s_3. The third row is therefore the pivot row. The unit vector $[0\ 0\ 1\ 0\ 0]^T$ must be transferred from the s_3 column to the s_2 column through row manipulations using the pivot row. This exercise results in Table 3.9.

The negative coefficient under the x_2 column suggests we are not finished yet. For the next iteration, EBV is x_2. LBV is s_1. The first row is the pivot row. The unit vector $[1\ 0\ 0\ 0\ 0\ ]^T$ must be constructed under the y column using row manipulation using the pivot row. This results in Table 3.10.

Table 3.8   Example 3.2: Simplex Table 1, Phase II (Canonical form)

x_1	x_2	s_1	s_2	s_3	s_4	b
0	0	1	1	0	0	1,360
1	1	0	−1	0	0	640
0	−1	0	1	1	0	160
0	1	0	0	0	1	1,440
0	0.4	0	1.4	0	0	$f - 18{,}224$

Table 3.9   Example 3.2: Simplex Table 2, Phase II (Canonical Form)

x_1	x_2	s_1	s_2	s_3	s_4	b
0	1	1	0	−1	0	1,200
1	0	0	0	1	0	800
0	−1	0	1	1	0	160
0	1	0	0	0	1	1,440
0	−1	0	0	1.4	0	$f - 18{,}000$

Table 3.10   Example 3.2: Simplex Table 3, Phase II (Final)

x_1	x_2	s_1	s_2	s_3	s_4	b
0	1	1	0	−1	0	1,200
1	0	0	0	1	0	800
0	0	1	1	0	0	1,360
0	0	−1	0	1	1	240
0	0	1	0	0.4	0	$f - 16{,}800$

There are no negative coefficients in the last row. The solution has been reached. From inspection of Table 3.10, the solution is

$$x_1 = 800, \quad x_2 = 1{,}200, \quad s_2 = 1{,}360, \quad s_4 = 240, \quad f = 16{,}800 \quad (3.32)$$

In this example, the same Simplex method was repeatedly applied although the problem was solved using *two phases*. The first phase was a pre-processing phase to move the surplus variables away from the initial set of basic variables. This was achieved by introducing additional variables for each surplus variable and an additional cost function that drove the iterations of the first phase.

To summarize, for (≥) and (=) constraints, artificial variables and an artificial cost function are introduced into the problem for convenience. Simplex method is applied in two phases. The first phase is terminated when the artificial variables can be eliminated from the problem.

Solution Using MATLAB Code: Like the Simplex method, the MATLAB solution is obtained by running two programs. The first one, **PhaseI_MATLABExample3_2.m**, executes the first phase. The last table is copied and edited to start **PhaseII_MATLABExample3_2.m**. The final table from running both pieces of code is included here.

```
-------------------
Phase I - Simplex Table-2
-------------------


    x1      x2     x3      x4      x5      x6      x7         b

 0.000   0.000  1.000   1.000  -1.000  0.000  0.000   1360.000

 1.000   1.000  0.000  -1.000   1.000  0.000  0.000    640.000

 0.000  -1.000  0.000   1.000  -1.000  1.000  0.000    160.000

 0.000   1.000  0.000   0.000   0.000  0.000  1.000   1440.000
```

```
0.000   0.400 0.000 -1.400   1.400 0.000 0.000-18224.000

0.000   0.000 0.000   0.000   1.000 0.000 0.000     0.000
```

```
------------------
Phase II:Simplex Table-3
------------------
```

x1	x2	x3	x4	x5	x6	b
0.000	1.000	1.000	0.000	-1.000	0.000	1200.000
1.000	0.000	0.000	0.000	1.000	0.000	800.000
0.000	0.000	1.000	1.000	0.000	0.000	1360.000
0.000	0.000	-1.000	0.000	1.000	1.000	240.000
0.000	0.000	1.000	0.000	0.400	0.000	-16800.000

```
--------------------------------
Further Iteration is Not Possible
Check for Solution
--------------------------------
```

```
 x( 2) = 1200.000
 x( 1) = 800.000
 x( 4) = 1360.000
 x( 6) = 240.000
```

Optimal value : 16800.000

3.4.2 Example 3.3 — Equality Constraints and Unrestricted Variables

Example 3.3 will include an equality constraint and illustrates the technique to handle variables that are unrestricted in sign (can have negative values). Variables like profit, temperature, and net income can be negative quite often. The equality constraint is handled by introducing an artificial variable corresponding to the equation and applying the Two-Phase Simplex technique illustrated in Example 3.2. When this corresponding artificial variable is zero (end of Phase I), the constraint has been met. This addition of an extra variable to a constraint that is already in the standard format must appear to be a waste of time. The idea is to set up the initial table without significant effort to handle large problems with many variables, since it will be difficult to relate the large matrices of numbers. Note that the two-phase approach still applies the same Simplex method with a different objective function called the augmented matrix. The negative value for a variable is simulated through the difference between two positive variables. Once again, these changes preserve the Simplex method in its original form.

The Problem: Today, a full-time student on campus is always a driven to maximize the grades. A strong influence for a favorable outcome is the amount of investment made in hitting the books versus the time spent in playing pinball. In a typical day, at least one hour is definitely extended to the pursuit of learning and pressing the sides of the coin-operated machine. Not more than five hours are available for such dispersion. Over the years, a fatigue threshold of four units, based on a combination of the two activities, has been established as a norm for acceptable performance. This combination is the sum of the hours spent at the pinball machine and twice the hours spent on homework, an acknowledgment that hitting the books is stressful. A negative hour of pinball playing will go to increase the time spent in studying, which will contribute positively to the grades. The overall grade is determined as a linear combination of twice the time spent academically and subtracting the time spent on playing pinball. The problem, therefore, is to distribute the time to maximize the grades obtained.

Problem Formulation: There are two original design variables in the problem. x_1 is the number of hours spent in studying, and x_2 is the time spent in enjoying the game of pinball when the same time can be advantageously spent in maintaining a high grade in the courses. The objective function can be expressed as follows:

$$\text{Maximize } f(x_1, x_2): 2x_1 - x_2$$

The constraints can be recognized from the statement of the problem directly:

$$g_1(x_1, x_2): \quad x_1 + x_2 \leq 5$$

$$g_2(x_1, x_2): \quad 2x_1 + x_2 = 4$$

$$g_3(x_1, x_2): \quad x_1 + x_2 \geq 1$$

$$x_1 \geq 0, x_2 \text{ is } \textbf{unrestricted in sign}$$

Figure 3.10, using the `Example3_3.m` m-file, is a graphical description of the problem and the solution. The solution is at $x_1 = 3$, and $x_2 = -2$ on the figure, where conveniently the objective function is also passing through.

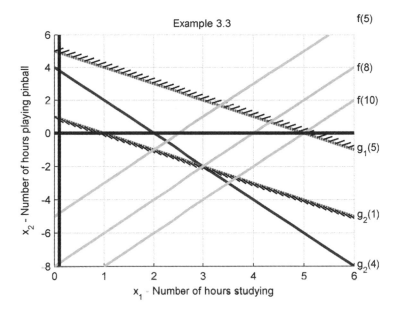

Figure 3.10 Graphical Solution, Example 3.3.

Standard Format: As the standard format expects only non-negative variables, the variable x_2 is replaced by a pair of variables in the formulation:

$$x_2 = x_{21} - x_{22} \tag{3.33}$$

The standard format after converting to a minimization problem, introducing slack, surplus, and artificial variables, and including equation (3.33) results in

$$\text{Minimize } f: \quad -2x_1 + x_{21} - x_{22} \tag{3.34a}$$

$$\text{Subject to:} \quad g_1: x_1 + x_{21} - x_{22} + s_1 = 5 \tag{3.34b}$$

$$g_2: 2x_1 + x_{21} - x_{22} + u_1 = 4 \tag{3.34c}$$

$$g_3: x_1 + x_{21} - x_{22} - s_2 + a_2 = 1 \tag{3.34d}$$

$$x_1, x_{21}, x_{22}, s_1, s_2, a_1, a_2 \geq 0 \tag{3.35}$$

The artificial function for the first phase of the two-phase Simplex method is

$$A_f = a_1 + a_2 \tag{3.36}$$

Table 3.11 captures the standard format in a table. It requires preprocessing with respect to the artificial variables so that a canonical form can be observed. Adding the second and third row and subtracting the result from the last row and replacing the last row with this computation will yield the Table 3.12.

Table 3.11 Example 3.3: Table 0, Phase I

x_1	x_{21}	x_{22}	s_1	a_1	s_2	a_2	b
1	1	−1	1	0	0	0	5
2	1	−1	0	1	0	0	4
1	1	−1	0	0	−1	1	1
−2	1	−1	0	0	0	0	f
0	0	0	0	1	0	1	A_f

Table 3.12 Example 3.3: Simplex Table 1, Phase I

x_1	x_{21}	x_{22}	s_1	a_1	s_2	a_2	b
1	1	−1	1	0	0	0	5
2	1	−1	0	1	0	0	4
1	1	−1	0	0	−1	1	1
−2	1	−1	0	0	0	0	f
−3	−2	2	0	0	1	0	$A_f - 5$

Phase 1: Simplex Table 1: Table 3.12 provides the start of the Simplex method. In Phase 1, the motivation is to drive the artificial function to 0, which has a value of 5. The EBV is x_1. The LBV is a_2. The third row is the pivot row. Row manipulations with the pivot row leads to Table 3.13.

Phase 1: Simplex Table 2: Table 3.13 is the second table under Phase 1. The value of the artificial function is 2, so Phase 1 is not yet over. The EBV is s_2. The LBV is a_1. The second row is the pivot row. This row will be used for row manipulations to yield the next table.

Phase 1: Simplex Table 3: Table 3.14 is the new table after the row operations. The value of the artificial function is 0 as both a_1 and a_2 are nonbasic

Table 3.13 Example 3.3: Simplex Table 2, Phase I

x_1	x_{21}	x_{22}	s_1	a_1	s_2	a_2	b
0	0	0	1	0	1	−1	4
0	−1	1	0	1	2	−2	2
1	1	−1	0	0	−1	1	1
0	3	−3	0	0	−2	2	$f + 2$
0	1	−1	0	0	−2	3	$A_f - 2$

Table 3.14   Example 3.3: Simplex Table 3, Phase I (Final)

x_1	x_{21}	x_{22}	s_1	a_1	s_2	a_2	b
0	0.5	−0.5	1	−0.5	0	0	3
0	−0.5	0.5	0	0.5	1	−1	1
1	0.5	−0.5	0	0.5	0	0	2
0	2	−2	0	1	0	0	$f+4$
0	0	0	0	1	0	1	A_f

Table 3.15   Example 3.3: Simplex Table 1, Phase II

x_1	x_{21}	x_{22}	s_1	s_2	b
0	0.5	−0.5	1	0	3
0	−0.5	0.5	0	1	1
1	0.5	−0.5	0	0	2
0	2	−2	0	0	$f+4$

variables with a value of 0 for this iteration. This signifies the end of Phase 1. Phase 2 will start with the last column (artificial function) and the two columns that represent the artificial variables removed from the current table.

Phase 2: Simplex Table 1:   Table 3.15 represents the table to start the iterations for Phase 2. The EBV is x_{22}. The LBV is s_2. The pivot row is the second row. Following the row operations, the Table 3.16 is developed.

Phase 2: Simplex Table 2:   Table 3.16 is the final table, as there are no negative coefficients in the last row.
    The solution is

$$x_1 = 3, x_{21} = 0, x_{22} = 2, s_1 = 4, s_2 = 0, f = -8$$

The value of x_2 is

$$x_2 = x_{21} - x_{22} = 0 - 2 = -2$$

which was identified graphically.

Table 3.16   Example 3.3: Simplex Table 2, Phase II (Final)

x_1	x_{21}	x_{22}	s_1	s_2	b
0	0	0	1	1	4
0	−1	1	0	2	2
1	0	0	0	1	3
0	0	0	0	4	$f+8$

This example illustrated the Simplex method for handling negative variables and equality constraints. Note the application of the Simplex technique itself did not change.

Solution Using MATLAB Code: Like the Simplex method, the MATLAB solution is obtained by running two programs. The first one, **PhaseI_ MATLABExample3_3.m**, executes the first phase. The last table is copied and edited to start **PhaseII_MATLABExample3_3.m**. The final table from running both pieces of code is included here.

```
------------------
Phase I - Simplex Table-3
------------------

   x1      x2       x3      x4      x5      x6      x7       b

 0.000   0.500  -0.500  1.000  -0.500  0.000   0.000  3.000

 0.000  -0.500   0.500  0.000   0.500  1.000  -1.000  1.000

 1.000   0.500  -0.500  0.000   0.500  0.000   0.000  2.000

 0.000   2.000  -2.000  0.000   1.000  0.000   0.000  4.000

 0.000   0.000   0.000  0.000   1.000  0.000   1.000  0.000


------------------------------------
Further Iteration is Not Possible
Check for Solution
------------------------------------
 x( 4) =    3.000
 x( 6) =    1.000
 x( 1) =    2.000
Optimal value :       0.000

------------------
Phase II:Simplex Table-2
------------------

        x1        x2        x3        x4        x5         b
      0.000     0.000     0.000     1.000     1.000     4.000
      0.000    -1.000     1.000     0.000     2.000     2.000
      1.000     0.000     0.000     0.000     1.000     3.000
      0.000     0.000     0.000     0.000     4.000     8.000
------------------------------------
Further Iteration is Not Possible
```

```
Check for Solution
----------------------------------
 x( 4) =    4.000
 x( 3) =    2.000
 x( 1) =    3.000

Optimal value :      -8.000
```

3.4.3   Example 3.4 – A Four-Variable Problem

Example 3.4 presents a problem with four variables. The procedure is identical to the one in Example 1, except that there is no graphical solution to the problem.

The Problem: The RIT student-run microelectronic fabrication facility is taking orders for four indigenously developed ASIC chips that can be used in touch sensors, LCD, pressure sensors, and controllers. There are several constraints on the production based on space, equipment availability, student hours, and the fact that the primary mission of the facility is student training. First, the handling time constraint, outside of processing, for all chips is 650 hours. Touch sensors require 4 hours, LCD 8 hours, pressure sensors 7 hours, and controllers require 10 hours. Second, the time available on the lithographic machines is about 500 hours. Touch sensors require 1 hour on the machine. An LCD requires 1.5 hours, pressure sensors require 2 hours, and controllers require 8 hours. Packaging considerations place the maximum at 1,200 volume units. Touch sensors require 25 volume units, LCD requires 40 volume units, pressure sensors require 18 units and controllers require 10 units because of their compact size. All the constraints are indicated per week of operation of the facility. The net revenue for each device is $6, $10, $9, and $20 for touch sensor, LCD, pressure sensor, and controller, respectively. The facility is interested in maximizing revenue per week and would like to determine the right mix of the four devices.

Problem Formulation: The formulation is straightforward and can be set up directly based on the problem description. Let x_1 represent the number of touch sensor chips per week, x_2, the number of LCD, x_3, the number of pressure sensors, and x_4, the number of controllers. The objective function is

$$\text{Maximize } f: \quad 6x_1 + 10x_2 + 9x_3 + 20x_4$$

The handling time constraint can be expressed as:

$$g_1: \quad 4x_1 + 8x_2 + 7x_3 + 10x_4 \le 650$$

The availability of the lithographic machine can be developed as:

$$g_2: \quad x_1 + 1.5x_2 + 2x_3 + 8x_4 \leq 500$$

The packaging constraint is

$$g_3: \quad 25x_1 + 40x_2 + 18x_3 + 10x_4 \leq 1{,}200$$

All design variables are expected to be greater than zero. As formulated, the problem displays a degree of incompleteness. There are *four* design variables and only *three* constraints. *The number of variables in the basis can only be three.* At least one of the original variables must have a value of zero at the solution—cannot be manufactured. Several useful additional constraints can be still included to define a valid optimization problem with a non-zero solution. This is now a modeling issue. For problems with a limited number of design variables, paying attention to the problem development allows anticipation of the solution, as well as the opportunity to troubleshoot decisions from practical consideration. The student can explore this problem further by extending the problem definition to generate various types of solutions.

Standard Format: The objective function requires a minimum formulation. Hence

$$\text{Minimize } f: \quad -6x_1 - 10x_2 - 9x_3 - 20x_4 \tag{3.37}$$

The symbol f has been retained for convenience even though the direction of optimization has changed. The constraints are set up using slack variables, $s_1, s_2,$ and s_3.

$$g_1: \quad 4x_1 + 8x_2 + 7x_3 + 10x_4 + s_1 = 650 \tag{3.38a}$$

$$g_2: \quad x_1 + 1.5x_2 + 2x_3 + 8x_4 + s_2 = 500 \tag{3.38b}$$

$$g_3: \quad 25x_1 + 40x_2 + 18x_3 + 10x_4 + s_3 = 1{,}200 \tag{3.38c}$$

All variables are semi positive.

$$x_1, x_2, x_3, x_4, s_1, s_2, s_3 \geq 0 \tag{3.39}$$

The standard Simplex method can be used to solve this problem.

Simplex Table 1: Table 3.17 is the initial table for Example 3.4. The canonical form is observable from the table and the initial basic feasible solution can be determined. s_1, s_2, s_3 are the basic variables. EBV is x_4 and the LBV is s_2. The pivot row is the second row used for the row manipulations to lead to Table 3.18.

Table 3.17   Example 3.4: Simplex Table 1

x_1	x_2	x_3	x_4	s_1	s_2	s_3	b
4.000	8.000	7.000	10.000	1.000	0.000	0.000	650.000
1.000	1.500	3.000	8.000	0.000	1.000	0.000	500.000
25.000	40.000	18.000	10.000	0.000	0.000	1.000	1200.000
−6	−10	−9	−20	0	0	0	f

Table 3.18   Example 3.4: Simplex Table 2

x_1	x_2	x_3	x_4	s_1	s_2	s_3	b
2.750	6.125	3.250	0.000	1.000	−1.250	0.000	25.000
0.125	0.188	0.375	1.000	0.000	0.125	0.000	62.500
23.750	38.125	14.250	0.000	0.000	−1.250	1.000	575.000
−3.500	−6.250	−1.500	0.000	0.000	2.500	0.000	$f + 1250.000$

Simplex Table 2:   Using the pivot row identified in the last table, the unit vector [0 0 1 0]' under the s_2 column needs to be transferred to the x_4 column through elementary row operations. Table 3.18 shows the canonical form after the completion of all the operations. The basic variables are x_4, s_1, s_3. The EBV is x_2 and the LBV is s_1. The pivot row is the first row. The objective of the row manipulations are to transfer the unit vector $[1\ 0\ 0\ 0]^T$ from the s_3 column to the x_2 column.

Simplex Table 3:   Table 3.19 denotes the reduced table with the canonical form after the required row operations are completed. The basis variables are x_2, x_4, s_3. The first row is the pivot row. The EBV is x_1 and the LBV is x_2. The pivot row is the first row.

Simplex Table 4:   Table 3.20 is the final Simplex table for this example. There are no negative coefficients in the last row, suggesting the solution has been

Table 3.19   Example 3.4: Simplex Table 3

x_1	x_2	x_3	x_4	s_1	s_2	s_3	b
0.449	1.000	0.531	0.000	0.163	−0.204	0.000	4.082
0.041	0.000	0.276	1.000	−0.031	0.163	0.000	61.735
6.633	0.000	−5.980	0.000	−6.224	6.531	1.000	419.388
−0.694	0.000	1.816	0.000	1.020	1.224	0.000	$f + 1275.510$

Table 3.20 Example 3.4: Simplex Table 4

x_1	x_2	x_3	x_4	s_1	s_2	s_3	b
1.000	2.227	1.182	0.000	0.364	−0.455	0.000	9.091
0.000	−0.091	0.227	1.000	−0.045	0.182	0.000	61.364
0.000	−14.773	−13.818	0.000	−8.636	9.545	1.000	359.091
0.000	1.545	2.636	0.000	1.273	0.909	0.000	$f + 1281.818$

obtained. From the table, the solution is:

$$x_1 = 9.091, \quad x_4 = 61.364, \quad x_2 = 0, x_3 = 51.61, \quad f = 1{,}281.82$$

This solution is not satisfactory because actual decision will involve integer values for the design variables. At this time, *integer programming* has not been introduced. Instead, the solution is adjusted to a neighboring integer is

$$x_1 = 9, x_2 = 0, x_3 = 0, x_4 = 61, \quad f = 1{,}274$$

The integer values should satisfy the constraints.

Solution Using MATLAB Code: The solution, including the various Tables, can be generated by running **MATLABExample3_4.m**. For the calculations, only the augmented matrix [A b] is needed to start the execution. The following is output to the Command window.

```
Example 3.4 - The Simplex Method
Calculations using MATLAB

Data Entry - Set up Augmented matrix Ab = [A b]
The values can be checked in Simplex Table 1

Number of variables [n]:    7
Number of constraints [m]:    3


------------------
Simplex Table-1
------------------


x1      x2      x3      x4      x5      x6      x7      b

4.000   8.000   7.000   10.000   1.000  0.000  0.000  650.000
```

```
1.000  1.500   3.000    8.000  0.000 1.000 0.000 500.000

25.000 40.000  18.000 10.000   0.000 0.000 1.000 1200.000

-6.000 -10.000 -9.000 -20.000 0.000 0.000 0.000 0.000
```

```
-----------------------------------
Another Iteration is Necessary !!!!
-----------------------------------
Entering basic Variable (EBV) = x4
Pivot Row:    2
Leaving Basic Variable (LBV) = x6
------------------
Simplex Table-2
------------------
```

```
x1     x2     x3     x4    x5    x6     x7     b

2.750  6.125  3.250  0.000 1.000 -1.250 0.000   25.000

0.125  0.188  0.375  1.000 0.000  0.125 0.000   62.500

23.750 38.125 14.250 0.000 0.000 -1.250 1.000   575.000

-3.500 -6.250 -1.500 0.000 0.000  2.500 0.000 1250.000
```

```
-----------------------------------
Another Iteration is Necessary !!!!
-----------------------------------
Entering basic variable (EBV) = x2
Pivot Row:    1
Leaving Basic Variable (LBV) = x5
------------------
Simplex Table-3
------------------
```

```
 x1     x2     x3     x4     x5    x6     x7      b

0.449  1.000  0.531  0.000  0.163 -0.204 0.000    4.082

0.041  0.000  0.276  1.000 -0.031  0.163 0.000   61.735
```

```
6.633   0.000  -5.980   0.000  -6.224   6.531 1.000   419.388

-0.694 0.000   1.816   0.000   1.020   1.224 0.000 1275.510



-----------------------------------
Another Iteration is Necessary !!!!
-----------------------------------
Entering basic Variable (EBV) = x1
Pivot Row:    1
Leaving Basic Variable (LBV) = x2
------------------
Simplex Table-4
------------------

   x1       x2      x3     x4     x5     x6     x7        b

 1.000    2.227   1.182 0.000   0.364 -0.455 0.000     9.091

 0.000   -0.091   0.227 1.000  -0.045  0.182 0.000    61.364

 0.000  -14.773 -13.818 0.000  -8.636  9.545 1.000   359.091

 0.000    1.545   2.636 0.000   1.273  0.909 0.000  1281.818



-----------------------------------
Further Iteration is Not Possible
Check for Solution
-----------------------------------

 x( 1) =    9.091
 x( 4) =   61.364
 x( 7) = 359.091

Optimal value :   -1281.818
```

Solution Using MATLAB's Optimization Toolbox: This section is useful for those that have access to the *Optimization Toolbox* from MATLAB (*ver 7.3*).

Start MATLAB. In the Command window, type *help linprog*. This will provide information on the use of the linear programming routine to solve the problem. The code for this example is available in **Toolbox_Example3_4.m**. Running the problem as formulated returns an unacceptable solution, with an error.

Example 3.4 Using the Optimization Toolbox:

```
Cost Coefficients:
  -6    -10     -9    -20
Constraint matrix A:
   4.0000     8.0000     7.0000    10.0000
   1.0000     1.5000     3.0000     8.0000
  25.0000    40.0000    18.0000    10.0000
Right hand side b:
          650             500          1200

Exiting: One or more of the residuals, duality gap, or
      total relative error has stalled:
         the dual appears to be infeasible (and the primal
             unbounded).
         (The primal residual < TolFun=1.00e-008.)
x =
  1.0c+008 *
   -1.0058
    1.7187
   -2.9222
    0.8993
fval =
 -2.8374e+008
```

The problem is not solved as formulated. It appears to be an unbounded problem. The lack of graphic solution makes it difficult to discover this fact. Placing (arbitrary) bounds on the design variable and using a different call statement (as indicated in the code) returns the solution, which matches the one obtained through the Simplex method.

Example 3.4 Using the Optimization Toolbox with Bounds:

```
Cost Coefficients:
  -6    -10     -9    -20
Constraint matrix A:
   4.0000     8.0000     7.0000    10.0000
   1.0000     1.5000     3.0000     8.0000
  25.0000    40.0000    18.0000    10.0000

Right hand side b:
          650             500          1200

Optimization terminated.
x =
```

```
    9.0909
    0.0000
    0.0000
   61.3636
fval =
  -1.2818e+003
```

3.5 ADDITIONAL TOPICS IN LINEAR PROGRAMMING

This section looks briefly at additional ideas associated with the LP problem. First, there is a discussion of the *dual* problem associated with an LP problem. Duality is important in the discussion of sensitivity analysis—*Variation in optimization solution is due to variation in the original parameters of the problem*. Then it discusses several pieces of information that can be extracted from the final table of the LP problem.

3.5.1 Primal and Dual Problem

Associated with every LP problem, which will be referred to a **primal problem**, there is a corresponding **dual problem**. In a certain way, the dual problem is a transposition of the primal problem. This is meant to imply that if the primal problem is a *minimization* problem the dual problem will be a *maximization* one. If the primal problem has n variables, the dual problem will have n- constraints. If the primal problem has m constraints, the dual problem will have m-variables. For the discussion of duality, the **standard primal problem** is generally defined as a *maximization* problem with $\leq$ *inequalities* (though the standard problem in the LP programming discussion was a minimization problem and the constraints were equalities). The standard primal problem in this section is defined as follows:

$$\text{Maximize } z: \quad c_1 x_1 + c_1 x_2 + \ldots + c_n x \tag{3.40}$$

$$\text{Subject to:} \quad a_{11} x_1 + a_{12} x_2 + \ldots + a_{1n} x_n \leq b_1 \tag{3.41a}$$

$$a_{21} x_1 + a_{22} x_2 + \ldots + a_{2n} x_n \leq b_2 \tag{3.41b}$$

$$\ldots \quad \ldots \quad \ldots \quad \ldots$$

$$a_{m1} x_1 + a_{m2} x_2 + \ldots + a_{mn} x_n \leq b_m \tag{3.41m}$$

$$x_j \geq 0, \quad j = 1, 2, \ldots, n \tag{3.42}$$

This is also referred to as a **normal maximum problem**. The *dual* of this problem is defined as:

$$\text{Minimize } w: \quad b_1 y_1 + b_1 y_2 + \ldots + b_m y_m \tag{3.43}$$

$$\text{Subject to:} \quad a_{11}y_1 + a_{21}y_2 + \ldots + a_{m1}y_m \geq c_1 \quad \text{(3.44a)}$$

$$a_{12}y_1 + a_{22}y_2 + \ldots + a_{m2}y_m \geq c_2 \quad \text{(3.44b)}$$

$$\ldots \quad \ldots \quad \ldots \quad \ldots$$

$$a_{1n}y_1 + a_{2n}y_2 + \ldots + a_{mn}y_m \geq c_n \quad \text{(3.44n)}$$

$$y_i \geq 0, \quad i = 1, 2, \ldots, m \quad \text{(3.45)}$$

Relations 3.42 to 3.45 describe a **normal minimum problem**. There is also an inverse relationship between the definition of the two problems above as far as the identification of the primal and dual problem. If the latter is considered a primal problem, then the former is the corresponding dual problem.

The following observations can be made with respect to the pair of problems just establishecd:

1. The number of dual variables is the same as the number of primal constraints.
2. The number of dual constraints is the same as the number of primal variables.
3. The coefficient matrix A of the primal problem is transposed to provide the coefficient matrix of the dual problem.
4. The inequalities are reversed in direction.
5. The maximization problem of the primal problem becomes a minimization problem in the dual problem.
6. The cost coefficients of the primal problem become the right-hand sides of the dual problem. The right-hand side values of the primal become the cost coefficients in the dual problem.
7. The primal and dual variables both satisfy the nonnegativity condition.

3.5.2 Example 3.5

The following example is used to illustrate the primal/dual versions of the same problem.[3] The problem is solved using techniques developed previously in this chapter.

The Problem: The school of American Craftsman at the Rochester Institute of Technology has decided to participate in a local charity event by manufacturing special commemorative desks, tables, and chairs. Each type of furniture requires lumber as well as two types of skilled labor: finishing and carpentry. Table 3.21 identifies the resources needed. The local lumber cooperative has donated 48 board feet of lumber. Faculty and staff have volunteered 20 finishing hours and 8 carpentry hours. The desk will be sold for $60, the table for $30, and the chair for $20. The school does not expect to sell more than five tables. The school would like to maximize the revenue collected.

Table 3.21   Example 3.5: Resources

Resources	Desk	Table	Chair
Lumber(board feet)	8	6	1
Finishing hours (hours)	4	2	1.5
Carpentry (hours)	2	1.5	0.5

Standard Format — Primal Problem: Defining x_1, as the number of desks produced; x_2, as the number of tables manufactured, and x_3, the number of chairs manufactured, the *primal* problem can be expressed in standard format:

$$\text{Maximize } z: \quad 60x_1 + 30x_2 + 20x_3 \tag{3.46}$$

$$\text{Subject to:} \quad g_1: 8x_1 + 6x_2 + x_3 \le 48 \tag{3.47a}$$

$$g_2: 4x_1 + 2x_2 + 1.5x_3 \le 20 \tag{3.47b}$$

$$g_3: 2x_1 + 1.5x_2 + 0.5x_3 \le 8 \tag{3.47c}$$

$$g_4: x_2 \le 5 \tag{3.47d}$$

$$x_1, x_2, x_3 \ge 0 \tag{3.48}$$

After the introduction of the slack variables (s_1, s_2, s_3), the solution to the primal problem is available in Tables 3.22 to 3.24. Table 3.22 is the initial table. The EBV is x_1 and LBV is s_3. The pivot row is the third row. The subsequent row operations determine Table 3.23. In this table, the EBV is x_3, the LBV is s_2, and the pivot row is the second row. Table 3.24 is the final table establishing the optimal values for the variables. x_1^*, the optimal number of desks, equals 2. x_3^*, the optimal number of chairs to be made, is 8. No tables will be made. The total revenue is $280.

Standard Format — Dual Problem: Although the *dual* problem can be expressed and set up mechanically, it is difficult to associate the design variables in a direct manner, as was done in the primal problem. Using y_1, y_2, y_3, and y_4

Table 3.22   Example 3.5: Primal Problem, Simplex Table 1

x_1	x_2	x_3	s_1	s_2	s_3	s_4	b
8	6	1	1	0	0	0	48
4	2	1.5	0	1	0	0	20
2	1.5	0.5	0	0	1	0	8
0	1	0	0	0	0	1	5
−60	−30	−20	0	0	0	0	$f + 0$

Table 3.23   Example 3.5: Primal Problem, Simplex Table 2

x_1	x_2	x_3	s_1	s_2	s_3	s_4	b
0	0	−1	1	0	−4	0	16
0	−1	0.5	0	1	−2	0	4
1	0.75	0.25	0	0	0.5	0	4
0	1	0	0	0	0	1	5
0	15	−5	0	0	30	0	$f + 240$

Table 3.24   Example 3.5: Primal Problem, Simplex Table 3

x_1	x_2	x_3	s_1	s_2	s_3	s_4	b
0	−2	0	1	2	−8	0	24
0	−2	1	0	2	−4	0	8
1	1.25	0	0	−0.5	1.5	0	2
0	1	0	0	0	0	1	5
0	5	0	0	10	10	0	$f + 280$

as the design variables, the problem can be formulated as in Equations (3.49) to (3.51) using the definition at the beginning of this section. Some associations are possible in light of transposition of the cost coefficients and the constraints. For example, the variable y_1 can be associated with the lumber constraint because in the dual problem its cost coefficient is the lumber constraint of the primal problem. In a similar manner, y_2 is associated with the finishing hours, y_3 with the carpentry hours, and y_4 with the special table constraint. The objective function of the dual problem is expressed in Equation (3.49). In view of the form of the expression of the objective function w, there is a possibility that the design variables in the dual problem can have an economic implication as follows:

y_1 unit price for a board-feet of lumber
y_2 unit price for a finishing hour
y_3 unit price for an hour of carpentry
y_4 this cannot be the unit price as the desk has a price defined in the primal problem—maybe a special price for the desk resource

The dual can be expressed as follows:

$$\text{Minimize } w: \quad 48y_1 + 20y_2 + 8y_3 + 5y_4 \tag{3.49}$$

$$\text{Subject to: } \quad h_1: 8y_1 + 4y_2 + 2y_3 + 0y_4 \geq 60 \tag{3.50a}$$

$$h_2: 6y_1 + 2y_2 + 1.5y_3 + 1y_4 \geq 30 \tag{3.50b}$$

$$h_3: 1y_1 + 1.5y_2 + 0.5y_3 + 0y_4 \geq 20 \tag{3.50c}$$

$$y_1, y_2, y_3 \geq 0 \tag{3.51}$$

Introducing surplus variables (s_1, s_2, s_3), and artificial variables (a_1, a_2, a_3), the dual problem is expressed as a standard LP problem:

$$\text{Minimize } w: \quad 48y_1 + 20y_2 + 8y_3 + 5y_4 \tag{3.52}$$

$$\text{Subject to:} \quad 8y_1 + 4y_2 + 2y_3 - s_1 + a_1 = 60 \tag{3.53a}$$

$$6y_1 + 2y_2 + 1.5y_3 + 1y_4 - s_2 + a_2 = 30 \tag{3.53b}$$

$$1y_1 + 1.5y_2 + 0.5y_3 + 0y_4 - s_3 + a_3 = 20 \tag{3.53c}$$

$$y_1, y_2, y_3, y_4, s_1, s_2, s_3, a_1, a_2, a_3 \geq 0 \tag{3.54}$$

In Phase 1 of the two-phase approach, the artificial variables are removed from the basis using an artificial cost function A, defined as follows:

$$\text{Minimize } A: \quad a_1 + a_2 + a_3 \tag{3.55}$$

The Phase 1 computations are displayed in Tables 3.25 to 3.28. Table 3.25 is identified as the Simplex Table 1, and it includes some preprocessing by which $a_1, a_2,$ and a_3 are made the **basis** variables. As a reminder, the last row represents the artificial cost function. From inspection of Table 3.25, y_1 is the EBV. The LBV is a_2 and the pivot row is the second row. Following the standard row operations, Table 3.26 is obtained. In this table, the EBV is y_2 (the largest negative coefficient). The LBV is a_3 (the third row contains the minimum of $\{[20/1.333], [5/0.3333], [15/1.6667]\}$. The pivot row is row 3. Since there is still some negative coefficient in the last row, Phase 1 is not yet complete and Table 3.27 needs to

Table 3.25   Example 3.5: Dual Problem, Phase I, Table 1

y_1	y_2	y_3	y_4	s_1	s_2	s_3	a_1	a_2	a_3	b
8	4	2	0	−	0	0	1	0	0	60
6	2	1.5	1	0	−1	0	0	1	0	30
1	1.5	0.5	0	0	0	−1	0	0	1	20
48	20	8	5	0	0	0	0	0	0	f
−15	−7.5	−4	−1	1	1	1	0	0	0	$A - 110$

Table 3.26   Example 3.5: Dual Problem, Phase I, Table 2

y_1	y_2	y_3	y_4	s_1	s_2	s_3	a_1	a_2	a_3	b
0	1.3333	0	−1.3331	−1	1.3333	0	1	−1.3333	0	20
1	0.3333	0.25	0.1666	0	−0.1666	0	0	0.1666	0	5
0	1.1666	0.25	−0.1667	0	0.1666	−1	0	−0.1666	1	15
0	4	−4	−3	0	8	0	0	−8	0	$f - 240$
0	−2.5	−0.25	1.5	1	−1.5	1	0	2.5	0	$A - 35$

Table 3.27 Example 3.5: Dual Problem, Phase I, Table 3

y_1	y_2	y_3	y_4	s_1	s_2	s_3	a_1	a_2	a_3	b
0	0	−0.2857	−1.1429	−1	1.1429	1.1429	1	−1.1429	−1.1429	2.8571
1	0	0.1786	0.2143	0	−0.2143	0.2857	0	0.2143	−0.2857	0.7143
0	1	0.2143	−0.1429	0	0.1429	−0.8571	0	−0.1429	0.8571	12.857
0	0	−4.8571	−2.4286	0	7.4286	3.4286	0	−7.4286	−3.4286	$f - 291.43$
0	0	0.2857	1.1429	1	−1.1429	−1.1429	0	2.1429	2.1429	$A - 2.857$

Table 3.28   Example 3.5: Dual Problem, Phase I, Table 4

y_1	y_2	y_3	y_4	s_1	s_2	s_3	a_1	a_2	a_3	b
0	0	−0.25	−1	−0.875	1	1	0.875	−1	−1	2.4999
1	0	0.125	−6E−06	−0.1875	6E−06	0.5	0.1875	−6E−06	−0.5	1.25
0	1	0.25	4E−05	0.125	−4E−05	−1	−0.125	4E−05	1	12.5
0	0	−3.0001	4.9998	6.4998	0.0002	−3.9998	−6.4998	−0.0002	3.9998	f − 310
0	0	0	0	0	0	0	1	1	1	A

be developed. The EBV is s_2. Note in this case that there are two candidates for EBV (s_2, s_3). s_2 is chosen arbitrarily. The LBV is a_1. The pivot row is the first row. After the required row manipulations, Table 3.28 is generated. Scanning the last row, there are only positive coefficients. Phase 1 is now complete. The artificial variables and the artificial cost function can be removed from the LP problem.

Eliminating the artificial variables and the artificial cost function from the problem, Phase 2 is started with Table 3.29. Please note that Table 3.29 contains no information that is not already available in Table 3.28. It only provides an uncluttered table for continuing the application of the Simplex technique, and therefore not really necessary. In Table 3.29, EBV is s_3. LBV is y_1 after a toss between y_1 and s_1. Pivot row is the second row. This row is used to obtain Table 3.30. In this table, note that the value of the basic variable s_1 is 0. This is called a **degenerate basic variable**. EBV for the next table is y_3. LBV is s_3. Pivot row is the second row (*Note*: The pivot row can be the row with the degenerate basic solution if the column of the EBV has a positive coefficient in that row). Table 3.31 is the final table and contains the optimal solution. The final solution of the dual problem is as follows:

$$y_1^* = 0, y_2^* = 10, y_3^* = 10, y_4^* = 0, s_2^* = 5, w^* = 280$$

To summarize, the solution of the primal problem is

$$x_1^* = 2, x_2^* = 0; \quad x_3^* = 8, \quad s_{1p}* = 24, \quad z^* = 280$$

Some Formal Observations: These observations are justified more formally in many books on Linear Programming, some of which have been included in the list of references at the end. Here, the formal results are elaborated using

Table 3.29   Example 3.5: Dual Problem, Phase II, Table 1

y_1	y_2	y_3	y_4	s_1	s_2	s_3	b
0	0	−0.25	−1	−0.875	1	1	2.5
1	0	0.125	0	−0.1875	0	0.5	1.25
0	1	0.25	0	0.125	0	−1	12.5
0	0	−3	5	6.5	0	−4	$w - 310$

Table 3.30   Example 3.5: Dual Problem, Phase II, Table 2

y_1	y_2	y_3	y_4	s_1	s_2	s_3	b
−2	0	−0.5	−1	−0.5	1	0	0
2	0	0.25	0	−0.375	0	1	2.5
2	1	0.5	0	−0.25	0	0	15
8	0	−2	5	5	0	0	$w - 300$

Table 3.31   Example 3.5: Dual Problem, Phase II, Table 3

y_1	y_2	y_3	y_4	s_1	s_2	s_3	b
2	0	0	−1	−1.25	1	2	5
8	0	1	0	−1.5	0	4	10
−2	1	0	0	0.5	0	−2	10
24	0	0	5	2	0	8	$w - 280$

the solution to the example discussed in this section. All nine observations may not be relevant to the Example 3.5; nevertheless, they are included here for completeness.

1. The *primal* and *dual* problem have the same value for the optimal objective function, $z^* = w^* = 280$. This is, however, true only if the problems have optimal solutions.

2. If x is any feasible solution to the *primal* problem, and y is any feasible solution to the *dual* problem, then $w(y) \geq z(x)$. This feature provides an estimate of the bounds of the *dual* optimal value if a feasible *primal* solution is known. These results also hold for the reverse case when a feasible *dual* is known.

3. If the *primal* problem is unbounded, the *dual* problem is infeasible. The *unbounded* problem implies that the objective function value can be pushed to infinite limits. This happens if the feasible domain is not closed. Of course, practical considerations will limit the objective function value to corresponding limits on the design variables. The inverse relationship holds, too. If the *dual* problem is unbounded, the *primal* is infeasible.

4. If the ith *primal constraint* is an equality constraint, the ith *dual variable* is unrestricted in sign. The reverse holds, too. If the *primal* variable is unrestricted in sign, then the *dual constraint* is an equality.

5. Obtaining primal solution from dual solution:

   i. If the ith *dual constraint* is a strict *inequality*, then the ith *primal variable* is *non-basic* (for optimum solutions only). From the dual solution, the second constraint (3.50b) is satisfied as an inequality. Therefore, the second primal variable x_2 is *non-basic*, and this is evident in Table 3.24.

   ii. If the ith *dual variable* is basic, then the ith *primal constraint* is a strict equality. From Table 3.31, y_2, y_3 are basic variables; therefore, 3.47b and 3.47c must be *equalities*. This is indeed true.

6. Recovering primal solution from final dual Table: When the primal and dual problems are in the standard form, the value of the ith *primal variable* equals the reduced coefficient (i.e., the coefficient in the last row of the final table) of the *slack/surplus* variable associated with the ith *dual constraint*. In Table 3.31, the values of the reduced cost coefficient corresponding

to the surplus variables s_1, s_2, s_3 are 2, 0, and 8, respectively. These are precisely the optimal values of the *primal variables*.

7. If an ith *dual variable* is non-basic, the value of its reduced cost coefficient is the value of the *slack/surplus* variable of the corresponding primal constraint. In Table 3.31, y_1, y_4 are non-basic variables with reduced cost coefficients of 24 and 5, respectively. These are the values of s_1 and s_4 in Table 3.24.

8. Obtaining dual solution from primal solution: When the primal and dual problems are in the standard form, the value of the ith *dual variable* equals the reduced coefficient (i.e., the coefficient in the last row of the final table) of the *slack/surplus* variable associated with the ith *primal constraint*. In Table 3.24, the values of the reduced-cost coefficient corresponding to the surplus variables s_1, s_2, s_3, s_4 are 0, 10, 10, and 0, respectively. These are precisely the optimal values of the *dual variables* in Table 3.31.

9. If an ith *primal variable* is non-basic, the value of its reduced cost coefficient is the value of the *slack/surplus* variable of the corresponding dual constraint. In Table 3.24, x_2 is a non-basic variable with reduced cost coefficients of 5. This is the value of s_2 in Table 3.31. Since x_1 and x_3 are *basic*, with the reduced coat coefficient of 0, the corresponding slack/surplus s_1, s_2 variables will be zero, as observed in Table 3.31.

This list referred to the primal and dual problems in standard form. A similar list can be obtained for non-standard form. There will be appropriate modifications for negative values of variables, as well as equality constraints. Please consult the listed reference.

Solution Using MATLAB Code: The solution, including the various tables can be generated by running **PRIMALMATLABExample3_5.m**, **DUALPhaseI_MATLABExample3_5.m**, and **DUALPHASEII_MATLABExample3_5.m**. For the calculations, only the augmented matrix [A b] is needed to start the execution. Only the output from the dual code is shown, since the tie for the selection of the LBV is the other variable. This results in two more tables but the same final solution. The following is output to the Command window.

Example 3.5 (Dual Problem):

```
Calculations using MATLAB


Data Entry - Set up Augmented matrix Ab = [A b]
Number of variables [n]:    7
Number of constraints [m]:    3
```

```
------------------
Starting Table
------------------
```

x1	x2	x3	x4	x5	x6	x7	b
0.000	0.000	-0.250	-1.000	-0.875	1.000	1.000	2.500
1.000	0.000	0.125	0.000	-0.188	0.000	0.500	1.250
0.000	1.000	0.250	0.000	0.125	0.000	-1.000	12.500
0.000	0.000	-3.000	5.000	6.500	0.000	-4.000	-310.000

```
------------------
Phase II - Simplex Table-1
------------------
```

x1	x2	x3	x4	x5	x6	x7	b
0.000	0.000	-0.250	-1.000	-0.875	1.000	1.000	2.500
1.000	0.000	0.125	0.000	-0.188	0.000	0.500	1.250
0.000	1.000	0.250	0.000	0.125	0.000	-1.000	12.500
0.000	0.000	-3.000	5.000	6.500	0.000	-4.000	-310.000

```
-----------------------------------
Another Iteration is Necessary !!!!
-----------------------------------
```

```
Entering basic Variable (EBV) = x7
Pivot Row:   1
Leaving Basic Variable (LBV) = x6
------------------
Phase II - Simplex Table-2
------------------
```

x1	x2	x3	x4	x5	x6	x7	b
0.000	0.000	-0.250	-1.000	-0.875	1.000	1.000	2.500

```
1.000 0.000  0.250  0.500  0.250 -0.500 0.000    0.000

0.000 1.000  0.000 -1.000 -0.750  1.000 0.000   15.000

0.000 0.000 -4.000  1.000  3.000  4.000 0.000 -300.000
```

```
-----------------------------------
Another Iteration is Necessary !!!!
-----------------------------------
```

```
Entering basic Variable (EBV) = x3
Pivot Row:    1
Leaving Basic Variable (LBV) = x7
------------------
Phase II - Simplex Table-3
------------------
```

x1	x2	x3	x4	x5	x6	x7	b
0.000	0.000	1.000	4.000	3.500	-4.000	-4.000	-10.000
1.000	0.000	0.000	-0.500	-0.625	0.500	1.000	2.500
0.000	1.000	0.000	-1.000	-0.750	1.000	0.000	15.000
0.000	0.000	0.000	17.000	17.000	-12.000	-16.000	-340.000

```
-----------------------------------
Another Iteration is Necessary !!!!
-----------------------------------
```

```
Entering basic Variable (EBV) = x7
Pivot Row:    2
Leaving Basic Variable (LBV) = x1
------------------
Phase II - Simplex Table-4
------------------
```

x1	x2	x3	x4	x5	x6	x7	b
4.000	0.000	1.000	2.000	0.998	-2.000	0.000	0.000

```
1.000 0.000 0.000 -0.500 -0.625  0.500 1.000    2.500

0.000 1.000 0.000 -1.000 -0.750  1.000 0.000   15.000

16.000 0.000 0.000  9.000  6.992 -4.000 0.000 -300.000
```

```
-----------------------------------
Another Iteration is Necessary !!!!
-----------------------------------
```

```
Entering basic Variable (EBV) = x6
Pivot Row:    2
Leaving Basic Variable (LBV) = x7
------------------
Phase II - Simplex Table-5
------------------
```

```
     x1     x2     x3     x4     x5     x6     x7      b

 8.000 0.000 1.000  0.000 -1.504 0.000  4.000   10.000

 2.000 0.000 0.000 -1.000 -1.251 1.000  2.000    5.000

-2.000 1.000 0.000  0.000  0.501 0.000 -2.000   10.000

24.000 0.000 0.000  5.000  1.988 0.000  8.000 -280.000
```

```
-----------------------------------
Further Iteration is Not Possible
Check for Solution
-----------------------------------
m =
      3
```

```
x( 3) =   10.000
x( 6) =    5.000
x( 2) =   10.000
```

```
Optimal value :    280.000
```

3.5.3 Sensitivity Analysis

The solution to the LP problem is dependent on the values for the coefficients, c, b, and A (also termed as **parameters**), involved in the problem. In many practical situations, these parameters are only known approximately and may change from their current value for any number of reasons, particularly after a solution has been established. Instead of recomputing a new solution, it is possible to obtain a fresh one based on the existing one and its final Simplex table. This issue is significant in practical problems with thousands of variables and constraints. The adaptation of a new solution without resolving the problem is called **sensitivity analysis**. In LP problems, *sensitivity analysis* refers to determining the range of parameters for which the optimal solution still has the same variables in the basis, even though values at the solution may change.

Example 3.1 is used for this discussion. Figures are used for illustration rather than a formal proof. Figures are not useful for more than two variables. Generally, most computer codes that solve linear programming problems also perform sensitivity analysis. Revisiting Example 3.1, it was necessary to identify the number of Component placement machines of type A and B. The objective is to maximize the number of boards to be manufactured. Constraint g_1 represents the acquisition dollars available. Constraint g_2 represents the floor space constraint. Constraint g_3 represents the number of operators available. The problem statement and the results are reproduced here (note that x_3, x_4, and x_5 are the slack variables).

$$\text{Minimize} \quad f(X): \ -990x_1 - 900x_2 - 5250 \tag{3.3}$$

$$\text{Subject to:} \quad g_1(X): 0.4x_1 + 0.6x_2 + x_3 = 8.5 \tag{3.4a}$$

$$g_2(X): 3x_1 - x_2 + x_4 = 25 \tag{3.4b}$$

$$g_3(X): 3x_1 + 6x_2 + x_5 = 70 \tag{3.4c}$$

$$x_1 \geq 0, x_2 \geq 0, x_3 \geq 0, \quad x_4 \geq 0, \quad x_5 \geq 0 \tag{3.5}$$

The solution is:

$$x_1{}^* = 10.4761, x_2{}^* = 6.4285, x_3{}^* = 0.4524, f^* = -21437.14$$

Changing Cost Coefficient Values (c): First, consider the effect of changing the cost coefficients of the design variables. In the following, we use c_1 as the example. Similar analysis can be performed for the other coefficients. An important question is if c_1, the coefficient of design variable x_1 is changed from -990, will the set of basis variables remain the same?

Changing the cost coefficient changes the slope of the line representing the objective function that may lead to a change in the solution. For example, Equation (3.3) can be written as follows:

$$x_2 = -(990/900)x_1 - ([5250 + f]/900)$$

If c_1 were to be made -1900, then the slope of the line will be $- (1900/900)$. The sensitivity due to c_1 can be experienced by running **Sensitivity_Cost_Example3_1.m**. The change with respect to original objective function is shown in animation, with the value of c_1 written to the figure. There is a pause of 5 seconds. The original solution is marked on the figure by a circle. The original solution is unchanged when the magnitude of the slope is raised by using c_1 as $-1,900$. However, if c_1 were to be made -200, the new objective function (in green) indicates the solution will change and will be on the x_2 axis, as shown by the green filled circle in Figure 3.11, after the final animation. The floor constraint is no longer binding. It can be established by simple calculations that infinitely many solutions are possible if c_1 has a value of -450. In this case, the cost function is parallel to the binding constraint g_3. From this discussion, it is apparent that to keep the location of the original solution unchanged, $\mathbf{c_1}$ must be greater than -450. Such analysis is possible for all the other cost coefficients involved in the problem. In particular, the analysis should indicate for what range of cost coefficients the solution would remain where it is. This is referred to as determining *sensitivity to cost coefficients*.

Change in the Resource Limits (b-vector):   When the values on the right-hand side changes, the constraint lines/planes/boundaries are moved parallel to themselves. This directly affects the feasible region. Therefore, the optimal values may also change due to the changes in the feasible region. The code **Sensitivity_ResourceLimits_Example3_1.m** provides an illustration by changing the value of the right-hand side for the first constraint (3.4a). There

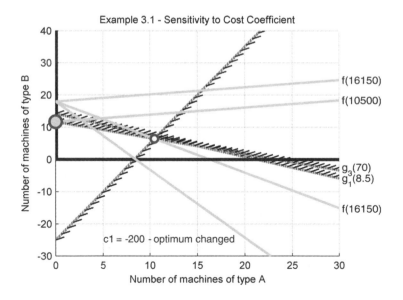

Figure 3.11   Sensitivity analysis, changing cost coefficient.

is a limited animation with two values varying around the original value. In the case where b_1 is 10.5, the solution does not change, while in the case of $b_1 = 6.5$, a new solution is determined due to the change in the feasible region. This makes g_1 an active constraint. For the original value of b_1, the constraint was not active. The new solution is shown in Figure 3.12. In *sensitivity analysis*, the problem is to discover the range of b_1 so that solution still has the same variables in the basis. If b_1 were to remain above 8.0475, then the solution would still be at A and g_1 will not be an active constraint.

Change in the Coefficient Matrix A: A change in the coefficient matrix is similar in effect to that due to the resource limits. In the discussion here, a change in either or both coefficients will change the slope of the first constraint. This directly impacts the problem by changing the feasible region. These changes also depend on whether the variable in the column is a basic variable or a non-basic variable. **Sensitivity_CoefficientA_Example3_1.m** illustrates the result of changing the coefficient a_{11}. An increase in the value moves the optimal solution, while a decrease leaves the solution unchanged. In both situations, x_1 and x_2 still are part of the basis. In the first case, g_1 becomes a binding/active constraint. This is illustrated in Figure 3.13. The coefficient is in the first column, which corresponds to a basic variable. Changing coefficient values in the first column in other constraints will yield similar information. You are encouraged to try changes in the other coefficients using the *m-files* available in the **code** that you downloaded from the publisher's website.

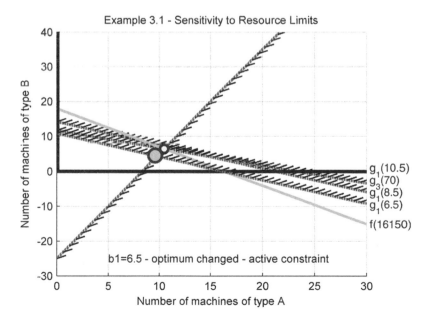

Figure 3.12 Sensitivity analysis, changing right-hand side.

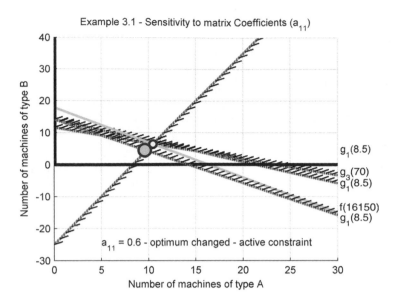

Figure 3.13 Sensitivity analysis, changing constraint coefficient.

REFERENCES

1. Dantzig, G.B. *Linear Programming and Extensions*. Princeton, NJ: Princeton University Press, 1963.

2. Luenberger, D.G. *Linear and Nonlinear Programming*, Reading, MA: Addison-Wesley, 1984.

3. Winston, W.L. *Introduction to Mathematical Programming, Applications and Algorithms*. CA: Duxbury Press, 1995.

4. Arora, J.S. *Introduction to Optimal Design*. New York: McGraw-Hill, Inc. 1989.

5. Williams, G. *Linear Algebra with Applications*. Wm,C. Brown Publishers, 1991.

6. Noble, Ben, and Daniel, J.W. *Applied Linear Algebra*. Englewood Cliffs, NJ: Prentice Hall, Inc., 1977.

PROBLEMS

All two-variable problems can be solved graphically. You can also use the Optimization Toolbox (Chapter 10) to verify the solution.

3.1 Solve the following linear programming problem:

$$Min \quad f(x_1, x_2): \qquad x_1 + x_2$$

$$Sub: \qquad 3x_1 - x_2 \le 3$$

$$x_1 + 2x_2 \le 5$$

$$x_1 + x_2 \leq 4$$
$$x_1 \geq 0; x_2 \geq 0$$

3.2 Solve the following linear programming problem:

Max $f(x_1, x_2)$:	$x_1 + x_2$
Sub:	$3x_1 - x_2 \leq 3$
	$x_1 + 2x_2 \leq 5$
	$x_1 + x_2 \leq 4$
	$x_1 \geq 0; x_2 \geq 0$

3.3 Solve the following linear programming problem:

Min $f(x_1, x_2)$:	$x_1 - x_2$
Sub:	$3x_1 - x_2 \leq 3$
	$x_1 + 2x_2 \leq 5$
	$x_1 + x_2 \leq 4$
	$x_1 \geq 0; x_2 \geq 0$

3.4 Solve the following linear programming problem:

Max $f(x_1, x_2)$:	$x_1 + x_2$
Sub:	$3x_1 - x_2 \geq 3$
	$x_1 + 2x_2 \leq 5$
	$x_1 + x_2 \leq 4$
	$x_1 \geq 0; x_2 \geq 0$

3.5 Solve the following linear programming problem:

Min $f(x_1, x_2)$:	$x_1 + x_2$
Sub:	$3x_1 - x_2 \leq 3$
	$x_1 + 2x_2 \leq 5$
	$x_1 + x_2 \leq 4$
	$x_1 \geq 0; x_2$ *unrestricted in sign*

3.6   The local bookstore must determine how many of each of the four new books on Photonics it must order to satisfy the new interest generated in the discipline. Book 1 costs $ 75, will provide a profit of $13, and requires 2 inches of shelf space. Book 2 costs $85, will provide a profit of $10, and requires 3 inches of shelf space. Book 3 costs $65, will provide a profit of $8, and requires 1 inches of shelf space. Book 4 costs $100, will provide a profit of $15, and requires 4 inches of shelf space. Find the number of each type that must be ordered to maximize profit. Total shelf space is 100 inches. Total amount available for ordering is $5,000. It has been decided to order at least 10 Books 2 and Books 4 (total).

3.7   The local community college is planning to grow the biotechnology offering through new federal and state grants. An ambitious program is being planned for recruiting at least 200 students from in and out of state. It is to recruit at least 40 out-of-state students. It will attempt to recruit at least 30 students who are in the top 20 percent of their graduating high school class. Current figures indicate that about 8 percent of the applicants from in state, and 6 percent of the applicants from out of state, belong to this pool. The college also plans to recruit at least 40 students who have AP courses in biology. The data suggest that 10 percent and 15 percent of in-state and out-of-state applicants, respectively, belong to this pool. The college anticipates that the additional cost per student is $800 for each in-state student and $1,200 for each out-of-state student. Find the actual enrollment needed to minimize cost and their actual cost.

3.8   In July, Venkat's Fruit Orchard usually sees a scheduling problem. Both cherries and blueberries ripen around the same time. A cartload of cherries sells for $2,750, while the cartload of blueberries fetches $1,750. The farm must deliver at least a half cartload of cherries and one cartload of blueberries to the local supermarket. Three persons can pick two cartloads of cherries in a day while one person will pick a cartload of blueberries in the same time. Only 46 people show up for work during the day. One cartload of cherries is contained in 4 pallets, while the same amount of blueberries occupies 6 pallets. There are 251 pallets. Each picker is paid $45 per day. The packaging and storage can handle, at most, 30 cartloads per day. Set up the LP problem for the orchard to maximize profits [Problem 1.22].

3.9   The energy crisis of 2008 is keeping the refineries very active trying to adjust the product mix due to demand and supply. Largely, the decision is about buying two varieties of crude oil and selling two kinds of product: gasoline and diesel. They can buy light crude at $135 per barrel, while heavy crude is discounted at $95 per barrel. Consider one barrel as 40 gallons. They can sell gasoline at $4.25/gallon and diesel at $4.85 a gallon. A barrel of light crude will yield 0.5 barrels of gasoline and 0.4 barrels of diesel. A barrel of heavy crude will yield 0.4 barrels of gasoline and 0.2 barrels of diesel. Totals supply of light crude is limited to 500,000 barrels, while heavy crude is unlimited in supply. In a day, the refinery can only handle 2 million gallons of gasoline and 3 million gallons of diesel. Set up the LP programming problem to maximize daily profits [Problem 1.23].

3.10 The ΦΧΨ fraternity is going to capitalize on the mouth-watering pizza sauce, accidentally created by one of its members by organizing a weekly fundraiser for its charities. It plans on two varieties of pizza based on two ingredients, sausage and cheese. Of course, each pizza will have the pie, base sauce, sausage, and cheese. The ingredients are in cups for each type of pizza. A cup of sausage costs $0.75, a cup of cheese $0.5, while a cup of sauce is $1.00. The pie costs $0.50 each.

Pizza	Sell Price	Sausage	Cheese	Sauce
Supergooey	$ 7.50	1.5	2.75	2.0
Justgooey	$ 7.50	2.5	1.5	1.5

The fraternity can obtain only 120 cups of sausage, 100 cups of cheese, and 120 cups of sauce, and any amount of pies. Find how many pizzas of the two kinds it must sell for maximum profit [Problem 1.34].

3.11 Fred wants to invest his $100,000 in four mutual funds whose holdings and return is shown in the following table. He wants his choice to reflect a growth strategy that requires that he has at least 60 invested in domestic stocks, at most 20 percent in international stock, and wants his exposure to real estate to be at most 10 percent. How should he invest his money for maximum return [Problem 1.37]?

Mutual Fund	Domestic Stocks	International Stocks	Bonds	Real Estate	Expected return
F1	0.2	0.6	0.2	0,0	6 %
F2	0.6	0.1	0.3	0.1	3 %
F3	0.4	0.5	0.0	0.1	4 %
F4	0.5	0.1	0.1	0.3	1 %

3.12 As the academic year approaches its end, the College Machine Shop sees a tremendous increase in demand because of the Capstone projects. To improve scheduling, the projects are divided into three broad groups. Project A requires no drill time but uses 30 minutes of milling and 60 minutes of machining on the lathe. Project B requires 30 minutes each of drilling, milling, and machining (lathe). Project C requires 15 minutes of drilling, 45 minutes of milling, and 15 minutes of machining. The machine shop has three drill presses. Each press takes 15 minutes to set up for a project. One operator can supervise the three drill presses. It has five milling machines. The machines take a half hour for set up, and two operators are required to supervise the five machines. The Shop has eight lathes, which require 45 minutes to set up and three supervisors for the eight machines. Maximize the number of projects that can be handled in a 10-hour day.

3.13   Solve the following LP problem using the table:

$$\textit{Minimize } f: \quad -2x_1 + 4x_2 - x_3 + 5x_4$$

$$\textit{Subject to}: \quad 4x_1 - x_2 + 2x_3 - x_4 = -2$$

$$x_1 + x_2 + 3x_3 - x_4 \leq 14$$

$$-2x_1 + 3x_2 - x_3 + 2x_4 \geq 2$$

$$x_1, x_2 \geq 0; \; x_3 \leq 0; \; x_4: \text{ unrestricted in sign}$$

3.14   Solve the following LP problem. Express its dual and verify its properties:

$$\textit{Minimize } f: \quad x_1 + 2x_2 + 3x_3 + 4x_4$$

$$x_1 + 2x_2 + 3x_3 + 3x_4 \leq 30$$

$$2x_1 + x_2 + 3x_3 + x_4 \leq 20$$

$$x_1, x_2, x_3, x_4 \geq 0$$

3.15   Solve the following LP problem. Express its dual and verify its properties:

$$\textit{Minimize } f: \quad 5x_1 + 4.5x_2$$

$$x_1 \leq 8$$

$$x_2 \leq 10$$

$$2x_1 + x_2 \geq 15$$

$$x_1, x_2 \geq 0$$

4

NONLINEAR
PROGRAMMING

Nonlinear Programming (NLP) defines a collection of numerical techniques for optimization problems whose mathematical models are characterized by nonlinear equations. *Mathematical Programming* is another name for this collection. In Chapter 2, several nonlinear optimization problems were examined graphically. The curvature and the gradient of the functions involved had a significant influence in determining the solution. This can be contrasted to problems in *Linear Programming* in Chapter 3, which required numerical techniques based on the linear nature of the problem definition. That technique will be very different from those required for NLP problems. We will center the discussion of the optimality conditions and numerical techniques around two variable problems because the ideas can be expressed graphically. Extension to more than two variables is quite straightforward, and is most simple when the presentation is made through the use of vector algebra.

Traditionally, there is a *bottom-up* presentation of material for nonlinear optimization. *Unconstrained* problems are discussed first, followed by *constrained* problems. For constrained problems, the *equality-constrained* problem is discussed first. A similar progression is observed with regard to the number of variables. A *single* variable problem is introduced, followed by *two* variables, which is then extended to a general problem involving *n*-variables. This order allows incremental introduction of new concepts, but primarily allows the creative use of existing rules to establish solutions to the extended problems.

An analytical foundation is essential to understand and establish the conditions that the optimal solution will have to satisfy. *Multivariable calculus* should be sufficient for aquiring this knowledge[1,4]. The numerical techniques will require *Linear Algebra*[2,4]. The mathematical foundation is an essential component of the development of the numerical technique. It will certainly be required when

203

creatively mating an existing numerical technique to the particular problem you want to solve. Essential mathematical definitions are reviewed and illustrated in this chapter. The suggested references will be useful for refreshing the calculus and other techniques essential to the development of NLP[1,5]. The books familiar to the reader should also do very admirably. This chapter also introduces the *symbolic computation* (computer algebra) resource available in MATLAB—*Symbolic Math Toolbox*[6,8]. Except for the exercise with symbolic programming all of the MATLAB code are available only in the identified files.

4.1 PROBLEM DEFINITION

In NLP it is not essential that *all* the functions involved be nonlinear. It is sufficient if just one of them is nonlinear. It is also not necessary that all functions be explicit functions of *all* the design variables. There are many examples in engineering in which only the objective function is a quadratic function while the constraints are linear. These problems are termed as *Linear Quadratic Problems (LQP)* and are recognized because they can be solved using numerical techniques that exploit their particular geometry. Optimization problems, for the most part, rely on experience and domain knowledge to frame the objective and the constraint functions in the mathematical model. The mathematical model, the collection of functions, is a set of relationships among the design variables. Expression of the problem in the standard format will assist in solving the problem using available software packages. For difficult optimization problems with large number of variables, implicit equations, or serious nonlinearity, the solution needs to be coaxed. This will require some understanding of the numerical techniques in the software package. These numerical technique themselves are all designed to solve the same mathematical problem. So, here is the question:

How does one establish the solution to the nonlinear optimization problem?

In mathematics (after all, at this stage there is a mathematical model for the problem) the *solution* is obtained by satisfying the **necessary** and **sufficient** conditions related to the class of problems. The **necessary conditions** are those relations that a candidate for the *optimum solution must satisfy*. If it does, and this is important, it *may* be an optimal solution. To qualify the design vector X (X represents the design vector) as an optimum, it must satisfy additional relations called the **sufficient** conditions. Therefore, an *optimum solution* must satisfy both *necessary* and *sufficient* conditions. This chapter establishes these conditions for the optimization problem.

First, the standard format is restated. Example 4.1 is then developed. The same example is restated in several ways in the chapter to define the different kinds of optimization problem and their corresponding conditions for the optimum.

Standard Format: The standard format of the NLP reproduced here for convenience

$$\text{Minimize} \quad f(x_1, x_2, \ldots x_n) \tag{4.1}$$

$$\text{Subject to:} \quad h_k(x_1, x_2, \ldots, x_n) = 0, \qquad k = 1, 2, \ldots, l \tag{4.2}$$

$$g_j(x_1, x_2, \ldots, x_n) \le 0, \qquad j = 1, 2, \ldots, m \tag{4.3}$$

$$x_i^l \le x_i \le x_i^u \qquad i = 1, 2, \ldots, n \tag{4.4}$$

In vector notation

$$\text{Minimize} \quad f(\mathbf{X}), \quad [\mathbf{X}]_n \tag{4.5}$$

$$\text{Subject to:} \quad [\mathbf{h}(\mathbf{X})]_l = \mathbf{0} \tag{4.6}$$

$$[g(X)]_m \le 0 \tag{4.7}$$

$$X^{low} \le X \le X^{up} \tag{4.8}$$

4.1.1 Problem Formulation–Example 4.1

The problem is restricted to two variables to draw graphical support for all of the discussions. There are two constraints, which, during the development of this chapter, may switch between equality and inequality constraints to illustrate the various problem types and the corresponding conditions that must be met by the optimum solution.

Problem: The fuel tank for the aircraft will be placed in the wing. It will be an elliptical tank and must be accommodated in the volume available. For design we will work with the airfoil of the wing, as shown in Figure 4.1. Since the tank will be aluminum we will expect it to assist in carrying some structural loads by having at least a perimeter of 1.75 m. Furthermore, we will assume that the airfoil rear surface can be approximated as a line in the area where the tank will be located. The optimization problem can be stated as follows:

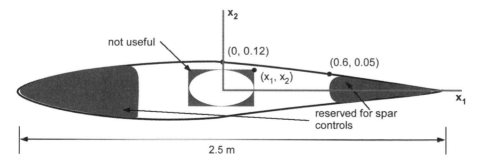

Figure 4.1 Constraints of Example 4.1.

Find the largest elliptical area, which must have a perimeter of at least 1.75 m that can be housed in the available area of the airfoil.

Mathematical Model: Figure 4.1 captures the essence of the problem. Several additional assumptions are made to keep the formulation simple. First, we will assume that the ellipse can be located symmetrically with respect to the origin shown in the figure. The size will be more influenced by the rear surface, since the thickness is higher forward of the origin. We can therefore work with the quarter of the ellipse in the positive quadrant. In Figure 4.1, it is shown that the elliptical area can be inscribed in a rectangle, with the excess area not useful in this investigation. The sides of the rectangle are the design variables, and they are also the semimajor and the semiminor axis of the ellipse that we are designing.

For the specific problem being discussed, and referring to Figure 4.1, the design variables are the coordinate values x_1 and x_2 of the rectangle. The optimization problem can be naturally formatted as follows:

$$\text{Maximize} \quad f(x_1, x_2) = \pi x_1 x_2 \tag{4.9}$$

$$\text{Subject to:} \quad g_1(x_1, x_2) = x_2 - 0.12 - \frac{(0.05 - 0.12)}{(0.6 - 0.0)} x_1 \leq 0 \tag{4.10}$$

$$g_2(x_1, x_2) : \pi \left[3(x_1 + x_2) - \sqrt{(3x_1 + x_2)(x_1 + 3x_2)} \right] \geq 1.75 \tag{4.11}$$

$$0 \leq x_1 \leq 0.6 \quad 0 \leq x_2 \leq 0.12 \tag{4.12}$$

The objective function is the area of the ellipse. The first inequality constraint is the size constraint imposed by the airfoil rear top surface. The second inequality constraint is an approximation to the perimeter of the ellipse suggested by Ramanujam[9].

Discussion of the Model: The natural formulation is a maximization problem. Instinctively, the two constraints in the formulation are inequalities. There are two nonlinear functions, one linear function, and two side constraints. Looking at Figure 4.1, the solution x_1^*, x_2^* will most likely lie on the linear inequality constraint, making it active. The perimeter constraint may or may not be active. Figure 4.2 is the graphical representation of the functions in (4.9) to (4.12). There is a scaling problem with the hash marks. The function $g_2(x_1, x_2)$ is steep, and the hash marks are thickened to show the forbidden region. It appears that the perimeter constraint will not be active. Figure 4.2 can be obtained by running **Solution_Example4_1.m**.[*]

[*]Files to be downloaded from the web site are indicated by boldface courier type.

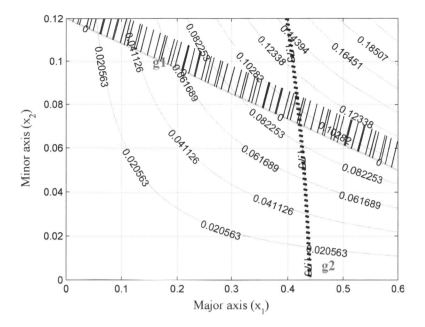

Figure 4.2 Graphical solution for Example 1.

4.1.2 Additional Optimization Problems

Using the relations (4.9) to (4.12) several additional classes of optimization problems can be described by including only a subset of the functions or changing the type of constraints. We explicitly set them up below so we can address the necessary and sufficient conditions associated with the particular problem before we generalize the conditions. All of the problems are expressed in standard format.

Unconstrained Problem: There are no functional constraints, although the side constraints are necessary to keep the solution finite. For this example:

$$\text{Minimize} \quad f(x_1, x_2) = -\pi x_1 x_2 \tag{4.13a}$$

$$0 \le x_1 \le 0.6 \quad 0 \le x_2 \le 0.12 \tag{4.13b}$$

In this problem, if the design variables are unbounded at the upper limit, no side constraint (4.12), then the solution would be at ∞, which is impractical. Therefore, a two-sided limit for the design variables is usually a necessary condition for the solution. The designer must define an acceptable design space.

Equality Constrained Problem 1: The functional constraints in this problem are only equalities. With reference to Example 4.1 the following problem can be set up:

$$\text{Minimize} \quad f(x_1, x_2) = -\pi x_1 x_2 \tag{4.14a}$$

Subject to: $h_1(x_1, x_2) = x_2 - 0.12 - \dfrac{(0.05 - 0.12)}{(0.6 - 0.0)} x_1 = 0$ (4.14b)

$$h_2(x_1, x_2) : \pi \left[3(x_1 + x_2) - \sqrt{(3x_1 + x_2)(x_1 + 3x_2)} \right] - 1.75 = 0$$

(4.14c)

$$0 \le x_1 \le 0.6 \quad 0 \le x_2 \le 0.12$$ (4.14d)

Intuitively, this problem cannot be optimized, since the two equality constraints by themselves should determine the values for the two design variables at the solution. The objective function in (4.14a) is immaterial. The arguments used in LP for acceptable problem definition are also valid here ($n > m$). There is always the possibility of multiple solutions – which is a strong feature of nonlinear problems. In such an event, the set of variables that yield the lowest value for the objective will be the optimal solution. Note, such a solution is obtained by scanning the acceptable solutions rather than through the applications of any *formal* conditions. This problem is not discussed further, as it is not a valid problem in optimization.

Equality Constrained Problem 2:   The problem with one constraint can be set up as follows:

Minimize $f(x_1, x_2) = -\pi x_1 x_2$ (4.15a)

Subject to: $h_1(x_1, x_2) = x_2 - 0.12 - \dfrac{(0.05 - 0.12)}{(0.6 - 0.0)} x_1 = 0$ (4.15b)

$$0 \le x_1 \le 0.6 \quad 0 \le x_2 \le 0.12$$ (4.15c)

Inequality Constrained Problem: This is the original problem.

Minimize $f(x_1, x_2) = -\pi x_1 x_2$ (4.9)

Subject to: $g_1(x_1, x_2) = x_2 - 0.12 - \dfrac{(0.05 - 0.12)}{(0.6 - 0.0)} x_1 \le 0$ (4.10)

$$g_2(x_1, x_2) : -\pi \left[3(x_1 + x_2) - \sqrt{(3x_1 + x_2)(x_1 + 3x_2)} \right] + 1.75 \le 0$$

(4.11)

$$0 \le x_1 \le 0.6 \quad 0 \le x_2 \le 0.12$$ (4.12)

Like its counterpart in linear programming, this is a valid optimization problem. There is no limit on the number of inequality constraints.

It is essential to understand both the nature and the number of constraints, as well as how they influence the problem. In general, equality constraints are easy to handle mathematically, difficult to satisfy numerically, and more restrictive on the search for the solution. Inequality constraints are difficult to resolve mathematically, are more flexible with respect to the search for the optimal solution, and define a larger feasible region. A well-posed optimization problem requires that some constraints are active (equality) at the optimal solution, as otherwise it would lead to a solution of an unconstrained optimization problem, suggesting that problem formulation needs to be revisited:

4.2 MATHEMATICAL CONCEPTS

Like LP, some mathematical definitions are necessary before the necessary and sufficient conditions for the NLP can be established. Definitions are needed both for the analytical discussion, as well as numerical techniques. MATLAB provides a *Symbolic Math Toolbox* and *Extended Symbolic Math Toolbox*, which permits symbolic computation to be integrated in the numerical environment of MATLAB. The symbolic computation are actually performed through the MAPLE engine/kernel[10]. This allows the user to explore problems in calculus, linear algebra, solutions of system of equations, and other areas. In fact, using symbolic computation students can easily recover the prerequisite information needed for design optimization. The student version does include some symbolic function handling in MATLAB. A short hands-on exercise to symbolic computation is provided in the following sections. This will allow us to revisit the mathematical prerequisites for the subject. These include ordinary derivatives, partial derivatives, matrices, derivatives of matrices, and solutions of linear and nonlinear equations.

4.2.1 Symbolic Computation Using MATLAB

One of the best ways to get familiar with symbolic computation is to take the quick online introduction to the Symbolic Math Toolbox available in the MATLAB Demos dialog box[8]. The following command locates the tutorial

```
>> demos
```

This will bring up the ***help browser***. In the toolbox list on the left, expand the **Symbolic Math**. Then you will see a choice of several demos. Click on them to view the demo. You can also run them in the Command window. The following hands-on exercise provides a small but important exposure to symbolic computation in MATLAB.

The computational engine executing the symbolic operations in MATLAB is the kernel of ***Maple*** marketed and supported by Waterloo Maple, Inc. If you are already familiar with Maple, then MATLAB provides a hook through which Maple commands can be executed in MATLAB. The symbolic computation in MATLAB is performed using a *symbolic object*, or ***sym***. This is another data type like the number and string data types used in earlier exercises. The Symbolic

Math Toolbox uses *sym* objects to represent symbolic variables, expressions, and matrices.

In the exercise that follows, f is a function of one variable x, g_1, and g_2 are functions of two variables x_1 and x_2. You will recognize the two variable functions from Example 4.1.

$$f = \sin(0.1 + x)/(0.1 + x)$$

$$g_1 = x_2 - x_1(0.05 - 0.12)/0.6 - 12$$

$$g_2 = \pi[3(x_1 + x_2) - \sqrt{(3x_1 + x_2)(x_1 + 3x_2)}]$$

For understanding, it is recommended that you treat this as a hands-on session. The commands are also available in **SymbolicExercise.m**. In the following the boldface words are commands that the reader types at the command line. The regular font is the response the user will see in the Command window.

```
x = sym('x')    % defining x as a symbolic object
x =
x


syms x1 x2      % defines multiple symbolic objects x1, x2

f  = 1*sin(0.1 + x)/(0.1 + x )

f =
sin(1/10+(x^2)^(1/2))/(1/10+(x^2)^(1/2))

g1 = x2 -x1*(0.5-0.12)/0.6-12
g1 =
x2-19/30*x1-12

g2 = pi*(3*(x1+x2)- sqrt((3*x1+x2)*(x1+3*x2)))
g2 =
pi*(3*x1+3*x2-((3*x1+x2)*(x1+3*x2))^(1/2))

whos   % display your variables
```

Name	Size	Bytes	Class	Attributes
f	1x1	204	sym	
g1	1x1	152	sym	
g2	1x1	208	sym	
x	1x1	126	sym	
x1	1x1	128	sym	
x2	1x1	128	sym	

```
%%% working with function of one variable
d1f = diff(f)  % first derivative of f wrt x
               % variable x is assumed
d1f =
cos(1/10+x)/(1/10+x)-sin(1/10+x)/(1/10+x)^2

d2f1 = diff(f,2)  % second derivative
d2f1 =
-sin(1/10+x)/(1/10+x)-2*cos(1/10+x)/
(1/10+x)^2+2*sin(1/10+x)/(1/10+x)^3

d2f2 = diff(d1f)  % same thing

d2f3 = diff(f,x,2) % same thing again

%%% the reader should see the same expression for second
%%% derivatives in the window

d1fval = subs(d1f,5) % evaluate first derivative at x = 5
d1fval =
    0.1097

d2fval = subs(d2f1,5) % evaluate second derivative at x = 5
d2fval =
    0.1385

%%% plot the function and the derivatives
%%% the plot is not shown here but should be clear

h1 = ezplot(f,[0 10])   % plot f between 0 and 10
hold on                 % multiple plots
h2 = ezplot(d1f,[0,10])  % plot first derivative of f
h3 = ezplot(d2f1,[0,10]) % plot second derivatives
hold off

% change plot styles
set(h1,'Color','r','LineStyle','-','LineWidth',2)
set(h2,'Color','b','LineStyle','--','LineWidth',2)
set(h3,'Color','g','LineStyle',':','LineWidth',2)
axis([0 10 -0.5 1.2])
grid
legend('function','first derivative','second derivative')
title('Plotting Symbolic Variables')

%%% working with function of two variables
```

```
%%% The derivatives are defined in the same way
%%% but they are partial derivatives
dg1_dx1 = diff(g1,x1)  % derivative of g1 wrt x1
dg1_dx1 =
7/60


dg1_dx2 = diff(g1,'x2') % derivative of g2 wrt x2
dg1_dx2 =
1
%%% Two functions with two variables
%%% The derivatives can be assembled as a Jacobian
g = [g1;g2]
g =
                                    x2+7/60*x1-12
 pi*(3*x1+3*x2-((3*x1+x2)*(x1+3*x2))^(1/2))
xx = [x1 x2]
xx =
[ x1, x2]
J = jacobian(g,xx)
[      7/60,                                               1]
[ pi*(3-1/2/((3*x1+x2)*(x1+3*x2))^(1/2)*(6*x1+10*x2)), pi*(3-
1/2/((3*x1+x2)*(x1+3*x2))^(1/2)*(10*x1+6*x2))]


%%% Gradient of function g2
grad_g2 = [diff(g2,x1); diff(g2,x2)]
grad_g2 =
 pi*(3-1/2/((3*x1+x2)*(x1+3*x2))^(1/2)*(6*x1+10*x2))
 pi*(3-1/2/((3*x1+x2)*(x1+3*x2))^(1/2)*(10*x1+6*x2))


%%% evaluate the gradient of g2 at x1 = 0.3, x2 = 0.2
grad_g2_val = subs(grad_g2,{x1,x2},{0.3,0.2})

grad_g2_val =
    3.4257
    2.7942


%%% Hessian of function g2
Hess_g2 = jacobian(grad_g2,xx)
%%% not shown as it is long

%%% evaluate Hessian at x1 = 0.3, x2 = 0.2
Hess_g2_val = subs(Hess_g2,{x1,x2},{0.3,0.2})
Hess_g2_val =
    2.0412   -3.0617
   -3.0617    4.5926
```

```
%%% draw contours of the two functions
%%% you need to use element by element operators
%%% you cannot label them

Figure    % use another figure window
h4 = ezcontour('x2 -x1*(0.05-0.12)/
0.6-12',[0 0.6 0 0.12],50)   %  one line
hold on
h5 = ezcontour('pi*(3*(x1+x2)- sqrt((3*x1+x2).
*(x1+3*x2)))',[0 0.6 0 0.12],50)
% h5   statement should be on one line
hold off
set(h4,'Color','r','LineStyle','-')
set(h5,'Color','b','LineStyle','--')
grid
xlabel('x_1')
ylabel('x_2')
title('Plot of g1 and g2')
%%%  Plot not shown here
```

Additional symbolic computations will be introduced as appropriate. Note that the result of both numeric and symbolic computations can be easily combined along with graphics to provide a powerful computing environment.

4.2.2 Basic Mathematical Concepts

The basic mathematical elements in the discussion of NLP are derivatives, partial derivatives, vectors, matrices, gradient, Jacobian, and Hessian. We have used the symbolic math toolbox in the previous section to calculate many of these quantities without defining them. These topics will have been extensively covered in the foundation courses on mathematics in most discipline. A brief review is offered in this section. The concepts are defined in a way that will be useful further on in the book. Graphical illustration is introduced wherever appropriate to sustain a geometrical connection to the definitions. Formal graphics are created through MATLAB. The opportunity to practice and extend the symbolic toolbox is exploited in this section.

Function of one variable: $f(x)$ identifies a function of one variable.

$$f(x) = \sin(0.1 + x)/(0.1 + x) \tag{4.16}$$

(4.16) is a specific example of such a function. The ***derivative*** of the function at any location x is written as follows (the derivative has a symbol delta):

$$\frac{df}{dx} = \lim_{\Delta x \to 0} \frac{f(x + \Delta x) - f(x)}{\Delta x} = \lim_{\Delta x \to 0} \frac{\Delta f}{\Delta x} \tag{4.17}$$

x is the point about which the derivative is computed. Δx is the distance to a neighboring point whose location therefore will be $x + \Delta x$. The value of the derivative is obtained as a limit of the ratio of the difference in the value of the function at the two points (Δf) to the distance separating the two points (Δx), as this separation is reduced to zero. The analytical computation of the derivative of a function usually follows several rules that are usually drilled into the students in the first course on calculus. The symbolic computation exercise indicates that the derivative of the function is as follows:

$$\frac{df}{dx} = \frac{\cos(0.1 + x)}{(0.1 + x)} - \frac{\sin(0.1 + x)}{(0.1 + x)^2}$$

which can also be established by calculus. Table 4.1 illustrates the limiting process for the derivative of the function in (4.16). The derivative is evaluated at the point $x = 3$. Several values of Δx, increasingly smaller, are used to illustrate the limiting process. The error in the derivative computation becomes insignificant as Δx is 0.0001. From these computations it can be expected that as Δx approaches 0, the ratio will reach the exact value of -0.32663.

Numerical Derivative Computation: Many numerical techniques in NLP require the computation of derivatives. Most code implementing these techniques do not use symbolic computation. Automation and ease of use require that these derivatives be computed numerically. The results in Table 4.1 justify the numerical computation of a derivative through a technique called the **first forward difference**[11]. Using a very small perturbation Δx, the derivative at a point is numerically calculated as the ratio of the change in the function value, Δf to the change in the displacement, Δx. For the example, the derivative at $x = 3$ with $\Delta x = 0.001$ is obtained as

$$\frac{df}{dx}\bigg|_{x=3} = \frac{f(3 + 0.001) - f(3)}{(3 + 0.001) - 3} = -0.32653 \qquad (4.18)$$

The derivative for the single variable function at any value x is also called the *slope* or the *gradient* of the function at that point. If a line is drawn tangent

Table 4.1   Limiting Process and Derivative

x	Δx	$x + \Delta x$	Δf	$\Delta f / \Delta x$	df/dx	error
3.00	1.0000	4.00000	-0.21299	-0.21299	-0.32663	34.79
3.00	0.1000	3.10000	-0.03166	-0.31655	-0.32663	3.09
3.00	0.0100	3.01000	-0.00326	-0.32564	-0.32663	0.30
3.00	0.0010	3.00100	-0.00033	-0.32653	-0.32663	0.03
3.00	0.0001	3.00010	-0.00003	-0.32662	-0.32663	0.00

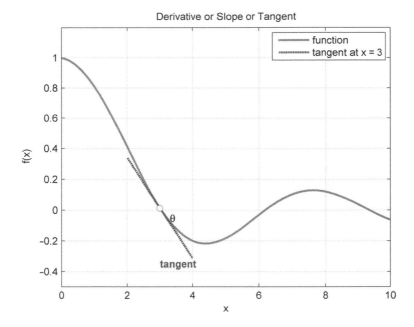

Figure 4.3 Illustration of the derivative or slope of the tangent.

to the function at the value x, the tangent of the angle that this line makes with the x-axis will have the same value as the derivative. If this angle is θ, then

$$\frac{df}{dx} = \tan\theta$$

Figure 4.3 illustrates the tangency property of the derivative. Figure 4.3 was created by the code **Tangent.m.**

Higher Derivatives: The derivative of the derivative is the called the *second derivative*. Formally, it is defined as follows:

$$\frac{d^2 f}{dx^2} = \frac{d}{dx}\left(\frac{df}{dx}\right) = \lim_{\Delta x \to 0} \frac{\Delta\left(\frac{df}{dx}\right)}{\Delta x} = \lim_{\Delta x \to 0} \frac{\left.\frac{df}{dx}\right|_{x+\Delta x} - \left.\frac{df}{dx}\right|_{x}}{\Delta x} \qquad (4.19)$$

Similarly the **third derivative** is defined as

$$\frac{d^3 f}{dx^3} = \frac{d}{dx}\left(\frac{d^2 f}{dx^2}\right) \qquad (4.20)$$

This can go on, provided the function has sufficient higher order dependence on the independent variable x.

Function of two variables: For illustration, we choose the nonlinear constraint g_2 and identify it as f for this section:

$$f(x_1, x_2) = \pi[3(x_1 + x_2) - \sqrt{(3x_1 + x_2)(x_1 + 3x_2)}] \tag{4.21}$$

The first important feature of two or more variables is that the derivatives are defined as *partial deriv*atives. Partial derivatives are defined for each independent variable. In (4.21) the independent variables are x_1 and x_2. The dependent variable is f. A *point* in this space refers to a particular combination of the values for the independent variables x_1 and x_2. The partial derivative is expressed with a curly $d(\partial)$. These derivatives are obtained in the same way as the ordinary derivatives, except that only one variable is changed while all the other variables are held at a constant value. In this example, when computing the partial derivative of x_1 the value of x_2 is not changed:

$$\left.\frac{\partial f}{\partial x_1}\right|_{(x_1, x_2)} = \lim_{\Delta x_1 \to 0} \frac{f(x_1 + \Delta x_1, x_2) - f(x_1, x_2)}{\Delta x_1} \tag{4.22}$$

(4.22) expresses the *partial derivative of the function f with respect to the variable x_1*. In (4.22) the subscript after the vertical line in the first expression, establishes the point about which the partial derivative with respect to x_1 is being evaluated. (x_1, x_2) represents any/all points. A similar expression can be written for *partial derivative of f with respect to x_2*. Most readers can evaluate these derivatives from the rules of calculus (or you can use symbolic programming) to verify that

$$\frac{\partial f}{\partial x_1} = \pi \left\{ 3 - \frac{1}{2} \frac{[3(x_1 + 3x_2) + (3x_1 + x_2)]}{\sqrt{(3x_1 + x_2)(x_1 + 3x_2)}} \right\} \tag{4.23a}$$

$$\frac{\partial f}{\partial x_2} = \pi \left\{ 3 - \frac{1}{2} \frac{[(x_1 + 3x_2) + 3(3x_1 + x_2)]}{\sqrt{(3x_1 + x_2)(x_1 + 3x_2)}} \right\} \tag{4.23b}$$

The value of the derivatives in (4.23) at $x_1 = 0.3$ and $x_2 = 0.4$ is

$$\left.\frac{\partial f}{\partial x_1}\right|_{(0.3, 0.4)} = 2.9005; \quad \left.\frac{\partial f}{\partial x_2}\right|_{(0.3, 0.4)} = 3.3505; \tag{4.24}$$

The value of $f(0.3, 0.4) = 2.2103$.

Until this point we have made use of symbols representing changes in values (functions, variables) without a formal definition. We have meant

$\Delta(\ )$: to represents total/finite/significant changes in the quantity $(\ )$

$d(\ ), \delta(\ )$: to represent differential/infinitesimal changes in $(\ )$

Changes in functions (dependent variables) occur due to changes in the independent variables. From calculus the differential change in $f(x_1, x_2)(df)$ due to the differential change in the variables $x_1(dx_1)$ and $x_2(dx_2)$ is expressed as follows[1,4]:

$$df = \frac{\partial f}{\partial x_1} dx_1 + \frac{\partial f}{\partial x_2} dx_2 \qquad (4.25)$$

The changes at any point in (4.25) would require evaluating the partial derivatives at the point. The definition of the partial derivative is apparent this expression as holding x_2 at a constant value implies that $dx_2 = 0$. Another interpretation for the *partial derivative* can be observed from the above definition: *change of the dependent variable (function) per unit change in the independent variables*

Gradient of the function: In the function of single variable, the *derivative* was associated with the *slope*. In two or more variables the slope is equivalent to the **gradient**. The *gradient* is a **vector**, and at any point represents *the direction in which the function will increase most rapidly*. Examining the conventional objective of NLP, *minimization of the objective function*, the reverse direction of the *gradient* has a natural part to play in the development of methods to solve the problem. The gradient is composed of the partial derivatives organized as a vector. Vectors in this book are column vectors unless otherwise noted. It has also a standard mathematical symbol. The definitions are captured in (4.26):

$$\nabla f = \begin{bmatrix} \dfrac{\partial f}{\partial x} \\ \dfrac{\partial f}{\partial y} \end{bmatrix} = \begin{bmatrix} \dfrac{\partial f}{\partial x} & \dfrac{\partial f}{\partial y} \end{bmatrix}^T \qquad (4.26)$$

The tangent to the function f at any point must be normal to the gradient. After all, if you are moving tangent to the function then, for small displacements, there should be no change in the function. Using (4.25) we can write

$$\Delta f = \frac{\partial f}{\partial x_1} \Delta x_1 + \frac{\partial f}{\partial x_2} \Delta x_2 = 0; \quad \Delta x_2 = -\frac{\dfrac{\partial f}{\partial x_1}}{\dfrac{\partial f}{\partial x_2}} \Delta x_1; \quad \Delta x_2 = -m_t \Delta x_1$$

$$(4.26)$$

At $x_1 = 0.3$ and $x_2 = 0.4$, and

$$\Delta x_2 = -\frac{2.9005}{3.3505} \Delta x_1 \qquad (4.27a)$$

We use this relation to draw a tangent line through the point $x_1 = 0.3$ and $x_2 = 0.4$. The Δx_1 and Δx_2 are distances measured from the selected point. Knowing the gradient is normal to the tangent, we can establish that the gradient

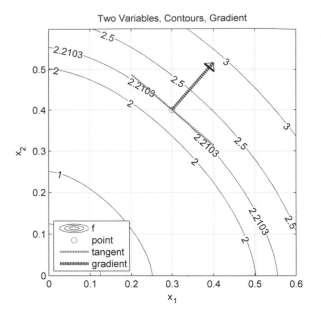

Figure 4.4 Two variables: contours, tangent, and gradient.

line at this point can be drawn using

$$\Delta x_2 = m_g \Delta x_1 = -\left(-\frac{3.3505}{2.9005}\right) \Delta x_1; \quad as \quad m_g m_t = -1 \qquad (4.27b)$$

MATLAB Code: `Gradient.m` will automatically generate Figure 4.4. First contours are drawn for several values of $f(x_1, x_2)$. The point $x_1 = 0.3$ and $x_2 = 0.4$ is identified on the figure. The contour passing through this point is also shown in the figure. The partial derivatives at the point are computed and the tangent and gradient lines through the point are drawn using two point established by (4.27). The direction of the gradient is drawn by a new function called **DrawArrow.m**. It draws a line and places an arrow on the second point so that the arrow indicates the direction from the first point to the second point. You might have to change the scaling in the function if the arrowhead is not visible or looks too big. A judicious mix of numerical and symbolic programming is utilized. The arrow on the gradient line, since it is a vector, should be to the right as the functions increases to the right. The tangent is, however, a line and is not a vector. Actually, for two independent variables, the tangent will be a plane in a three-dimensional plot. Three-dimensional plot of this information is quite messy and is suggested as a problem at the end of the chapter.

You can actually see the tangent and gradient lines are normal to each other in Figure 4.4. In order to display this geometric property it was first necessary to make the plot square. For aesthetic and page size reasons, most plotting software will usually have the x-axis longer than the y-axis. The second important trick was to make the range of the variables the same, which is not true in the definition if Example 4.1. If you were to retain original limits on the variables, the lines are no longer at 90 degrees. You can easily verify it by running **Gradient.m** with the original limits.

Scaling: What if we wish to stick with the original side constraint limits for the design variables? Is there a way to display the perpendicularity of the normal and the tangent lines on the figure? The answer is in the affirmative if we scale the function. Scaling is very important in applied optimization. It is introduced so that all functional constraints are equally effective in discovering the solution. Consider the design of the I beam in Example 1.2. The inequality constraints included one on stress and another one on displacement. The typical value of stress is about 10^8 times the typical value of the displacement. If they were incorporated in the problem directly, then the displacement constraint would be overwhelmed by the stress constraint in the search for the optimum solution.

In this situation, scaling the variables will create a design region that will have the same range in the two variables. We can easily accomplish this by defining the following:

$$\tilde{x}_1 = \frac{x_1}{0.6}; \quad x_1 = 0.6\tilde{x}_1; \quad \tilde{x}_2 = \frac{x_2}{0.12}; \quad x_2 = 0.12\tilde{x}_2 \tag{4.28a}$$

$$f(x_1, x_2) = \pi[3(0.6\tilde{x}_1 + 0.12\tilde{x}_2) - \sqrt{(3 \times 0.6\tilde{x}_1 + 0.12\tilde{x}_2)(0.6\tilde{x}_1 + 3 \times 0.12\tilde{x}_2)}] \tag{4.28b}$$

$$0 \le \tilde{x}_1 \le 1; \quad 0 \le \tilde{x}_2 \le 1 \tag{4.28c}$$

Keeping the axis square we can display the function contours, the tangent line, and the gradient line in Figure 4.5. You can see the geometric property between the gradient and the tangent is displayed. The code is available in **ScaledGradient.m**. As you can see the effort is still the same once the pre-processing is taken care off. Scaling should not affect the value of the optimum but should improve the search. In this particular instant you can witness the visual improvement.

Jacobian: The Jacobian $[J]$ defines a useful way to organize the gradients of several functions. For example, using two inequality constraints in Example 4.1, $g_1(x_1, x_2)$ and $g_2(x_1, x_2)$, the definition of the Jacobian is

$$g_1 = x_2 - x_1(0.05 - 0.12)/0.6 - 12$$

$$g_2 = \pi[3(x_1 + x_2) - \sqrt{(3x_1 + x_2)(x_1 + 3x_2)}]$$

$$[J] = \begin{bmatrix} \dfrac{\partial g_1}{\partial x_1} & \dfrac{\partial g_1}{\partial x_2} \\ \dfrac{\partial g_2}{\partial x_1} & \dfrac{\partial g_2}{\partial x_2} \end{bmatrix} \tag{4.29}$$

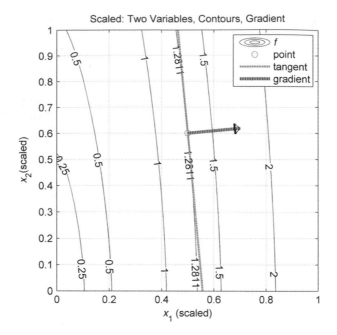

Figure 4.5   Scaled function: Two variables, contours, tangent, and gradient.

$$[J] = \begin{bmatrix} -\dfrac{(0.05 - 0.12)}{6} & 1 \\ \pi\left\{3 - \dfrac{1}{2}\dfrac{[3(x_1 + 3x_2) + (3x_1 + x_2)]}{\sqrt{(3x_1 + x_2)(x_1 + 3x_2)}}\right\} & \pi\left\{3 - \dfrac{1}{2}\dfrac{[(x_1 + 3x_2) + 3(3x_1 + x_2)]}{\sqrt{(3x_1 + x_2)(x_1 + 3x_2)}}\right\} \end{bmatrix}$$

In equation (4.29) the gradients of the function appear in the same row. The first row is the gradient of g_1, while the second row is the gradient of g_2. If the two functions are collected into a column vector, the differential changes $[dg_1 \; dg_2]^{\mathrm{T}}$ in the functions, due to the differential change in the variables $[dx_1 \; dx_2]$ can be expressed as a matrix multiplication using the Jacobian

$$\begin{bmatrix} dg_1 \\ dg_2 \end{bmatrix} = [J]\begin{bmatrix} dx_1 \\ dx_2 \end{bmatrix} \tag{4.30}$$

which is similar to equation (4.25) for a single function.

Hessian: The Hessian matrix $[H]$ is the same as the matrix of second derivatives of a single function of several variables. For $f(x_1, x_2)$:

$$[H] = \begin{bmatrix} \dfrac{\partial^2 f}{\partial x_1^2} & \dfrac{\partial^2 f}{\partial x_1 \partial x_2} \\ \dfrac{\partial^2 f}{\partial x_2 \partial x_1} & \dfrac{\partial^2 f}{\partial x_2^2} \end{bmatrix} \tag{4.31}$$

For the example, for space reasons, the individual terms of the Hessian matrix are given here:

$$f(x_1, x_2) = \pi[3(x_1 + x_2) - \sqrt{(3x_1 + x_2)(x_1 + 3x_2)}]$$

$$\frac{\partial^2 f}{\partial x_1^2} = \frac{-3\pi}{\sqrt{(3x_1 + x_2)(x_1 + 3x_2)}} + \frac{1}{4}\frac{\pi(6x_1 + 10x_2)^2}{[(3x_1 + x_2)(x_1 + 3x_2)]^{3/2}}$$

$$\frac{\partial^2 f}{\partial x_1 \partial x_2} = \frac{-5\pi}{\sqrt{(3x_1 + x_2)(x_1 + 3x_2)}} + \frac{1}{4}\frac{\pi(6x_1 + 10x_2)(6x_2 + 10x_1)}{[(3x_1 + x_2)(x_1 + 3x_2)]^{3/2}} = \frac{\partial^2 f}{\partial x_2 \partial x_1}$$

$$\frac{\partial^2 f}{\partial x_2^2} = \frac{-3\pi}{\sqrt{(3x_1 + x_2)(x_1 + 3x_2)}} + \frac{1}{4}\frac{\pi(6x_2 + 10x_1)^2}{[(3x_1 + x_2)(x_1 + 3x_2)]^{3/2}}$$

The Hessian matrix is symmetric. For the example

$$[H]|_{(0.3, 0.2)} = \begin{bmatrix} 2.0412 & -3.0617 \\ -3.0617 & 4.5926 \end{bmatrix}$$

Function of n variables: For $f(X)$, where $[X] = [x_1, x_2, \ldots, x_n]^T$, the gradient is

$$\nabla f = \begin{bmatrix} \dfrac{\partial f}{\partial x_1} & \dfrac{\partial f}{\partial x_2} & \cdots & \dfrac{\partial f}{\partial x_n} \end{bmatrix}^T \tag{4.32}$$

The Hessian is

$$[H] = \begin{bmatrix} \dfrac{\partial^2 f}{\partial x_1^2} & \dfrac{\partial^2 f}{\partial x_1 \partial x_2} & \cdot & \dfrac{\partial^2 f}{\partial x_1 \partial x_n} \\ \dfrac{\partial^2 f}{\partial x_2 \partial x_1} & \dfrac{\partial^2 f}{\partial x_2^2} & & \\ & & \cdot & \\ \dfrac{\partial^2 f}{\partial x_n \partial x_1} & & & \dfrac{\partial^2 f}{\partial x_n^2} \end{bmatrix} \tag{4.33}$$

Equations (4.32) and (4.33) will appear quite often in succeeding chapters. A few minutes of familiarization with these concepts will provide sustained comprehension in later discussions.

4.2.3 Taylor's Theorem/Series

Single Variable: The Taylor's series is a useful mechanism to approximate the value of the function $f(x)$ at the point $(x_p + \Delta x)$ if the function is completely known (its value and the value of all its derivatives) at the point x_p. The expansion is

$$
f(x_p + \Delta x) \cong f(x_p) + \frac{df}{dx}\bigg|_{x_p} (\Delta x) + \frac{1}{2!} \frac{d^2 f}{dx^2}\bigg|_{x_p}
$$

$$
\times (\Delta x)^2 + \cdots + \frac{1}{n!} \frac{d^n f}{dx^n}\bigg|_{x_p} (\Delta x)^n \qquad (4.34)
$$

The series is widely used in most disciplines to establish continuous mathematical models. It is the mainstay of many numerical techniques, including those in optimization. Equation (4.34) is usually truncated to the first two or three terms, with the understanding that the approximation will suffer some error whose order depends on the term that is being truncated. The representation in (4.35) suggests that the expression is truncated at the fourth term, with the error being of the order of $(\Delta x)^3$.

$$
f(x_p + \Delta x) \cong f(x_p) + \frac{df}{dx}\bigg|_{x_p} (\Delta x) + \frac{1}{2!} \frac{d^2 f}{dx^2}\bigg|_{x_p} (\Delta x)^2 + E_{rr}(\Delta x)^3 \quad (4.35)
$$

If the first term is brought to the left and the error term discarded, the expression can be written as

$$
\Delta f = f(x_p + \Delta x) - f(x_p) \cong \frac{df}{dx}\bigg|_{x_p} (\Delta x) + \frac{1}{2!} \frac{d^2 f}{dx^2}\bigg|_{x_p} (\Delta x)^2
$$

$$
\Delta f = f(x_p + \Delta x) - f(x_p) \cong \delta f + \delta^2 f
$$

(4.36)

In Equation 4.36, the first term on the right is called the first-order/linear variation (δf) and the second term is second-order/quadratic variation $(\delta^2 f)$. For the example

$$
f(x) = \sin(0.1 + x)/(0.1 + x)
$$

and about the point $x = 5$, the Taylor's quadratic expansion can be written as follows:

$$
f(5 + \Delta x) = -0.1815 + 0.1097(\Delta x) + \frac{1}{2} 0.1385(\Delta x)^2
$$

Figure 4.6 illustrates the approximations using Taylor's series of various order at the point $x = 5$. In using the Taylor's expansion, it is important to focus on the neighborhood of the chosen point rather than the entire range of the function. This is clearly demonstrated in Figure 4.6, where the approximation over the range leaves a lot to be desired. In the rectangle, however, clearly the quadratic expansion is decent and the quintic approximation is excellent. Figure 4.6 can be created using the code **Taylor.m**. MATLAB provides a symbolic function *taylor* to create the approximations which are then plotted and styled.

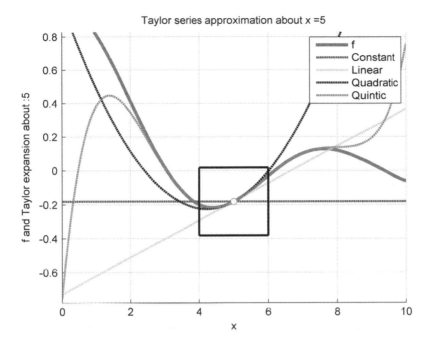

Figure 4.6 Taylor's expansion : one variable.

Two or More Variables: The series is only expanded to the quadratic term in the expression below. The truncation error is ignored, and can be controlled by choosing a small Δx. The two variable Taylor expansion is first shown in detail and then expressed in terms of vectors and matrices. The first-order expansion can be expressed in terms of the gradient. The second order expansion involves the *Hessian* matrix

$$f(x_{1p} + \Delta x_1, x_{2p} + \Delta x_2) \cong f(x_{1p}, x_{2p})$$

$$+ \left[\left. \frac{\partial f}{\partial x_1} \right|_{(x_{1p}, x_{2p})} \Delta x_1 + \left. \frac{\partial f}{\partial x_2} \right|_{(x_{1p}, x_{2p})} \Delta x_2 \right]$$

$$+ \frac{1}{2} \left[\left. \frac{\partial^2 f}{\partial x_1^2} \right|_{(x_{1p}, x_{2p})} (\Delta x_1)^2 \right.$$

$$+ 2 \left. \frac{\partial^2 f}{\partial x_1 \partial x_2} \right|_{(x_{1p}, x_{2p})} \Delta x_1 \Delta x_2$$

$$\left. + \left. \frac{\partial^2 f}{\partial x_2^2} \right|_{(x_{1p}, x_{2p})} (\Delta x_2)^2 \right] \tag{4.37}$$

If the displacements are organized as a column vector $[\Delta x_1 \ \Delta x_2]^T$, the expansion in 4.37 can be expressed in a condensed manner as

$$f(x_{1p} + \Delta x_1, x_{2_p} + \Delta x_2) \cong f(x_{1p}, x_{2p}) + \nabla f|_{(x_{1p}, x_{2p})}^T \begin{bmatrix} \Delta x_1 \\ \Delta x_2 \end{bmatrix}$$

$$+ \frac{1}{2} [\Delta x_1 \ \Delta x_1]^T [H(x_{1p}, x_{2_p})] \begin{bmatrix} \Delta x_1 \\ \Delta x_2 \end{bmatrix} \quad (4.38)$$

This can also be expressed as first-and second-order variations as

$$f(x_{1p} + \Delta x_1, x_{2_p} + \Delta x_2) - f(x_{1p}, x_{2p}) = \Delta f \cong \delta f + \delta^2 f$$

For n-variables, with $\mathbf{X_p}$ as the current point and $\mathbf{\Delta X}$ the displacement vector,

$$f(\mathbf{X_p} + \mathbf{\Delta X}) - f(\mathbf{X_p}) = \Delta f \cong \nabla f(\mathbf{X_p})^T \mathbf{\Delta X} + \frac{1}{2} \mathbf{\Delta X}^T H(\mathbf{X_p}) \mathbf{\Delta X} \quad (4.38)$$

For the example

$$f(x_1, x_2) = \pi [3(x_1 + x_2) - \sqrt{(3x_1 + x_2)(x_1 + 3x_2)}]$$

About the point $x_1 = 0.3$, $x_2 = 0.2$

$$f(0.3 + \Delta x_1, 0.2 + \Delta x_2) = 1.5865 + [3.4257 \ 2.7942] \begin{bmatrix} \mathbf{\Delta x_1} \\ \mathbf{\Delta x_2} \end{bmatrix}$$

$$+ \frac{1}{2} [\Delta x_1 \ \Delta x_2] \begin{bmatrix} 2.0412 & -3.0617 \\ -3.0617 & 4.5296 \end{bmatrix} \begin{bmatrix} \mathbf{\Delta x_1} \\ \mathbf{\Delta x_2} \end{bmatrix}$$

Figure 4.7, generated through `Taylor2Var.m`, illustrates the Taylor's expansion for two variables. This is best seen in color by running the code. Once more, we have defined the neighborhood about the point $x_1 = 0.3$, $x_2 = 0.2$ where the approximation of the function f is considered. The linear expansion about the point in consideration indicates error in the approximation. The quadratic expansion, on the other hand, is generally good in the neighborhood.

4.3 ANALYTICAL CONDITIONS

Analytical conditions refer to the necessary and sufficient conditions that will permit the recognition of the solution to the optimal design problem. The conditions developed here empower the numerical techniques to follow later. We develop the conditions starting with the unconstrained problem. We then look at the equality constraints and then the inequality constraints. Instead of formal mathematical details, the conditions are established less formally from the geometrical description of the problem, graphical solution, and/or through intuitive

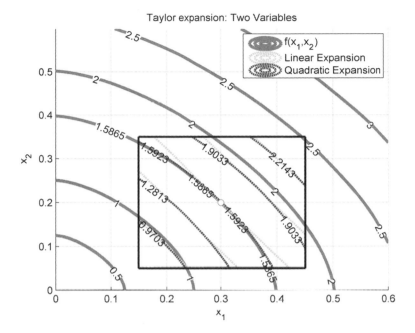

Figure 4.7 Taylor's expansion: Two variables.

reasoning. A more formal development can be found in[11-16] . While establishing the conditions it is expected that the *solution is in the interior of the feasible region*, and *there is only one minimum in the design space*.

Earlier, we had introduced the different optimization problems based on Example 4.1. In order to provide a geometric perspective in the graphical solutions, we will transform all the problems to include the simple variable scaling introduced in the previous section as

$$\tilde{x}_1 = \frac{x_1}{0.6} \quad or \quad x_1 = 0.6\tilde{x}_1; \quad \tilde{x}_2 = \frac{x_2}{0.12} \quad or \quad x_2 = 0.12\tilde{x}_2$$

In the following discussion, we drop the tilde notation for convenience. The reference to the scaled variable can be seen through the side constraints.

4.3.1 Unconstrained Problem

$$\text{Minimize} \quad f(x_1, x_2) = -\pi(0.6)(0.12)x_1 x_2 \tag{4.39a}$$

$$0 \le x_1 \le 1 \qquad 0 \le x_2 \le 1 \tag{4.39b}$$

Figure 4.8 provides the graphical solution of the problem. The side constraints define the design space enclosed by the black rectangle. The solution is at the point (1, 1). This is on the boundary of the region, and f^* is -0.22619. This

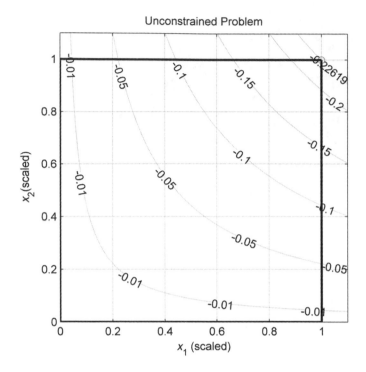

Figure 4.8 Solution for original unconstrained optimization problem.

solution on the boundary does not permit the development of the analytical condition for the unconstrained minimum problem. We will construct a different unconstrained optimization problem that will move the solution to the interior. Adding two quadratic terms should do the trick.

$$\text{Minimize} \quad f(x_1, x_2) = -\pi(0.6)(0.12)x_1x_2 + (x_1 - 0.5)^2 + (x_2 - 0.3)^2 \tag{4.40a}$$

$$0 \le x_1 \le 1 \quad 0 \le x_2 \le 1 \tag{4.40b}$$

Graphical Solution: Figure 4.9 provides a three-dimensional plot of the refined problem, along with the contours of the objective function on the lower plane. The solution is identified on the plot, along with the feasible region. From the figure we can observe a single minimum and also note that the tangent plane at the solution is horizontal.

Necessary Conditions: The tangent plane at the solution, being horizontal, or in the plane of the independent variables x_1 and x_2 allows us to develop the mathematical conditions for the solution. We recognize at the solution (x_1^*, x_2^*) of the minimization problem, for differential changes in the independent variables

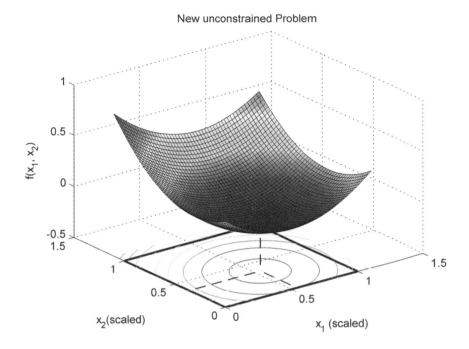

Figure 4.9 3D plot of modified unconstrained problem.

dx_1 and dx_2 *(in the limit that these changes go to zero)*, we will be moving on the plane tangent to this point. On this plane the objective function f cannot change. We can state that df must be 0. Therefore,

$$df = \frac{\partial f}{\partial x_1}dx_1 + \frac{\partial f}{\partial x_2}dx_2 = 0 \qquad (4.41)$$

Since (4.41) must be valid for all kind of changes dx_1 and dx_2, the best way this requirement can be met is to require that the partial derivatives at the solution must be zero. This requires that *the gradient of the objective function f at the optimum must be zero* or

$$\left.\frac{\partial f}{\partial x_1}\right|_{(x_1^*,x_2^*)} = 0; \quad \left.\frac{\partial f}{\partial x_2}\right|_{(x_1^*,x_2^*)} = 0; \quad \textbf{\textit{OR}} \quad \nabla f|_{(x_1^*,x_2^*)} = \begin{bmatrix} \dfrac{\partial f}{\partial x_1} \\[2mm] \dfrac{\partial f}{\partial x_2} \end{bmatrix} = \begin{bmatrix} \mathbf{0} \\ \mathbf{0} \end{bmatrix}$$

$$(4.42)$$

For the example in (4.40) two equations will be required to establish the values of the optimum design variables and they are provided by (4.42). These are also called **first-order conditions** (FOC)because of the first derivative of the objective function f involved in the conditions. For the problem in (4.40) the first-order conditions are

$$\frac{\partial f}{\partial x_1} = -\pi(0.6)(0.12)x_2 + 2(x_1 - 0.5) = 0$$

$$\frac{\partial f}{\partial x_2} = -\pi(0.6)(0.12)x_1 + 2(x_2 - 0.3) = 0$$

(4.43)

And the solution for x_1^* and x_2^* is

$$x_1^* = 0.5408; \quad x_2^* = 0.3612; \quad f^* = -0.0388$$

Now consider the problem in (4.39) where the constraint is on the boundary. The first order conditions (4.42) will determine that

$$x_1^* = 0.0; \quad x_2^* = 0.0; \quad f^* = 0$$

While in fact the actual solution can be seen in Figure 4.8 as

$$x_1^* = 1.0; \quad x_2^* = 1.0; \quad f^* = -0.2262$$

This suggests that the conditions (4.42) apply for solution *in the interior of the feasible region only*. Solution on the boundary will have to be determined in other ways. Note that the same conditions hold if you are looking for a maximum. At the peak of the function the tangent plane will again be horizontal, giving the same conditions (4.42) for a *maximum*. Therefore the term necessary conditions implies that these conditions must be met for being considered as a solution To establish the solution itself additional conditions are required, which we develop as **second-order conditions** (SOC). Meanwhile for the general unconstrained problem in n-variables, the necessary conditions can be stated as

$$\nabla f(X^*) = 0 \tag{4.44}$$

The equation in 4.43 is used to establish the value of the design variables X^* both analytically and numerically

Sufficient Conditions: These conditions are usually regarded as *second-order conditions*. It can be inferred that these conditions will involve *second derivatives* of the function. The SOC is obtained through the Taylor's expansion of the function to second order. If X^* is the solution, and ΔX represents the change of the variables from the optimal value which will yield a change Δf, then

$$\Delta f = f(X^* + \Delta X) - f(X^*) = \nabla f(X^*)^T \Delta X + \frac{1}{2}\Delta X^T H(X^*)\Delta X \tag{4.45}$$

This is similar to Equation (4.38) except the expansion is about the solution (X^*). Δf must be greater than zero if the minimum is at X^*. Employing the necessary conditions (4.43) the first term on the right of Equation (4.45) is zero. This leaves the following inequality

$$\Delta f = \frac{1}{2}\Delta X^T H(X^*)\Delta X > 0 \tag{4.46}$$

where $H(X^*)$ is the *Hessian* matrix (the matrix of second derivatives) of the function f at the possible optimum value X^*. For the relations in 4.46 to hold, the matrix $H(X^*)$ must be positive definite. There are three ways to establish that H is positive definite.

   i. For all possible ΔX, $\Delta X^T H(X^*)\Delta X > 0$
  ii. The eigenvalues of $H(X^*)$ must all be positive
 iii. The determinant of all lower order (sub matrices) of $H(X^*)$ that include the terms on the main diagonal are all positive

Of the three only (ii) and (iii) can be practically applied. It is illustrated below At the solution (actually everywhere)

$$H = \begin{bmatrix} 2 & -0.2262 \\ -0.2262 & 2 \end{bmatrix}$$

Is it positive definite?

  ii. To calculate the eigenvalues of H in condition (ii) above,

$$\begin{vmatrix} 2 - \lambda & -0.2262 \\ -0.2262 & 2 - \lambda \end{vmatrix} = (2 \quad \lambda)(2 \quad \lambda) \quad 0.0512 = 0$$

The eigenvalues are $\lambda = 2.2262$, $\lambda = 1.7738$ and therefore the matrix is ***positive definite***.

 iii. Condition (iii) requires calculating determinants of increasing orders that includes the main diagonal elements. For a 2×2 matrix there are two determinants to be considered:

$$|2| > 0; \quad \begin{vmatrix} 2 & -0.2262 \\ -0.2262 & 2 \end{vmatrix} = 4 - (0.2262)^2 = 3.9488 > 0$$

The determinants are positive and the matrix is *positive definite*
This establishes that.

$$x_1^* = 0.5408; \quad x_2^* = 0.3612; \quad f^* = -0.0388$$

is a minimum, as both the *necessary* and *sufficient* conditions are satisfied.

MATLAB Code: `UnconstrainedProblem.m` contains the code that will create Figure 4.8 and Figure 4.9. It also calculates the solution using the first-order conditions. The Hessian at the solution is then calculated and the eigenvalues are obtained using a standard MATLAB function. Once again, we have used a combination of symbolic and numeric calculations.

Summary: For an unconstrained problem, for a solution in the interior of the feasible region, the necessary conditions for optimum are

$$\nabla f(X^*) = 0 \tag{4.44}$$

The sufficient conditions for the optimum are

$$H(X^*) \quad is \ positive \ definite \tag{4.47}$$

4.3.2 Equality-Constrained Problem 2

Its scaled form is as follows:

Minimize $f(x_1, x_2) = -\pi(0.6)(0.12)x_1x_2$ \hfill (4.48a)

Subject to: $h_1(x_1, x_2) = 0.12x_2 - 0.12 - \dfrac{(0.05 - 0.12)}{(0.6 - 0.0)}0.6x_1 = 0$ \hfill (4.48b)

$0 \le x_1 \le 1 \quad 0 \le x_2 \le 1$ \hfill (4.48c)

Its general description is

Minimize $f(x_1, x_2)$ \hfill (4.49a)

Subject to $h_1(x_1, x_2) = 0$ \hfill (4.49b)

$x_1^l \le x_1 \le x_1^u; \quad x_2^l \le x_2 \le x_2^u$ \hfill (4.49c)

Graphical Solution: Figure 4.10 is the graphical solution to the problem in (4.48). It is created by running `EqualityConstrainedProblem2.m.` You can see the contours of the objective function, the equality constraint, the solution, the tangent to the function at the solution that overlays the constraint, the gradient of the function and the constraint at the solution, and the feasible region. On the screen you will see an enhanced figure with color.

From Figure 4.10, at the *solution*, the gradient of the objective function and the gradient of the constraint are parallel and oppositely directed. Also, at the solution the constraint is tangent to the objective function. Examining other feasible points in Figure 4.10, which must lie on the equality constraint, it can be seen that the above special geometrical relationship is only possible at the solution. If the

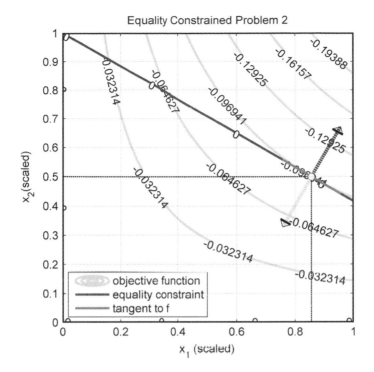

Figure 4.10 Graphical solution: Equality-constrained problem 2.

constant of proportionality is λ_1 then the relationship between the gradients can be expressed as follows:

$$\nabla f = -\lambda_1 \nabla h_1$$

or

$$\nabla f + \lambda_1 \nabla h_1 = 0 \tag{4.50}$$

Equation 4.50 is obtained in a more formal way using the method of Lagrange multiplier.

Method of Lagrange: In this method, the problem is transformed into an *unconstrained minimization* problem of an augmented function called the **Lagrangian (F)**. The **Lagrangian** is defined as the sum of the original objective function and a *linear combination* of the constraints. The coefficients of this linear combination are known as the **Lagrange multipliers**. The unconstrained problem in three variables x_1, x_2, λ_1 is

$$\text{Minimize} \quad F(x_1, x_2, \lambda_1) = f(x_1, x_2) + \lambda_1 h_1(x_1, x_2) \tag{4.51a}$$

$$x_1^l \le x_1 \le x_1^u; \qquad x_2^l \le x_2 \le x_2^u \tag{4.51b}$$

Substituting from (4.48) the Lagrangian in detail is written as

$$\text{Minimize} \quad F(x_1, x_2, \lambda_1) = -\pi(0.6)(0.12)x_1 x_2$$

$$+ \lambda_1 \left(0.12x_2 - \frac{(0.05 - 0.12)}{(0.6 - 0.0)} 0.6x_1 - 0.12 \right) \qquad (4.52a)$$

$$0 \le x_1 \le 1 \quad 0 \le x_2 \le 1 \qquad (4.52b)$$

The equity constant must be satisfied. Is the solution to the transformed problem (4.51) same as the solution to the original problem (4.48)? If the design is feasible, then most definitely yes. For feasible designs, $h_1(x_1, x_2) = 0$, the objective functions in equations (4.51) and (4.48) are the same. If design is not feasible, then by definition there is no solution anyway.

Necessary Conditions: Applying the FOC for an unconstrained optimization problem (4.44) in the variables x_1, x_2, λ_1, the three relations to solve for $x_1^*, x_2^*, \lambda_1^*$ are

$$\frac{\partial F}{\partial x_1} = \frac{\partial f}{\partial x_1} + \lambda_1 \frac{\partial h_1}{\partial x_1} = 0$$

$$\frac{\partial F}{\partial x_2} = \frac{\partial f}{\partial x_2} + \lambda_1 \frac{\partial h_1}{\partial x_2} = 0 \qquad (4.53)$$

$$\frac{\partial F}{\partial \lambda_1} = h_1 = 0$$

Equation 4.53 expresses the FOC, or necessary conditions, for an equality constrained problem in two variables. The first two equations can be assembled in vector form to yield the same information expressed by (4.50), which was noted in the graphical solution Figure 4.10. Therefore, (4.53) can be also expressed as

$$\nabla F = \nabla f + \lambda_1 \nabla h_1 = 0 \qquad (4.54a)$$

$$h_1 = 0 \qquad (4.54b)$$

where the gradient of the augmented function or Lagrangian is defined only with respect to the design variables x_1 and x_2. Applying Equations 4.54 to the example of this section:

$$\frac{\partial F}{\partial x_1} = -\pi(0.6)(0.12)x_2 - \lambda_1 \frac{(0.05 - 0.12)}{(0.6 - 0.0)} 0.6 = 0$$

$$\frac{\partial F}{\partial x_2} = -\pi(0.6)(0.12)x_1 + 0.12\lambda_1 = 0 \qquad (4.55)$$

$$h_1 = 0.12x_2 - \frac{(0.05 - 0.12)}{(0.6 - 0.0)} 0.6x_1 - 0.12 = 0$$

Equations (4.55) represent three equations in three variables which should determine the values for $x_1^*, x_2^*, \lambda_1^*$. Note that (4.54) only define the necessary conditions, which means the solution could be a maximum also. (4.55) is a set of three linear equations and can be solved as

$$x_1^* = 0.8571; \quad x_2^* = 0.5; \quad \lambda_1^* = 1.6157; \quad f^* = -0.0969; \quad h^* = 0 \quad (4.56)$$

In this particular example the side constraints were ignored in the conditions in (4.54). After obtaining the solution we can verify that the solution does not violate the side constraints. In many problems (4.54) may establish a nonlinear set of equations. Most pre-requisite courses on numerical methods do not attempt to solve a nonlinear system of equations. Usually, they only handle a single one through Newton-Raphson or the bisection method. In fact, NLP or design optimization, is primarily about techniques for solving a system of nonlinear equation, albeit of specific forms. For simple examples we can solve the equations using symbolic programming in MATLAB. Often, symbolic solution may be clumsy and we may have to use numerical techniques before we have discussed them. Fortunately, having a tool like MATLAB obviates this difficulty, and provides a strong justification of using a numerical/symbolic tool for supporting the development of the course.*

Sufficient Conditions: The Lagrangian allows us to transform the problem into an unconstrained problem. One can then attempt to use the Hessian of the Lagrangian at the solution to be positive definite as sufficient conditions for the minimum (4.47). For this problem, the eigenvalues of the Hessian are -0.2262 and 0.2262. This does not provide the sufficient conditions as both the eigenvalues are required to be positive. This approach does not appear useful, as we can see graphically in Figure 4.10 that a minimum exists.

At the solution determined by the FOC (4.54), the sufficient conditions will require that the function should increase for all changes in ΔX *(neighborhood of X^*)*. Moreover the changes in ΔX cannot be arbitrary. They have to satisfy the equality constraint. For small changes we can require that they move tangentially to the constraint. It is becoming clear that the sufficient conditions are usually applied in a small neighborhood of the optimum, thereby establishing *local optimums*. Also, changes are contemplated only with respect to X and not with respect to the Lagrange multiplier. The Lagrange method is often called as the **method of undetermined coefficient** – indicating it is not a bonafide variable. In the analytical derivation of the FOC for this problem the Lagrangian F was considered unconstrained. Borrowing from the previous section (unconstrained minimization), the SOC can be expected to satisfy the following relations:

$$\Delta F = F(X^* + \Delta X) - F(X^*)$$

$$= \nabla F(X^*) + \frac{1}{2}\Delta X^T[\nabla^2 F(X^*)]\Delta X > 0 \quad (4.57a)$$

*There are other tools beside MATLAB like Mathcad, Excel, etc.

$$\nabla h_1{}^T \Delta X = 0 \tag{4.57b}$$

$[\nabla^2 F(X^*)]$ is similar to the *Hessian* of the Lagrangian, but includes the derivatives with respect to the design variables only. Also, the FOC require that $\nabla F(X^*) = 0$. With reference to two variables and one constraint:

$$\Delta F = \frac{1}{2} \left. \left\{ \frac{\partial^2 F}{\partial x_1^2}(\Delta x_1)^2 + 2\frac{\partial^2 F}{\partial x_1 \partial x_2}(\Delta x_1)(\Delta x_2) + \frac{\partial^2 F}{\partial x_2^2}(\Delta x_2)^2 \right\} \right|_{x_1^*, x_2^*} > 0 \tag{4.58}$$

$$\Delta F = \frac{1}{2} \left. \left\{ \frac{\partial^2 F}{\partial x_1^2}\left(\frac{\Delta x_1}{\Delta x_2}\right)^2 + 2\frac{\partial^2 F}{\partial x_1 \partial x_2}\left(\frac{\Delta x_1}{\Delta x_2}\right) + \frac{\partial^2 F}{\partial x_2^2} \right\} \right|_{x_1^*, x_2^*} (\Delta x_2)^2 > 0 \tag{4.59a}$$

$$\frac{\Delta x_1}{\Delta x_2} = - \left. \frac{\dfrac{\partial h_1}{\partial x_2}}{\dfrac{\partial h_1}{\partial x_1}} \right|_{x_1^*, x_2^*} \tag{4.59b}$$

$$\text{SOC} = \left. \left\{ \frac{\partial^2 F}{\partial x_1^2}\left(\frac{\Delta x_1}{\Delta x_2}\right)^2 + 2\frac{\partial^2 F}{\partial x_1 \partial x_2}\left(\frac{\Delta x_1}{\Delta x_2}\right) + \frac{\partial^2 F}{\partial x_2^2} \right\} \right|_{x_1^*, x_2^*} > 0 \tag{4.60}$$

The SOC requires expression in the curly brackets must be positive. For this example it has a value of 0.7755. All the derivatives are evaluated at the solution. Compared to the SOC for the unconstrained minimization problem, the conditions in (4.59b and 4.60a) require a lot of extra effort, particularly with many variables and several constraints.

One important issue that arises in this discussion is whether it is necessary to consider the Lagrangian (F) at all in (4.57). Since we are concerned with the minimum of the objective function can we use the objective function (f) (4.57a) and its subsequent simplification? In that case our sufficient conditions become

$$\text{SOC} = \left. \left\{ \frac{\partial^2 f}{\partial x_1^2}\left(\frac{\Delta x_1}{\Delta x_2}\right)^2 + 2\frac{\partial^2 f}{\partial x_1 \partial x_2}\left(\frac{\Delta x_1}{\Delta x_2}\right) + \frac{\partial^2 f}{\partial x_2^2} \right\} \right|_{x_1^*, x_2^*} > 0 \tag{4.61a}$$

$$\frac{\Delta x_1}{\Delta x_2} = - \left. \frac{\dfrac{\partial h_1}{\partial x_2}}{\dfrac{\partial h_1}{\partial x_1}} \right|_{x_1^*, x_2^*} \tag{4.61b}$$

For this example, (4.61) yields a value of 0.7755.

It is possible to develop an intuitive requirement on the Lagrange multiplier λ_1 as SOC. The value and sign for λ was ignored in the (4.56) and is also immaterial in most derivations of the sufficient conditions in the various books on optimization. But, for a well-posed equality constrained problem, the sign of the Lagrange multiplier at the solution should be positive, since increasing the constraint value is useful only if we are trying to identify a lower value for the minimum. This forces h and f to move in opposite directions (gradients). Increasing constraint value can be associated with enlarging the feasible domain, which may yield a better design. So, as SOC we can require that, for a minimization problem,

$$\lambda_1^* > 0 \qquad (4.62)$$

which is true in (4.56).

Discussion on Lagrange Multiplier: The Lagrange multiplier method is an elegant formulation to obtain the solution to a constrained problem. In overview, it seems strange that we have to introduce an additional unknown (λ_1) to solve the constrained problem. This violates the conventional rule for NLP that the fewer the variables, the better the chances of obtaining the solution. Also indicated in the discussion earlier, the Lagrangian allows the transformation of a *constrained problem* into an *unconstrained problem*. The Lagrange multiplier also has a physical significance. At the solution it expresses the ratio of the change in the objective function to the change in the constraint value. To illustrate this, consider:

$$F = f + \lambda_1 h_1$$
$$dF = df + \lambda_1 dh$$
$$dF = \frac{\partial F}{\partial x_1} dx_1 + \frac{\partial F}{\partial x_2} dx_2 + \frac{\partial F}{\partial \lambda_1} d\lambda_1$$

At the solution, the FOC deems that $\boldsymbol{dF} = \boldsymbol{0}$. Hence

$$\lambda_1 = -\frac{df}{dh} = -\frac{\Delta f}{\Delta h} \qquad (4.63)$$

This dependence does not affect the establishment of the optimal design. It does have an important role in the discussion of *design sensitivity* in NLP problems.

MATLAB Code: `EqualityConstrainedProblem2.m` which was mentioned earlier for generating Figure 4.10 also computes the optimum values in (4.56). It computes the second order conditions in (4.60, 4.61a). Symbolic programming is mostly used for all the derivative computation. Equations 4.55 are solved symbolically. Several of the values are used for generating Figure 4.10.

4.3.3 Equality-Constrained Problem 3

This problem was not identified earlier but has very interesting features that are different from the previous problem and well worth the investigation. Here the other constraint is used to define a valid equality constrained problem. Its scaled form, after the scaling factors are absorbed in the expression is

$$\text{Minimize} \quad f(x_1, x_2) = -0.2262x_1x_2 \tag{4.64a}$$

$$\text{Subject to:} \quad h_1(x_1, x_2) : 5.6549x_1 + 1.131x_2$$

$$- \pi\sqrt{(0.18x_1 + 0.12x_2)(0.6x_1 + 0.36x_2)} - 1.75 = 0 \tag{4.64b}$$

$$0 \le x_1 \le 1 \quad 0 \le x_2 \le 1 \tag{4.64c}$$

Graphical Solution: Figure 4.11 presents the graphical solution to the problem. The solution is on the boundary $x_2 = 1$. The analytical conditions (4.54) do not yield a solution to the problem within the feasible region. This implies that the conditions discussed in the previous section apply only to the solution in the interior of the region. The solution is at the intersection of the upper bound for x_2 and the equality constraint. Once again the gradients are drawn using the **DrawArrow.m** function. The gradients of the objective function and the constraint

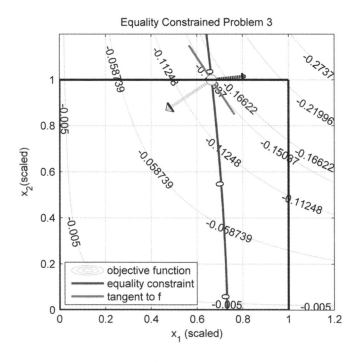

Figure 4.11 Graphical solution: Equality-constrained problem 3.

are not parallel at the solution this time. The graphics are produced by running
EqualityConstrainedProblem3.m.

Necessary condition: In order to determine the optimum, the Lagrangian is
first defined as follows:

$$F = -0.2262x_1x_2 + \lambda_1\Big(5.6549x_1 + 1.131x_2$$

$$- \pi\sqrt{(0.18x_1 + 0.12x_2)(0.6x_1 + 0.36x_2)]} - 1.75\Big)$$

To save on space because of the length of the algebraic expression and its
derivatives, we define the following quantity:

$$define\ A = (0.18x_1 + 0.12x_2)(0.6x_1 + 0.36x_2)$$

$$\frac{\partial A}{\partial x_1} = 0.18(0.6x_1 + 0.36x_2) + 0.6(0.18x_1 + 0.12x_2)$$

$$\frac{\partial A}{\partial x_2} = 0.12(0.6x_1 + 0.36x_2) + 0.36(0.18x_1 + 0.12x_2)$$

The first order conditions can be expressed as follows:

$$\frac{\partial F}{\partial x_1} = -0.2262x_2 + 5.6549\lambda_1 - \frac{1}{2}\lambda_1\pi A^{-\frac{1}{2}}\frac{\partial A}{\partial x_1} = 0$$

$$\frac{\partial F}{\partial x_2} = -0.2262x_1 + 1.131\lambda_1\frac{1}{2}\pi\lambda_1 A^{-\frac{1}{2}}\frac{\partial A}{\partial x_2} = 0$$

$$h_1(x_1, x_2) = 5.6549x_1 + 1.131x_2 - \pi A^{\frac{1}{2}} = 0 \qquad (4.65)$$

Equation (4.65) gives three equations to solve for the variables and the
Lagrange multiplier at the solution. This is not easy to solve analytically because
of the nonlinear terms. The solution can be seen in the Command window by
running **EqualityConstrainedProblem3.m**. The solutions to the FOC are
given in Table 4.2 and are obtained using symbolic programming. There are
no solutions to the FOC in the feasible region. In other words, *the FOC is not
helpful for this equality-constrained problem.*

Table 4.2 Solutions for Equality-Constrained Problem 3

x_1	x_2	λ	f	h
0.4642	2.3210	0.2785	−0.2437	−0.0000
0.6341	−0.8495	−0.1393	0.1219	−0.0000
−0.1699	3.1706	−0.1393	0.1219	−0.0000

There is definitely a solution to the problem. The solution must lie on the equality constraint. It must be feasible. Therefore, we can look for it at the intersection of the equality constraint and the boundary of the feasible region. In Figure 4.11 the gradient of the objective function and the gradient of the constraint at the solution are not parallel—violating the necessary condition (4.50). The solution to this problem is

$$x_1^* = 0.6648; \quad x_2^* = 1; \quad f^* = -0.1504; \quad h^* = 0 \tag{4.66}$$

Because the solution is not on the interior we cannot apply the second order or sufficient conditions for this example.

MATLAB Code: `EqualityConstrainedProblem3.m` which generates the graphical solution also computes the optimum values in (4.66). Symbolic programming is mostly used for all the derivative computation. The equations based on the FOC are solved symbolically. Several of the values are used for generating Figure 4.11 itself. The author strongly encourages the reader to at least walk through the code to see how computations are being carried out and how the graphics are constructed. The graphics are generated automatically without user intervention.

Sufficient Conditions: Since the solution does not satisfy the necessary conditions, the sufficient conditions do not apply.

General Equality Constrained Problem: Remembering $n - l > 0$,

$$\text{Minimize} \quad f(\mathbf{X}), [\mathbf{X}]_n \tag{4.5}$$

$$\text{Subject to:} \quad [\mathbf{h}(\mathbf{X})]_l = \mathbf{0} \tag{4.6}$$

$$\mathbf{X}^{low} \le \mathbf{X} \le \mathbf{X}^{up} \tag{4.8}$$

In terms of the augmented function or the Lagrangian:

$$\text{Minimize} \quad F(X, \lambda) = f(X) + \sum_{k=1}^{l} \lambda_k h_k(X)$$

$$= f(X) + \lambda \cdot h; \quad [\lambda]_l; [h]_l$$

$$= f(X) + \lambda^T h \tag{4.67}$$

$$\text{Subject to:} \quad [\mathbf{h}(\mathbf{X})]_l = \mathbf{0} \tag{4.6}$$

$$\mathbf{X}^{low} \le \mathbf{X} \le \mathbf{X}^{up} \tag{4.8}$$

Necessary Conditions: In (4.67), three equivalent representations for the Lagrangian are shown. The necessary conditions (FOC) are

$$\frac{\partial F}{\partial x_i} = \frac{\partial f}{\partial x_i} + \sum_{k=1}^{l} \lambda_k \frac{\partial h_k}{\partial x_i} = 0; \quad i = 1, 2, \ldots n \qquad (4.68)$$

Subject to: $\quad [\mathbf{h}(\mathbf{X})]_l = \mathbf{0}$ (4.6)

$$\mathbf{X}^{low} \le \mathbf{X} \le \mathbf{X}^{up}$$ (4.8)

Equations (4.6) and (4.68) provide the $n + l$ relations to determine the $n + l$ unknowns X^*, λ^*. Equation (4.8) is used after the solution is obtained, if necessary. Equation (4.72) is also expressed as

$$\nabla F = \nabla f + [\nabla h_1 \ \nabla h_2 \ \ldots \ \nabla h_l] \begin{bmatrix} \lambda_1 \\ \lambda_2 \\ \cdot \\ \lambda_l \end{bmatrix} = \nabla f + [\lambda_1 \ \lambda_2 \ \cdot \ \lambda_l] \begin{bmatrix} \nabla h_1 \\ \nabla h_2 \\ \cdot \\ \nabla h_l \end{bmatrix} = 0$$
(4.69)

Sufficient Conditions: The sufficient conditions for the general problem are quite cumbersome and are not developed here. We can insist that the λ^*s must be positive though we have not formally established that. A simple but practical way of confirming the minimum is to make a set of random perturbations in the small neighborhood of the optimum and verify that the solution is feasible and is a local minimum.

4.3.4 Inequality-Constrained Optimization

The original problem (4.9) to (4.12) is scaled and rewritten as

Minimize $\quad f(x_1, x_2) = -0.2262 x_1 x_2$ (4.70a)

Subject to: $\quad g_1(x_1, x_2) = 0.12 x_2 - 0.07 x_1 - 0.12 \le 0$ (4.70b)

$\quad g_1(x_1, x_2) : -5.6549 x_1 - 1.131 x_2$

$\quad\quad + \pi \sqrt{(0.18 x_1 + 0.12 x_2)(0.6 x_1 + 0.36 x_2)} + 1.75 \le 0$ (4.70c)

$\quad 0 \le x_1 \le 1 \quad\quad 0 \le x_2 \le 1$ (4.70d)

The number of variables ($n = 2$) and the number of inequality constraints ($m = 2$) are not constrained by each other. In fact an inequality constrained problem in a single variable can be usefully designed and solved. Before reading further, please revisit Figure 4.2 to examine the graphical solution the problem. The solution is not marked on the figure but you should be able to mark the solution and verify later if the instinct was correct. Section 4.3.2 solved the equality-constrained problem and established the necessary and sufficient conditions. If the *inequality-constrained problem* can be transformed to an equivalent *equality- constrained problem* then we have the means to establish the solution. The standard transformation involves introduces a slack variable z_j for each inequality constraint g_j. Unlike LP problem, the slack variable for NLP is not restricted in sign. Therefore, the square of the slack variable is added to the left-hand side of the corresponding inequality constraint to transform it into an equality constraint. This adds a positive value to the left hand side to bring up the constraint to zero as the constraints are required to be less than or equal to zero. Of course, a zero value should be added if the constraint is already zero.

Transformation to an Equality-constrained problem

$$\text{Minimize} \quad f(x_1, x_2) = -0.2262 x_1 x_2 \tag{4.71a}$$

$$\text{Subject to:} \quad g_1(x_1, x_2) + z_1^2 = 0.12 x_2 - 0.07 x_1 - 0.12 + z_1^2 = 0 \tag{4.71b}$$

$$g_1(x_1, x_2) + z_2^2 = -5.6549 x_1 - 1.131 x_2$$
$$+ \pi \sqrt{(0.18 x_1 + 0.12 x_2)(0.6 x_1 + 0.36 x_2)} + 1.75 + z_2^2 = 0 \tag{4.71c}$$

$$0 \le x_1 \le 1 \quad 0 \le x_2 \le 1 \tag{4.71d}$$

Technically we should recognize the equality constraint in (4.71) as **h** but we will retain **g** for clarity. There are four variables (x_1, x_2, z_1, z_2) and two equality constraints. *It is a valid equality-constrained problem*. The Lagrange multiplier method can be applied to this transformation. Although the method is the same we will distinguish the multipliers using the variable $\boldsymbol{\beta}$, once again for clarity and association with the type of optimization problem.

Method of Lagrange: The unconstrained optimization problem can be defined using the Lagrangian for (4.71). The variable dependence is retained in the following expression below so that the connection to the partial derivatives is clear.

$$\text{Minimize} \quad F(x_1, x_2, z_1, z_2, \beta_1, \beta_2) = f(x_1, x_2)$$
$$+ \beta_1 \left[g_1(x_1, x_2) + z_1^2 \right] + \beta_2 \left[g_2(x_1, x_2) + z_2^2 \right] \tag{4.72}$$

The FOC or necessary conditions are:

$$\frac{\partial F}{\partial x_1} = \frac{\partial f}{\partial x_1} + \beta_1 \frac{\partial g_1}{\partial x_1} + \beta_2 \frac{\partial g_2}{\partial x_1} = 0 \tag{4.73a}$$

$$\frac{\partial F}{\partial x_2} = \frac{\partial f}{\partial x_2} + \beta_1 \frac{\partial g_1}{\partial x_2} + \beta_2 \frac{\partial g_2}{\partial x_2} = 0 \tag{4.73b}$$

$$\frac{\partial F}{\partial z_1} = 2\beta_1 z_1 = 0 \tag{4.73c}$$

$$\frac{\partial F}{\partial z_2} = 2\beta_2 z_2 = 0 \tag{4.73d}$$

$$\frac{\partial F}{\partial \beta_1} = g_1 + z_1^2 = 0 \tag{4.73e}$$

$$\frac{\partial F}{\partial \beta_2} = g_2 + z_2^2 = 0 \tag{4.73f}$$

Equations (4.73e, 4.73f) are the transformed equality constraints. Equations (4.73) provides six equations to solve for $x_1^*, x_2^*, z_1^*, z_2^*, \beta_1^*, \beta_2^*$.

Graphical Solution: The graphical solution is available in Figure 4.12. It is seen that only g_1 is an active constraint for this problem. The solution is identified on the figure. The gradient of the objective function and the constraints are drawn using the **DrawArrow** function. The gradient of the objective function and the active constrain g_1 are parallel. The inactive constraint passing through the solution is shown as a dotted line. The gradient of the inactive constraint is not parallel to the two other gradients.

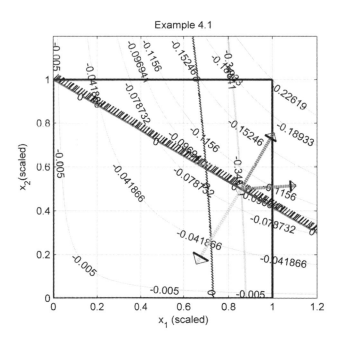

Figure 4.12 Graphical solution: Inequality constrained problem.

Necessary Conditions: Working with (4.71c to 4.71f) the following sequence of operations is possible:

$$0 = 2\beta_1 z_1 = \beta_1 z_1^2 = \beta_1(-g_1) = \boxed{\beta_1 g_1 = 0;} \qquad (4.74a)$$

$$0 = 2\beta_2 z_2 = \beta_2 z_2^2 = \beta_2(-g_2) = \boxed{\beta_2 g_2 = 0;} \qquad (4.74b)$$

The FOC can be reduced to

$$\frac{\partial F}{\partial x_1} = \frac{\partial f}{\partial x_1} + \beta_1 \frac{\partial g_1}{\partial x_1} + \beta_2 \frac{\partial g_2}{\partial x_1} = 0 \qquad (4.75a)$$

$$\frac{\partial F}{\partial x_2} = \frac{\partial f}{\partial x_2} + \beta_1 \frac{\partial g_1}{\partial x_2} + \beta_2 \frac{\partial g_2}{\partial x_2} = 0 \qquad (4.75b)$$

$$\beta_1 g_1 = 0 \qquad (4.75c)$$

$$\beta_2 g_2 = 0 \qquad (4.75d)$$

These 4 equations have to be solved for $x_1^*, x_2^* \beta_1^*, \beta_2$. Note that z_1^*, z_2 are not being determined –which suggests that they can be discarded from the problem. Also, if the constraints are zero they are considered ***active constraints***.

For nontrivial solutions, where both g and β are not simultaneously zero, Equations (4.75a, 4.75b) force the consideration of: *either β_i is zero (and corresponding $g_i \neq 0$) or g_i is zero (and corresponding $\beta_i \neq 0$).* An acceptable solution requires $g_i \leq 0$. Since simultaneous equations are being solved, the conditions on the multipliers and constraints must be satisfied simultaneously. For the equation set in (4.75), this translates into the following four cases:

$$\text{Case (a):} \beta_1 = 0[g_1 < 0]; \ \beta_2 = 0[g_2 < 0]; \qquad (4.76a)$$

$$\text{Case (b):} \beta_1 = 0[g_1 < 0]; \ \beta_2 \neq 0[g_2 = 0]; \qquad (4.76b)$$

$$\text{Case (c):} \beta_1 \neq 0[g_1 = 0]; \ \beta_2 = 0[g_2 < 0]; \qquad (4.76c)$$

$$\text{Case (d):} \beta_1 \neq 0[g_1 = 0]; \ \beta_2 \neq 0[g_2 = 0]; \qquad (4.76d)$$

In Equation (4.76a), if $\boldsymbol{\beta_i \neq 0}$, (or corresponding $\boldsymbol{g_i = 0}$), then the corresponding constraint is an *equality*. In the previous section a simple reasoning was used to show that the sign of the multiplier must be positive ($> \mathbf{0}$) for a well-formulated problem. The sign of the multiplier was unimportant for the equality-constrained problem, but it is included as part of the FOC for the inequality-constrained problem. Before restating (4.76), the Lagrangian is reformulated after discarding the slack variables:

$$\text{Minimize} \quad F(x_1, x_2, \beta_1, \beta_2) = f(x_1, x_2) + \beta_1 \big[g_1(x_1, x_2) \big]$$

$$+ \beta_2 \big[g_2(x_1, x_2) \big] \qquad (4.77)$$

This is the same formulation in (4.67). The slack variable was introduced to provide the transformation to an equality constraint. It is also evident that the construction of the Lagrangian function is insensitive to the type of constraint. Since the multipliers tied to the inequality constraint are required to be positive, while those corresponding to the equality constraints are not checked for sign, the book will continue to distinguish between the multiplier symbols to clarify the type of constraint. The FOC for the problem are as follows:

$$\frac{\partial F}{\partial x_1} = \frac{\partial f}{\partial x_1} + \beta_1 \frac{\partial g_1}{\partial x_1} + \beta_2 \frac{\partial g_2}{\partial x_1} = 0 \tag{4.78a}$$

$$\frac{\partial F}{\partial x_2} = \frac{\partial f}{\partial x_2} + \beta_1 \frac{\partial g_1}{\partial x_2} + \beta_2 \frac{\partial g_2}{\partial x_2} = 0 \tag{4.78b}$$

And one of the following cases

$$\text{Case (a)}: \beta_1 = 0[g_1 < 0]; \; \beta_2 = 0[g_2 < 0]; \tag{4.79a}$$

$$\text{Case (b)}: \beta_1 = 0[g_1 < 0]; \; \beta_2 > 0[g_2 = 0]; \tag{4.79b}$$

$$\text{Case (c)}: \beta_1 > 0[g_1 = 0]; \; \beta_2 = 0[g_2 < 0]; \tag{4.79c}$$

$$\text{Case (d)}: \beta_1 > 0[g_1 = 0]; \; \beta_2 > 0[g_2 = 0]; \tag{4.79d}$$

The equations in (4.78) and any single case in (4.79) provide one set of equations to solve for the four unknowns of the problem. These are the *necessary conditions* for this problem. The four cases identify four sets to be examined to obtain the solution. The best design is decided by scanning the several acceptable solutions from the four sets.

The Lagrangian for the example is defined as follows:

$$F(x_1, x_2, \beta_1, \beta_2) = -0.2262x_1x_2 + \beta_1(0.12x_2 - 0.07x_1 - 0.12)$$

$$+ \beta_2(-5.6549x_1 - 1.131x_2$$

$$+ \pi \sqrt{(0.18x_1 + 0.12x_2)(0.6x_1 + 0.36x_2)} + 1.75) \tag{4.80}$$

$$define \; A = (0.18x_1 + 0.12x_2)(0.6x_1 + 0.36x_2)$$

$$\frac{\partial A}{\partial x_1} = 0.18(0.6x_1 + 0.36x_2) + 0.6(0.18x_1 + 0.12x_2)$$

$$\frac{\partial A}{\partial x_2} = 0.12(0.6x_1 + 0.36x_2) + 0.36(0.18x_1 + 0.12x_2)$$

$$F(x_1, x_2, \beta_1, \beta_2) = -0.2262x_1x_2 + \beta_1(0.12x_2 - 0.07x_1 - 0.12)$$

$$+ \beta_2(-5.6549x_1 - 1.131x_2 + \pi A^{1/2} + 1.75) \tag{4.81}$$

The FOC are

$$\frac{\partial F}{\partial x_1} = -0.2262x_2 - 0.07\beta_1 - 5.6549\beta_2 + \frac{\pi}{2}\beta_2 A^{-1/2}\frac{\partial A}{\partial x_1} = 0 \qquad (4.82a)$$

$$\frac{\partial F}{\partial x_1} = -0.2262x_1 + 0.12\beta_1 - 1.131\beta_2 + \frac{\pi}{2}\beta_2 A^{-1/2}\frac{\partial A}{\partial x_2} = 0 \qquad (4.82b)$$

$$\beta_1(0.12x_2 - 0.07x_1 - 0.12) = 0 \qquad (4.82c)$$

$$\beta_2(-5.6549x_1 - 1.131x_2 + \pi A^{1/2} + 1.75) = 0 \qquad (4.82d)$$

The side constraints are

$$0 \le x_1 \le 1 \quad 0 \le x_2 \le 1$$

Case(a) $:\beta_1 = 0[g_1 < 0]; \beta_2 = 0[g_2 < 0]$

The solution is trivial and infeasible, as g_2 is violated:

$$x_1 = 0; x_2 = 0; \beta_1 = 0; \beta_2 = 0; f = 0; g_1 = -0.1200 \ g_2 = 1.7500 \qquad (4.83)$$

Case(b) $:\beta_1 = 0[g_1 < 0]; \beta_2 > 0[g_2 = 0]$

There are three solutions and none of them are feasible

$$x_1 = 0.4642; \quad x_2 = 2.3210; \quad \beta_1 = 0; \quad \beta_2 = -0.2785;$$

$$f = -0.2437; \quad g_1 = 0.1910; \quad g_2 = 0;$$

$$x_1 = 0.6341 \ x_2 = -0.8495; \quad \beta_1 = 0; \quad \beta_2 = 0.1393;$$

$$f = 0.1219; \quad g_1 = -0.1776; \quad g_2 = 0;$$

$$x_1 = -0.1699 \ x_2 = 3.1706; \quad \beta_1 = 0; \quad \beta_2 = 0.1393;$$

$$f = 0.1219; \quad g_1 = 0.2486; \quad g_2 = 0; \qquad (4.84)$$

Case(c) $:\beta_1 > 0[g_1 = 0]; \quad \beta_2 = 0[g_2 < 0]$

There is one solution. The nonzero Lagrange multiplier is positive. The inactive constraint is negative. All of the necessary conditions are satisfied. This is a possible optimum solution.

$$x_1 = 0.8571; x_2 = 0.5000; \beta_1 = 1.6157; \beta_2 = 0; f = -0.0969;$$

$$g_1 = 0; g_2 = -0.3484; \qquad (4.85)$$

Case(d) $:\beta_1 > 0[g_1 = 0]; \beta_2 > 0[g_2 = 0]$

There is one solution. It is not a minimum, as the multiplier for the second constraint is negative.

$$x_1 = 0.7027; \quad x_2 = 0.5901; \quad \beta_1 = 1.2972; \beta_2 = -0.0182;$$

$$f = -0.0938; g_1 = 0; \quad g_2 = 0; \tag{4.86}$$

MATLAB Code: The code is available in `InEqualityConstrainedProblem.m`. All of the four cases are solved. The solution for each case is reported to the Command window, along with the in-between results to understand the code. The best solution is also tracked through all of the cases. The necessary conditions are applied using symbolic calculations. The final results are used to create Figure 4.12 without user intervention.

You are strongly recommended to walk through the code. Suddenly the length of the code has increased. However, you will notice it except for a couple of lines setting up the cases; most of the code is copy and paste. The code for creating the graphics, the contour plots, the tangents, the gradients, and the arrows are also copied from previous effort. By now you should start to get the feeling that an initial investment in making a piece of code robust can pay back very well because of its reuse. The reuse ensures less debugging, and more importantly, brings a sophisticated level of consistency to the application.

Sufficient Conditions: The sign of the Lagrange multiplier is *not a sufficient* condition for the inequality constrained problem. The value is unimportant for optimization but may be relevant for sensitivity analysis. Generally a positive value of the multiplier indicates that the solution is not a **local maxima**. Formally verifying a minimum solution requires consideration of the second derivative of the Lagrangian. In practical situations, if the problem is well defined, the positive value of the multiplier usually suggests a minimum solution.

One can even verify through computations that the function increases in a small neighborhood of the point in contention. The effort for checking the sufficient conditions is very significant and is similar to the ones illustrated for the equality-constrained problem. Most practical approaches eschew these calculations, as they are considered expensive. Computing second derivatives, for real problems, is considered a stretch especially when the justification of the first derivatives is itself a concern.

General inequality constrained problem: Remembering n and m can be unrelated

$$\text{Minimize} \quad f(\mathbf{X}), [\mathbf{X}]_n \tag{4.5}$$

$$\text{Subject to:} \quad [g]_m \leq 0 \tag{4.7}$$

$$\mathbf{X}^{low} \leq \mathbf{X} \leq \mathbf{X}^{up} \tag{4.8}$$

In terms of the augmented function or the Lagrangian:

$$\text{Minimize} \quad F(X, \lambda) = f(X) + \sum_{j=1}^{m} \beta_j g_j(X)$$

$$= f(X) + \beta \cdot g; \quad [\beta]_m; [g]_m \tag{4.87}$$

$$= f(X) + \beta^T g$$

$$X^{low} \leq X \leq X^{up} \tag{4.8}$$

Necessary Conditions: The necessary conditions (FOC) are

$$\frac{\partial F}{\partial x_i} = \frac{\partial f}{\partial x_i} + \sum_{j=1}^{m} \beta_k \frac{\partial g_j}{\partial x_i} = 0; \quad i = 1, 2, \ldots n \tag{4.88}$$

along with 2^m cases/conditions where

$$\beta_j g_j = 0 \rightarrow if \ \beta_j = 0 \quad then \quad g_j < 0$$

$$if \ g_j = 0 \quad then \quad \beta_j > 0 \tag{4.89}$$

$$X^{low} \leq X \leq X^{up} \tag{4.8}$$

Each case involves m choices in (4.89) and involves verifying that the corresponding solution is feasible or the multipliers are positive.

Equations 4.88 and 4.89 provide the $n + m$ relations to determine the $n + m$ unknowns X^*, β^*. Equation (4.8) is used after the solution is obtained, if necessary.

Sufficient Conditions: The sufficient conditions for the general problem are quite cumbersome and are not developed here. We can insist that the β^*s must be positive or zero. A simple but practical way of confirming the minimum is to make a set of random perturbations in the small neighborhood of the optimum and verify that the solution is feasible and is a local minimum.

4.3.5 A General Optimization Problem

We will create another problem based on Example 4.1 that was not included before. Let us also create another solution to the problem by making the perimeter constraint binding while relaxing the linear surface constraint. This will create an optimization problem with both the equality and an inequality constraint. This is the general optimization problem described by the set of equations (4.1−4.4). The specific problem is as follows:

$$\text{Minimize} \quad f(x_1, x_2) = -0.2262 x_1 x_2 \tag{4.90a}$$

$$\text{Subject to:} \quad g_1(x_1, x_2) = 0.12 x_2 - 0.07 x_1 - 0.12 \leq 0 \tag{4.90b}$$

$$h_1(x_1, x_2) : -5.6549x_1 - 1.131x_2$$

$$+ \pi\sqrt{(0.18x_1 + 0.12x_2)(0.6x_1 + 0.36x_2)} + 1.75 = 0 \quad (4.90c)$$

$$0 \le x_1 \le 1 \quad 0 \le x_2 \le 1 \tag{4.90d}$$

The FOC for this problem is a combination of the conditions in Sections 4.3.2 and 4.3.3. No new concepts are required. The Lagrange multiplier method is again utilized to set up the FOC. In the following development, the multipliers are kept distinct for comprehension.

Lagrange Multiplier Method: The problem is transformed by minimizing the Lagrangian:

$$F(x_1, x_2, \beta_1, \beta_2) = -0.2262x_1x_2 + \beta_1(0.12x_2 - 0.07x_1 - 0.12)$$

$$+ \lambda_1(-5.6549x_1 - 1.131x_2$$

$$| \pi\sqrt{(0.18x_1 + 0.12x_2)(0.6x_1 + 0.36x_2)} + 1.75)$$

$$define\ A = (0.18x_1 + 0.12x_2)(0.6x_1 + 0.36x_2)$$

$$\frac{\partial A}{\partial x_1} = 0.18(0.6x_1 + 0.36x_2) + 0.6(0.18x_1 + 0.12x_2)$$

$$\frac{\partial A}{\partial x_2} = 0.12(0.6x_1 + 0.36x_2) + 0.36(0.18x_1 + 0.12x_2)$$

$$F(x_1, x_2, \beta_1, \beta_2) = -0.2262x_1x_2 + \beta_1(0.12x_2 - 0.07x_1 - 0.12)$$

$$+ \lambda_1(-5.6549x_1 - 1.131x_2 + \pi A^{1/2} + 1.75) \tag{4.91}$$

Necessary Conditions: The FOC are

$$\frac{\partial F}{\partial x_1} = -0.2262x_2 - 0.07\beta_1 - 5.6549\lambda_1 + \frac{\pi}{2}\lambda_1 A^{-1/2}\frac{\partial A}{\partial x_1} = 0 \tag{4.92a}$$

$$\frac{\partial F}{\partial x_1} = -0.2262x_1 + 0.12\beta_1 - 1.131\lambda_1 + \frac{\pi}{2}\lambda_1 A^{-1/2}\frac{\partial A}{\partial x_2} = 0 \tag{4.92b}$$

$$- 5.6549x_1 - 1.131x_2 + \pi\sqrt{(0.18x_1 + 0.12x_2)(0.6x_1 + 0.36x_2)}$$

$$+ 1.75 = 0 \tag{4.92c}$$

$$\beta_1(0.12x_2 - 0.07x_1 - 0.12) = 0 \tag{4.92d}$$

$$\textbf{Case(a)} : \beta_1 = 0;\ g_1 < 0 \tag{4.93a}$$

$$\textbf{Case(b)} : \beta_1 > 0;\ g_1 = 0 \tag{4.93b}$$

Two solutions must be examined. The first requires the solution of a system of three equations in three unknowns λ_1, x_1, x_2 using equations (4.92a, 4.92b, 4.92c). The second is a system of four equations in the four unknowns $\lambda_1, \beta_1, x_1, x_2$ using (4.92) without the multiplier term in (4.92d). The solution of this optimization problem is one of the problems at the end of the chapter.

Kuhn-Tucker Conditions: The FOC associated the general optimization problem in Equations (4.1 to 4.4) or (4.5 to 4.8) is termed as the Kuhn–Tucker conditions. These conditions are established in the same manner as in the previous section. The general optimization is problem is repeated here for sake of completeness:

$$\text{Minimize} \quad f(x_1, x_2, \ldots x_n) \tag{4.1}$$

$$\text{Subject to:} \quad h_k(x_1, x_2, \ldots, x_n) = 0, \qquad k = 1, 2, \ldots, l \tag{4.2}$$

$$g_j(x_1, x_2, \ldots, x_n) \le 0, \qquad j = 1, 2, \ldots, m \tag{4.3}$$

$$x_i^l \le x_i \le x_i^u \qquad i = 1, 2, \ldots, n \tag{4.4}$$

The Lagrangian:

$$\text{Minimize} \quad F(x_1 \ldots x_n, \lambda_1 \ldots \lambda_l, \beta_1 \ldots \beta_m) = f(x_1, \ldots, x_n) + \lambda_1 h_1 + \cdots$$
$$+ \lambda_l h_l + \beta_1 g_1 + \cdots + \beta_m g_m \tag{4.94}$$

There are $n + l + m$ unknowns. The same number of equations are required to solve the problem. These are provided by the FOC or the Kuhn-Tucker conditions[17]:

n equations are obtained as follows:

$$\frac{\partial F}{\partial x_i} = \frac{\partial f}{\partial x_i} + \lambda_1 \frac{\partial h_1}{\partial x_i} + \cdots + \lambda_l \frac{\partial h_l}{\partial x_i} + \beta_1 \frac{\partial g_1}{\partial x_i} + \cdots + \beta_m \frac{\partial g_m}{\partial x_i} = 0;$$
$$i = 1, 2, \ldots, n \tag{4.95}$$

l equations are obtained directly through the equality constraints:

$$h_k(x_1, x_2, \ldots, x_n) = 0; \quad k = 1, 2, \ldots, l \tag{4.96}$$

m equations are applied through the 2^m cases. This implies that there are 2^m possible solutions. These solution must include (4.95) and (4.96). Each case sets the multiplier β_j or the corresponding inequality constraint g_j to zero. If the multiplier is set to zero, then the corresponding constraint must be feasible for an acceptable solution. If the constraint is set to zero (active constraint) then the corresponding multiplier must be positive for a minimum. With this in mind the m equations can be expressed as follows:

$$\beta_j g_j = 0 \rightarrow if \; \beta_j = 0 \quad then \quad g_j < 0$$

$$if \ g_j = 0 \quad then \quad \beta_j > 0 \tag{4.97}$$

If conditions in (4.97) are not met, the design is not acceptable. In implementing (4.97), for each case a simultaneous choice of m values must be made. Once these FOC conditions determine a possible solution the side constraints have to be checked. As evidenced in the examples earlier, this is not built into the FOC. It is only confirmed after a possible solution has been identified. Equations (4.95) to (4.97) are referred to as the Kuhn-Tucker conditions. Note: There is a caveat with respect to these conditions. These conditions are applicable at regular points, which we will introduce in the discussion following Example 4.3 (Section 4.4.2).

4.4 EXAMPLES

Three examples are presented in this section. The first is an unconstrained problem that has significant use in data reduction. Specifically, it illustrates the problem of curve fitting or regression. The second is a problem with equality constraints. The third is the beam design problem in Chapters 1 and 2.

4.4.1 Example 4.2–Curve Fitting

Problem: Given a set of y, z data find the best polynomial that will provide an error of at least 1.0e-06 between the data and the polynomial.

The input to the polynomial is the set of y_i and z_i data. Let us generate the data using the function we have used for illustration before

$$z(y) = 1^* \sin(0.1 + y)/(0.1 + y) \tag{4.98}$$

We will choose 51 linearly space points for y between 0 and 10. These are organized data. Data do not have to be organized for this example. You can also create random data to check out the program. This exercise is termed as *curve fitting* or *data reduction* or *regression*.

Least Squared Error: The objective function that drives a large number of curve fitting/regression methods is the minimization of the sum of the **squared error** between the original data and the value evaluated using a representative function. The construction of the objective requires two entities—the data and the type mathematical relationship between the data. In this example it is an mth order polynomial. Generally it can be any function for that manner, but polynomials are neat and convenient. Functions other than polynomials are used in correlation equations determined in experimental fluid dynamics and heat transfer. For this example the data are the collection of points (y_i, z_i), $i = 1, 2, \ldots, n$. The expected polynomial can be characterized as:

$$z_p = x_1 y^m + x_2 y^{m-1} + \ldots + x_m y + x_{m+1}$$

where x_i are the design variables. For a cubic polynomial, there will be four design variables. If z_{pi} is the fitted value for the independent variable y_i, the objective function, which is the minimum of the square of the error over all the data points can be expressed as

Minimize

$$f(X) = \sum_{i=1}^{n} \left[z_i - z_{pi} \right]^2 = \sum_{i=1}^{n} \left[z_i - \left(x_1 y_i^m + x_2 y_i^{m-1} + \ldots + x_m y_i + x_{m+1} \right) \right]^2$$

(4.99)

It is possible to complete the formulation with side constraints, though it is not necessary.

Necessary Conditions:

$$\frac{\partial f}{\partial x_1} = \sum_{i=1}^{n} 2 \left[z_i - \left(x_1 y_i^m + x_2 y_i^{m-1} + \ldots + x_m y_i + x_{m+1} \right) \right] \left(-y_i^m \right) = 0$$

$$\frac{\partial f}{\partial x_2} = \sum_{i=1}^{n} 2 \left[z_i - \left(x_1 y_i^m + x_2 y_i^{m-1} + \ldots + x_m y_i + x_{m+1} \right) \right] \left(-y_i^{m-1} \right) = 0$$

$$\cdots\cdots\cdots\cdots\cdots\cdots\cdots\cdots\cdots\cdots\cdots$$

$$\frac{\partial f}{\partial x_{m+1}} = \sum_{i=1}^{n} 2 \left[z_i - \left(x_1 y_i^m + x_2 y_i^{m-1} + \ldots + x_m y_i + x_{m+1} \right) \right] (-1) = 0$$

(4.100)

Dropping the factor of 2 and reorganizing the terms, we can set up a matrix equation

$$\begin{bmatrix} \sum_{i=1}^{n} y_i^{2m} & \sum_{i=1}^{n} y_i^{2m-1} & \cdot & \sum_{i=1}^{n} y_i^{m} \\ \sum_{i=1}^{n} y_i^{2m-1} & \sum_{i=1}^{n} y_i^{2m-2} & \cdot & \sum_{i=1}^{n} y_i^{m-1} \\ \cdot & \cdot & \cdot & \cdot \\ \sum_{i=1}^{n} y_i^{m} & \sum_{i=1}^{n} y_i^{m-1} & \cdot & \sum_{i=1}^{n} 1 \end{bmatrix} \begin{bmatrix} x_1 \\ x_2 \\ \cdot \\ x_{m+1} \end{bmatrix} = \begin{bmatrix} \sum_{i=1}^{n} z_i y_i^{m} \\ \sum_{i=1}^{n} z_i y_i^{m-1} \\ \cdot \\ \sum_{i=1}^{n} z_i \end{bmatrix}$$

(4.101)

Note that the matrices can be set up easily and solved using MATLAB. The Hessian matrix is the square matrix on the left in (4.101). The SOC requires that it be positive definite.

MATLAB Code: The code for Example 4.2 is available in **Example4_2.m.** Compare this code to the one in Chapter 1. Figure 4.13 illustrates that an order of $m = 9$ keeps the error under 1.0e-06. The results for the least squared error and

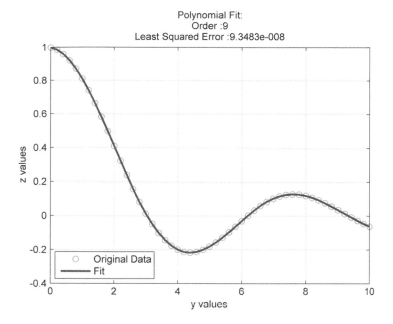

Figure 4.13 Graphical solution: Example 4.2.

the order are displayed in the plot title. The sufficient conditions are the Eigenvalues that are computed and presented in the Command window. The code is mainly numerical. The symbolic part is only used to generate the original data.

4.4.2 Example 4.3–Flagpole Problem

This example is the same as Example 2.3, the flagpole problem. It is traditionally considered difficult to solve. It is chosen here to reveal additional features in applying the Kuhn-Tucker conditions that are not easily resolved. In these situations the designer can bring his experience, intuition, and common sense into the solution process. The problem was fully developed in Section 2.3.2 and is not repeated here. The graphical solution is available in Figures 2.8 and 2.9. These figures are exploited in this section to determine the correct combination of β's for applying the Kuhn-Tucker conditions. If all the parameters are substituted in the various functions the relations can be expressed with numerical coefficients as:

$$\text{Minimize} \quad f(x_1, x_2) = 6.0559\,E\,05\,(x_1^2 - x_2^2) \tag{4.102a}$$

$$\text{Subject to} \quad g_1(x_1, x_2): 2399x_1^2 + 128x_1 - 31174(x_1^4 - x_2^4) \le 0 \tag{4.102b}$$

$$g_2(x_1, x_2): (250 + 7497\,x_1)(x_1^2 + x_1x_2 + x_2^2)$$

$$- 854140(x_1^4 - x_2^4) \le 0 \tag{4.102c}$$

$$g_3(x_1, x_2) : 2500x_1 + 800 - 65450(x_1^4 - x_2^4) \leq 0 \qquad (4.102\text{d})$$

$$g_4(x_1, x_2) : x_2 - x_1 + 0.001 \leq 0 \qquad (4.102\text{e})$$

$$0.02 \leq x_1 \leq 1.0; 0.02 \leq x_2 \leq 1.0 \qquad (4.102\text{f})$$

The coefficients in the equations were obtained by running the code with a variable precision arithmetic (vpa) of five digits in the symbolic calculations. The actual computation using MATLAB does not truncate the values, so the expressions in (4.102) are approximate but can be included in this page neatly.

Kuhn–Tucker Conditions: Four Lagrange multipliers are introduced for the four inequality constraints and the Lagrangian is

$$F(X, \beta) = 6.0559E\ 05(x_1^2 - x_2^2)$$

$$+ \beta_1(2399x_1{}^2 + 128x_1 - 31174(x_1^4 - x_2^4))$$

$$+ \beta_2((250 + 7497x_1)(x_1^2 + x_1x_2 + x_2^2) - 854140(x_1^4 - x_2^4))$$

$$+ \beta_3(2500x_1 + 800 - 65450(x_1^4 - x_2^4))$$

$$+ \beta_4(x_2 - x_1 + 0.001) \qquad (4.103)$$

where $X = [x_1, x_2]$; and $\beta = [\beta_1, \beta_2, \beta_3, \beta_4]^{\text{T}}$. The Kuhn-Tucker conditions require two equations using the gradients of the Lagrangian with respect to the design variables

$$\frac{\partial F}{\partial x_1} = 0; \quad \frac{\partial F}{\partial x_2} = 0 \qquad (4.104)$$

The actual expression for the gradients obtained from Equation (4.103) is left as an exercise for the student. The Kuhn-Tucker conditions are applied by identifying the various cases based on the activeness of various sets of constraints. For this problem $n = 2$; and $m = 4$. There are $(2)^m = 2^4 = 16$ cases that must be investigated as part of the Kuhn-Tucker conditions given by (4.97). While some of these cases are trivial, nevertheless it is a formidable task. A graphical solution to this problem is available in Chapter 2. The graphical solution, which is one of the cases, can be exploited to identify the particular case that needs to be solved for the solution. Returning to Figure 2.9, it can be identified that constraints g_1 and g_3 are active. The solution is around $x_1^* = 0.68m$ and $x_2^* = 0.65m$

If g_1 and g_3 are active constraints then the multipliers β_1 and β_3 must be positive. By the same reasoning the multipliers associated with the inactive constraints g_2 and g_4, that is β_2 and β_4 must be set to zero. The active constraints can be used to solve for x_1^*, x_2^* as this is a system of two equations in two unknowns. This does not however complete the satisfaction of the Kuhn-Tucker conditions. β_1 and β_3 must be solved and verified that they are positive. g_2 and g_4 must be evaluated and verified they are less than zero.

MATLAB Code: The code is available in **Solution_Example4_3.m**. All of the calculations are symbolic. All of the functions are generated using the parameter list identified at the beginning. The two constraints g_1 and g_3 are solved for x_1 and x_2. There are eight solutions for the design variables of which four are imaginary which should be visible in the Command window; *for* and *if* conditions are used to filter the solution. The best solution is then determined. The Lagrange multipliers are then obtained at the best solution. The lines of code that filter the best acceptable solution are shown next. The outer loop is over all the solution pairs. Note first the filter of real numbers through **isreal** function. If the solution is real, we convert it into a number (from a symbolic object). Then we check the side constraints. We compute the functions at the current value of the variables. Then we check if the inequality constraints g_2 and g_4 are satisfied. We update the best solution if the objective function is less than the previous best.

```
%%% capture the best solution in x1b x2b f1b
x1b = 0;
x2b = 0;
fb =1.0e10;
%%% Collect the best solution
for i = 1:length(xsol.x1)
    x1v - xsol.x1(i);
    x2v = xsol.x2(i);

    if isreal(x1v) & isreal(x2v)
       % real numbers ???
       x1vv = double(x1v)
       x2vv = double(x2v)
       if x1vv > 0 & x2vv > 0 & x1vv <=1 & x2vv <= 1
       % meets side    constraint ???
           fv = subs(f,{x1,x2},{x1vv,x2vv})
           g1v - subs(g1,{x1,x2},{x1vv,x2vv})
           g3v = subs(g3,{x1,x2},{x1vv,x2vv})
           g2v = subs(g2,{x1,x2},{x1vv,x2vv})
           g4v = subs(g4,{x1,x2},{x1vv,x2vv})
           if g2v <= 0 & g4v <= 0
           % satisfies other inequality constraints
               if fv < fb
                   x1b = x1vv;
                   x2b = x2vv;
                   fb = fv;
               end
           end
       end
    end
end
```

The values of the design variables, the functions, and the multipliers are printed to the Command window.

Example 4.3: Flag Pole Design:

```
        x1* =   0.6773
        x2* =   0.6443

Functions
  f*    = 26391.4035
  g1*   = -0.0000
  g2*   = -45552734.5497
  g3*   = -0.0000
  g4*   = -0.0320

Multipliers
  beta1*  =   0.0003
  beta2*  =   0.0000
  beta3*  = -76906.7141
  beta4*  =   0.0000
```

Referring to the solution of this case, above the values for the design variables are $x_1^* = 0.6773\,m$ and $x_2^* = 0.6443\,m$. This is very close to the graphical solution. Before we conclude that the solution is achieved, take note of the multipliers. $\beta_1^* = 0.0003$, $\beta_3^* = -76906.7141$. We expect β_3^* to be positive, since the graphical solution is quite clear that g_3 is an active constraint. If the multipliers are negative then it is not a minimum, at least according to the Kuhn-Tucker conditions we noted earlier. The values of the constraints at the optimal values of the design are $g_2 = -4.5553e + 007$ and $g_4 = -0.0320$. Both these values suggest that the constraints are inactive. At least the solution is feasible—all constraints are satisfied. The value of β_1^* is almost zero, while it is also an active constraint ($g_1 = 0$). This corresponds to a trivial case in the Kuhn-Tucker conditions.

Examining other cases that involve intersection of different pairs of inequalities (so that we can solve for x_1 and x_2) to make them active constraints, the trivial case shows up often (left to the students as an to explore). For the only case when the multipliers are positive is when g_2 and g_3 are considered active. The solution however violates the side constraints.

One important question that must be asked is whether the graphic solution in Chapter 2, which was used here, is actually an optimal solution. In Chapter 2 it was selected because it was feasible with two active constraints. The previous discussion suggests that it is definitely not an optimum solution based on the necessary conditions. Is it possible to find the best solution to the problem? The answer is yes and the way to accomplish it is by scanning the design space for feasibility and recording the best values. Today, everybody's desktop and laptop has incredible processing power that is never used to potential.

Table 4.3 Direct Solution of Flagpole with Different Increments

N	x_1^*	x_2^*	f^*	g_1^*	G_2^*	G_3^*	G_4^*
11	0.5	0.4	54503	−4.2e7	−5.4e7	−0.0076	−0.099
101	0.74	0.71	26343	−1.23e6	−4.4e7	−0.0058	−0.029
1001	0.9970	0.9760	25051	−7.85e04	−3.9e7	−0.0018	−0.020
10001	0.9963	0.9753	25073	−3118.6	−3.9e7	−0.0188	−0.020
100001	0.9997	0.9788	25064	−101.87	−3.9e7	−0.0189	−0.0199

DirectSolution_Example4_3.m is the code that will scan the design space in prescribed increments in the design space and store the best five solutions. This is a numerical scan. Table 4.3 is the best solution recorded for different increments in the design variables that effected through the value for n.

```
x_1 = linspace(0,1,n);
x_2 = linspace(0,1,n);
```

The CPU time on a current day laptop for $n = 10001$ is around 2230 seconds. That is about less than an hour of computer time. There is a hundredfold increases in CPU time for a decade increase in the number of points (n). One can always narrow the range while keeping the large number of points. This was the way the last row in Table 4.3 was obtained. Table 4.3 suggests that the solution is possibly determined by the *side constraint* once again. If none of the constraints are active, then the problem is essentially an unconstrained one. The FOC will determine that the optimum values for the design variables are zero contrary to Table 4.3. It appears that the Kuhn-Tucker conditions are not applicable to this problem. The solution is based on the direct investigation of the design space with $n = 10001$ for the original range of the design variables (row 4 in Table 4.3).

Example 4.3: Flagpole Design - Scanning the Design Space:

```
Design Variables:
        x1* =   0.9963
        x2* =   0.9753

Functions
f*    = 25073.7643
g1*   = -3118.5630
g2*   = -39011079.3125
g3*   = -0.0188
g4*   = -0.0200
```

The solution is feasible, and none of the constraints are active. There are no conditions available for determining a solution that is at the boundary.

4.4.3 Additional Topics

Regular Points: Regular points [14] arise when equality constraints, *h(X)*, are present in the problem. They are also extended to ***active*** inequality constraints (pseudo equality constraints). *Kuhn-Tucker conditions are valid only for regular points*. There are two essential features of a regular point X^*:

1. The point X^* is feasible (satisfies all constraints).
2. The gradients of the equality constraints as well as the active inequality constraints at X^* must form a linear independent set of vectors

Consider the graphical solution in Example 4.3, where constraints g_1 and g_3 are active. The solution at $x_1^* = 0.6773$, and $x_2^* = 0.6443$ is feasible. The following can be checked by running **RegularPoint_Example4_3.m**. The gradient of the constraints are

$$\nabla g_1(x_1^*, x_2^*) = 1.0e + 009 \begin{bmatrix} -2.9514 \\ 2.7835 \end{bmatrix}; \quad \nabla g_2(x_1^*, x_2^*) = \begin{bmatrix} -1.5792 \\ 1.4026 \end{bmatrix}$$

The included angle between the gradient vectors is

$$\theta = 1.7137°$$

It appears that the two gradients are almost parallel to each other. This is also evident in the graphical solution. The gradients therefore are not (*highly*) linearly independent—*Kuhn-Tucker conditions (probably) cannot be applied at this point*. The magnitude of the gradients differs by large orders of magnitude of the constraints. This affects the magnitude of the multipliers. In the code the simple form of the constraints from (4.102) is also used. While the angles are the same, the gradients of the two constraints are of the same order, suggesting that the simple form (4.102) is better scaled.

Kuhn-Tucker conditions are the only formal conditions for the recognition of the optimum values for constrained optimization problems in the literature. In practice it is applied without regard to *regularity*. Some additional considerations should be kept in mind[15].

If equality constraints are present and all the inequality constraints are inactive then the points satisfying the Kuhn-Tucker conditions may be minimum, maximum, or a saddle point. Higher order conditions are necessary to identify the type of solution

If the multipliers of the active inequality constraints are positive, the points cannot be a local maxima. They may not be local minima either. A point may be a maximum if the multiplier of an active inequality constraint is zero.

Scaling: We have implemented scaling of the variables in dealing with different versions of Example 4.1. For Example 4.3, all of the functions differ by orders of magnitude. Numerical calculations are driven by larger magnitudes. Inequality (4.102e) will be ignored in relation to the other functions in the calculations. This is a frequent concern in most numerical techniques in all disciplines. The standard approach to minimize the impact of large variations in magnitudes among different equations is to *normalize* the relations. In practice this is also extended to the variables. This is referred to as *scaling the variables* and *scaling the functions*. Many of the current optimization software will scale the problem without user invocation.

Scaling Variables: The presence of side constraints in problem formulation allows a natural definition of scaled variables. The user defined upper and lower bounds are used to scale each variable between 0 and 1. Therefore

$$\tilde{x}_i = \frac{x_i - x_i^l}{x_i^u - x_i^l}; \quad \tilde{x}_i - \text{ scaled } i^{\text{th}} \text{variable} \tag{4.105a}$$

$$x_i = \tilde{x}_i(x_i^u - x_i^l) + x_i^l \tag{4.105b}$$

Using (4.105b), the original problem in (4.102) can be restated in terms of the scaled variables. This is left as an exercise for the student since using symbolic functions this can be easily incorporated.

Scaling the Constraints: Scaling of the functions in the problem is usually critical for a successful solution. Numerical techniques used in optimization are iterative. During the iteration the gradient of the functions at the current value of the design variables is involved in the calculations. These gradients, expressed as a matrix, are called the *Jacobian* matrix or simply the *Jacobian*. Sophisticated scaling techniques[12,13] employ the diagonal entries of this matrix as metrics to scale the respective functions. These entries are evaluated at the starting values of the design variables. Function can also be scaled in the same manner as (4.105). In this case an upper and lower bound for the functions will be necessary. A simpler approach is to divide the each function by its maximum value. Typically, maximum values of functions are not available and determining them is usually considered an expensive effort. Remember, these discussions apply to n-variable problems. Scanning the function space in n-dimensions is not trivial. The whole idea of optimization techniques is to discover the solution without this effort. In this case the value of the functions at the initial guess is used for scaling.

It is important to understand tha the discussion on scaling is associated with the numerical techniques. So far, we have not introduced or invoked any. It starts with the next chapter. However, to bring some closure to this discussion, a practical method of obtaining the scaling factor for each function is to use the starting guess for the design variables and calculates the function values. These can serve as the scaling factors for the respective functions. Since the graphical solution is available, we can use it as the initial guess for two variable problems. Constructing the function scaled optimization problem of the examples is left as an exercise for the reader.

Side Constraints: In all of the examples in this chapter the side constraints were checked after a solution was found. It is possible to incorporate the side constraints through linear inequality constraints so they become part of solution discovery. This will be accompanied by increase in the computing effort. For example

$$x_i^l \leq x_i \leq x_i^u \tag{4.106}$$

can be replaced by

$$g_{j+i} : x_i^l - x_i \leq 0$$

$$g_{j+i+1} : x_i - x_i^u \leq 0 \tag{4.107}$$

Remember each inequality constraint contributes two additional cases that have to be solved for the Kuhn-Tucker conditions.

Summary: This chapter discusses the condition that is needed to recognize the optimum solution for different types of optimization problem. These conditions will be an essential part of the numerical techniques for optimization. In this chapter, it is shown through several examples that these conditions apply to problems that have a solution in the interior of the feasible domain. Most often it is taken for granted that such is the case. A solution on the boundary requires the designer to be vigilant and use extra procedures to confirm the solution. These have not been formalized successfully and are difficult to implement. This is where creativity, intuition, expertise, and common sense become valuable.

REFERENCES

1. Thomas, G. B., L. F. Ross, *Calculus and Analytical Geometry*. Reading, MA., Addison-Wesley, 1998.
2. Lay, C. *Linear Algebra and its Applications*. Reading, MA. Addison-Wesley, 2003.
3. Banchoff, T., Wermer, J., *Linear Algebra through Geometry*. New York: Springer, 1992.
4. Kreyszig, E. *Advanced Engineering Mathematics*, John Wiley & Sons, Inc. Hoboken, NJ: 2006.
5. Hostetler, G.H., Santina, M.S., and Montalvo, P.D., *Analytical, Numerical, and Computational Methods for Science and Engineering*. Prentice Hall, Englewood Cliffs, NJ.
6. Higham, D. J., and N.J. Higham. "MATLAB Guide", *SIAM*, 2006.
7. Venkataraman, P., *MATLAB 6*, http://www.rit.edu/~pnveme/Matlab6/, 2004.

8. *Matlab Demo—Symbolic Tool Box—Introduction*. The Mathworks Inc., MA.

9. Ramanujan, S. "Modular Equations and Approximations to π", *Quarterly Journal of Pure and Applied Mathematics*, vol. 45 (1913–1914).

10. *Maple 10* is developed and distributed by Waterloo Maple Software, Inc.

11. Burden, R.L., and J.D. Faires. *Numerical Analysis, 4th ed.* Boston, MA: PWS-KENT Publishing Company.

12. Fox, R.L., *Optimization Methods for Engineering Design.* Reading, MA: Addison-Wesley.

13. Vanderplaats, G.N., *Numerical Optimization Techniques for Engineering Design.* New York: McGraw-Hill.

14. Arora, J.S., *Introduction to Optimal Design.* New York: McGraw-Hill.

15. Ravindran, A., K.M. Ragsdell, and G.V. Relaitis, *Engineering Optimization, Methods and Applications.* Hoboken, NJ: John Wiley & Sons, 2006.

16. Rao, S. S., *Engineering Optimization, Theory and Practice.* Hoboken, NJ: John Wiley & Sons, 1996.

17. Kuhn, H.W., and Tucker, A.W., "Nonlinear Programming." In *Proceedings of the Second Berkeley Symposium on Mathematical Statistics and Probability*, J. Neyman ed., University of California Press, CA.

PROBLEMS

In many of the following problems, you are required to obtain numerical solution. For analytical solution, use symbolic calculations as much as you can after identifying the KT conditions.

4.1 Define two nonlinear functions in two variables. Find their solution through contour plots.

4.2 For the functions in Problem 4.1 obtain the gradients of the function and the Jacobian. Confirm them using the Symbolic Math Toolbox.

4.3 Define the design space (chooses side constraints) for a two variable problem. Define two nonlinear functions in two variables that do not have a solution within this space. Graphically confirm the result.

4.4 Define a design space for a two-variable problem. Define two non linear functions in two variables that have at least two solutions within the space.

4.5 Define a nonlinear function of two variables. Choose a contour value and draw the contour. Identify a point on the contour. Calculate the value of the gradient at that point. Draw the gradient at the point using the computed value. Calculate the Hessian at the above point

4.6 Using the relation in Equation (4.23) establish that the gradient is normal to the tangent (This is a physical interpretation for Equation (4.23).

4.7 Define a nonlinear function of three variables. Choose a point in the design space. Find the gradient of the function at the point. Calculate the Hessian matrix at the same point.

4.8   Express the Taylor series expansion (quadratic) of the function $f(x) = (2 - 3x\, x^2)\, sin\, x$ about the point $x = 0.707$. Confirm your results through the Symbolic Math Toolbox. Plot the original function and the approximation.

4.9   Expand the function $f(x, y) = 10\, (1 - x^2)^2 + (y - 2)^2$ quadratically about the point $(1,1)$. How will you display the information? Draw the contours of the original function and the approximation.

4.10   Obtain the solution to the optimization problem in Section 4.3.5 using MATLAB.

4.11   Use an example to show that optimum values for the design variables do not change if the objective function is multiplied by a constant. Prove the same if a constant is added to the function.

4.12   How does scaling of the objective function affect the value of the multipliers? Use an example to infer the result. How does scaling the constraint affect the multipliers? Check with an example.

4.13   Solve Problem 1.11.

4.14   Solve Problem 1.12.

4.15   Solve Problem 1.13.

4.16   Solve Problem 1.14.

4.17   Solve Problem 1.16.

4.18   Solve Problem 1.17.

4.19   Solve Problem 1.18.

4.20   Set up the analytical conditions for Problem 1.19.

4.21   Solve Problem 1.20.

4.22   Solve Problem 1.21.

4.23   Solve Problem 1.23.

4.24   Solve Problem 1.25.

4.25   Solve Problem 1.26.

4.26   Setup the analytical conditions for Problem 1.27.

4.27   Solve Problem 1.35.

4.28   Solve Problem 1.36.

4.29   Solve Problem 1.39.

4.30   Solve Problem 1.40.

5

NUMERICAL TECHNIQUES – THE ONE-DIMENSIONAL PROBLEM

A real optimal design problem will involve several variables (or many dimensions). We will see that the numerical techniques for these problems will involve the solution to a single variable (or one-dimension) problem in the step size. Readers who have attended a course in numerical techniques, which in several curriculums is a prerequisite for the optimization course, will recognize that they have dealt with the one-dimensional unconstrained optimization problem. It sometimes appears indirectly, notably as finding the root of a nonlinear equation, which is the same as satisfying the FOC for an unconstrained optimization problem. Two of the popular techniques for this problem are the Newton–Raphson and the bisection technique. These are termed as **root-finding algorithms**. They are introduced early in Section 5.2 to provide a comparison to the methods currently used in optimization.

In engineering design, the single variable optimization problem is probably a triviality, but one-dimensional optimization is a critical component of multivariable design, as discussed in Section 5.3. Meanwhile, the classical root-finding methods do not play any significant part in numerical optimization techniques, as they are considered computationally expensive and not robust enough over a large class of problems. An important consideration in nonlinear design optimization is that during the earlier iterations, when you are far away from the solution, accuracy can be traded for quickness and the success of moving through to the next iteration, instead of faltering at the current one. These criteria have required that the one-dimensional techniques be simple in concept, as well as easily implementable. Two different techniques, the polynomial approximation and the golden section, provide popular support for most optimization software. Very often, the two techniques are implemented in combination.

5.1 PROBLEM DEFINITION

In order to connect with the discussions in later chapters where one-dimensional optimization is used, the one-dimensional variable will be identified as α instead of x or x_1. The variable α is usually called the **step size**.

Example 5.1 We will use an example similar to the one used in Chapter 4.

$$\text{Minimize} \quad f(\alpha) = \sin(0.1 + 2\alpha)/(0.1 + \alpha) \tag{5.1a}$$

$$\text{Subject to:} \quad 0 \leq \alpha \leq 10 \tag{5.1b}$$

The problem does not represent any real design. It was constructed to have multiple local minimums within the area of interest. Equations (5.1a) and (5.1b) represent an unconstrained problem with the ubiquitous but necessary side constraints. The presence of side constraints really imply that there are no truly unconstrained problems. The side constraints also define an acceptable design region for the problems, and must be respected by the designer at all times. In Chapter 4, we have shown that these constraints determined the solution for some of the examples. Section 5.2 will use this example to demonstrate the various numerical methods for one-dimensional problems.

5.1.1 Constrained One-Dimensional Problem

The only constraint that 5.1, a single-variable problem, can accommodate is *inequality* constraints. There can also be more than one of them. A single-variable problem cannot have an equality constraint (see Chapter 4). In constrained multivariable optimization problems it will often be necessary to find the step size α such that a constraint changes its status from an inactive constraint to an active constraint. We define the constrained problem as 5.2.

Example 5.2

$$\text{Minimize} \quad f(\alpha) = \sin(0.1 + 2\alpha)/(0.1 + \alpha) \tag{5.2a}$$

$$\text{Subject to:} \quad g(\alpha) : \alpha^2 - 7\alpha - 0.5 \leq 0 \tag{5.2b}$$

$$0 \leq \alpha \leq 10 \tag{5.2c}$$

In actual applications, it is more likely that (5.2) will really define two problems for identifying two step sizes. The first one is the solution to (5.1). The second one is the solution to

$$g(\alpha) : \alpha^2 - 7\alpha + 0.5 = 0 \tag{5.3a}$$

$$0 \leq \alpha \leq 10 \tag{5.3b}$$

which makes (5.3a) an active constraint. A graphical investigation in the next section should definitely provide more light on this idea.

5.1.2 Necessary and Sufficient Conditions

The necessary and sufficient conditions are borrowed from Chapter 4. For the unconstrained optimization problem in (5.1) the necessary, or FOC, is

$$\frac{df}{d\alpha} = 0 \qquad (5.4a)$$

The sufficient condition is

$$\left.\frac{d^2 f}{d\alpha^2}\right|_{\alpha*} > 0 \qquad (5.4b)$$

For the constrained problem in (5.2) we first construct the Lagrangian F as

$$F(\alpha, \beta) = f(\alpha) + \beta g(\alpha)$$

and the FOC are

$$\frac{\partial F}{\partial \alpha} = \frac{df}{d\alpha} + \beta \frac{dg}{d\alpha} = 0 \qquad (5.5a)$$

with

$$\boldsymbol{Case(a)} : g(\alpha) = 0 \quad and \ \beta > 0$$
$$\boldsymbol{Case(b)} : \beta = 0 \quad and \ g(\alpha) < 0 \qquad (5.5b)$$

5.1.3 Solution to the Examples

There are two ways we can determine the solution without using any special techniques. The first way is to scan the design space for a minimum. The second is to identify the solution graphically by plotting the objective function and the constraints

Solution through Scanning: You can control the accuracy of the solution by dividing the interval dictated by the side constraint into appropriate number of intervals. This can be run in the background or during the time the computer is not being used. This procedure is quite simple to program, and, unlike the more sophisticated numerical techniques in the next section, will determine the best solution or the global minimum.

MATLAB Code: The code **ScanningExample5_1.m*** will scan the design space for the solution to 5.1 and 5.2. It will find the lowest value of the objective

*Files to be downloaded from the web site are indicated by boldface courier type.

function while checking for feasibility. It will find the very best solution. The solution to both examples is the same, and is (output in Command window)

$$\alpha^* = 2.1991$$

$$f^* = -0.4250 \tag{5.6}$$

$$g^* = -10.0577$$

For 100,000 points, the CPU time was about 744 seconds on the author's laptop.

Graphical Solution: The graphical solution to both Example 5.1 and 5.2 is shown in Figure 5.1. The unconstrained problem has three local minimums within the design region and they are marked as local minimum. The best solution is also marked as Global minimum and is verified by the direct solution above. For the constrained problem, the feasible region shrinks the design space. There are only two local minimums in this space. The solution is the interior of the feasible region with the constraint being inactive. Note the special nature of the feasible region for a one variable problem where it is the intersection of constraint and the axis. In many variable problems, the constraint itself may be part of the solution.

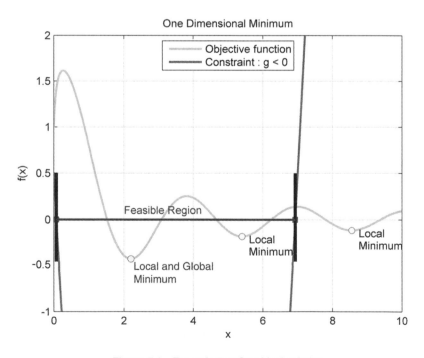

Figure 5.1 Example 5.1: Graphical solution.

MATLAB   Code: The graphics are created by running the code GraphicalSolution_Example5_1.m. All of the notations in Figure 5.1 are generated in the code. Many interesting observations can be seen in the code. The code initially uses symbolic computation. The optimum value determined by applying the FOC (5.4a), through symbolic calculations, will identify the point at $\alpha = 0.3$, which is the maximum. There is no other way to identify the actual minimum. This brings the need for numerical solution using the fsolve function to handle the FOC. The fsolve function requires a starting point for the iterations. Using different starting points, the solution converges to the closest minimum or maximum in the design region. A simple scan of all the solutions that satisfy the SOC will identify local minimums. The constrained problem will require the satisfaction of the constraints. The graphical solution confirms the solution obtained through the simple scanning of the design space (5.6).

It can be concluded that symbolic computation has only limited value in optimization. Numerical computation is necessary for discovering the solution. Numerical techniques are iterative and require an intelligent guess to start the search for the optimal solution. Second-order conditions will be necessary for identifying a minimum. As we look at Figure 5.1, again consider the difficulty of identifying the minimum if it is on the constraint boundary. For example, if the left constraint limit is moved to about 0.3, then the FOC and SOC do not apply. Another situation is when the constraint has segmented feasible regions. It is necessary for the designer to understand the problem, the mathematical model, and the solution space, and must have some instinct about the problem in order to obtain a solution.

5.2 NUMERICAL TECHNIQUES

Several techniques for solving the one-dimensional unconstrained optimization problem are developed in this section. The first two are traditional root-finding techniques that are typically not used in optimization. If you are developing a unique one-time approach for a particular class of problem, then there are essentially no limitations on what you can use to make your application work, including the methods espoused in the root-finding methods.

5.2.1 Features of the Numerical Techniques

Root-finding problems involve identifying the value of the variable for which the function will have a value of zero. Typically, they are postulated as the solution to

$$\phi(\alpha) = 0 \qquad (5.7)$$

where $\phi(\alpha)$ is a nonlinear equation in the variable α. We can identify that (5.4a) and (5.5a) belongs to this class of problem. In general, nonlinear equations are solved iteratively. This means that they repeat a set of similar calculations until they arrive at the solution. This repeating framework is called an **algorithm**. For example, a simple algorithm for solving (5.7) can be set up as follows:

Generic Algorithm (A5.1):

Step 1: Assume α

Step 2: Calculate $\Delta\alpha$

Step 3: Update $\tilde{\alpha} = \alpha + \Delta\alpha$
        If converged $(\phi(\tilde{\alpha}) = 0)$ *then* exit
        If not converged $(\phi(\tilde{\alpha}) \neq 0)$ *then* $\alpha \leftarrow \tilde{\alpha}$
        Go to Step 1

This sequence is a standard procedure in iterative techniques for many classes of problems. It also captures the essence for most of the techniques to follow in this book. Step 1 indicates the primary feature of iterative methods—the *starting solution*. This is an initial guess provided by the user to start the iterative process. The iterative process is continued to the solution by calculating changes in the variable ($\Delta\alpha$). The value of the variable is updated. Convergence is checked. If Equation (5.7) is satisfied, then the solution has been found. If convergence is not achieved, then the updated variable is used to proceed with the next iteration. $\Delta\alpha$ can be calculated in many ways and *Newton–Raphson* is just one of the ways to establish the change in the variable. As we develop the various techniques in this chapter, a couple of assumptions are made about the functions:

1. These functions are explicitly defined, and their derivatives can also be obtained.
2. The solution is expected to be in the interior of the design space—that is, it is not on the boundary.

5.2.2 Newton–Raphson Technique

The **Newton – Raphson technique**, also referred as Newton's technique, is a gradient-based solution to finding the root of a single nonlinear equation:

$$\phi(\alpha) = 0 \tag{5.7}$$

where $\phi(\alpha)$ is a nonlinear equation in the variable α.

The technique has the following features:

It can be explained geometrically.

It uses the Taylor series expanded linearly.

It has the property of quadratic convergence; that is, as it gets closer to the solution, the change in $\Delta\alpha$ is quadratic.

Calculation of Δα: Let α be the current value of the variable. It is assumed that Equation (5.7) is not satisfied at this value of the variable (the reason for us to continue to iterate). Let $\tilde{\alpha}$ be a neighboring value. Ideally, we would like to achieve convergence at this value (even as we are aware it might take us several iterations to achieve convergence). Using Taylor's theorem expanded to the linear term only

$$\phi(\tilde{\alpha}) = \phi(\alpha + \Delta\alpha) = \phi(\alpha) + \frac{d\phi}{d\alpha}\Delta\alpha = 0 \qquad (5.8)$$

$$\Delta\alpha = -\frac{\phi(\alpha)}{\dfrac{d\phi}{d\alpha}} = -\left[\frac{d\phi}{d\alpha}\right]^{-1}\phi(\alpha) \qquad (5.9)$$

For Equation (5.9) to be effective, the gradient of the function should not be zero. It is also inevitable that changes in α will be large where φ is flat, and small when the slope is large. For the Newton–Raphson method to be effective, the iterations should avoid regions where the slope is small, or be more creative in moving through this region. This is a disadvantage of this method.

Newton–Raphson Algorithm (A5.2):

Step 1: Assume α

Step 2: Calculate $\Delta\alpha = -\dfrac{\phi(\alpha)}{\dfrac{d\phi}{d\alpha}} = -\left[\dfrac{d\phi}{d\alpha}\right]^{-1}\phi(\alpha)$

Step 3: Update $\tilde{\alpha} = \alpha + \Delta\alpha$
If converged ($\phi(\tilde{\alpha}) = 0$) *then* exit
If not converged ($\phi(\tilde{\alpha}) \neq 0$) *then* $\alpha \leftarrow \tilde{\alpha}$
Go to Step 1

Example 5.1 The Newton–Raphson technique is applied to FOC of Example 5.1.

$$\text{Minimize} \quad f(\alpha) = \sin(0.1 + 2\alpha)/(0.1 + \alpha) \qquad (5.1a)$$

$$\text{Subject to:} \quad 0 \leq \alpha \leq 10 \qquad (5.1b)$$

The FOC is

$$\phi(\alpha) = \frac{df}{d\alpha} = \frac{2\cos(0.1 + 2\alpha)}{(0.1 + \alpha)} - \frac{\sin(0.1 + 2\alpha)}{(0.1 + \alpha)^2} = 0 \qquad (5.10)$$

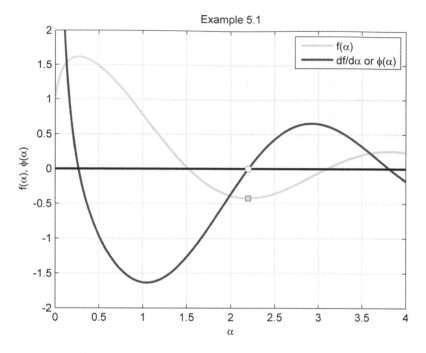

Figure 5.2   Newton–Raphson Technique: Example 5.1.

MATLAB Code: The code is available in **NewtonRaphson_Example5_1.m**. The code will create the graphics in Figure 5.2 and the Table 5.1. Figure 5.2 is the final figure in a series of animation that follows the current value of α and places a marker on the function and the derivative. The markers move to the current value of the variable during the iteration. Five iterations are necessary for convergence. The code however does not test for SOC, or check if the second derivative of f is close to zero, which will stall the progress to the solution.

The starting value is $\alpha = 1.5$. The final values match the one in (5.6). Table 5.1 shows that α changes by orders of magnitude as the solution converges. It suggests that the biggest change occurs at the start of the iterations and solution converges quickly. The convergence criteria, or the stopping criteria, or the zero

Table 5.1   Newton-Raphson Technique for Example 5.1

Iterations(i)	α_i	$\phi(\alpha_i)$	$\Delta\alpha_i$	ϕ_{i+1}	$f(\alpha)$
1	1.5000	−1.2652	0.8563	−1.2652	0.0260
2	2.3563	0.2463	−0.1735	0.2463	−0.4051
3	2.1828	−0.0280	0.0162	−0.0280	−0.4248
4	2.1990	−0.0002	0.0001	−0.0002	−0.4250
5	2.1991	−0.0000	0.0000	−0.0000	−0.4250

value for f, was set to 1.0e-08. The acceleration of convergence or rate of convergence to the solution is formally defined in many different ways. One of the ways is to monitor the ratio of $\phi(\alpha + \Delta\alpha)/\phi(\alpha)$. Another is to track $\Delta\alpha$. In applying and understanding numerical techniques, it is important to develop an instinct and feel for the numbers corresponding to various functions, as the iterations progress. The speed or the quality of convergence can be gauged by the shift in the decimal point in the quantity of interest through the iterations in Table 5.1.

The Newton–Raphson technique can be directly applied to the (5.3)—after all the method is designed to operate to find the zero of functions. It is left as an exercise for the reader. The reader is encouraged to try running the code from different starting points. A starting value of $\alpha = 3$, leads to a solution of $\alpha = 6.9833$ in six iterations. As part of the criteria, we can also check second-order conditions. Most iterative numerical techniques are designed to converge to solutions that are close to where they start from.

5.2.3 Bisection Technique

This is a popular technique to find the root of a function. It is also called a *binary search* or *interval halving* procedure. Unlike the Newton–Raphson method, this procedure does not need the evaluation of the gradient of the function whose root is being sought. The method can be compared to the *Golden Section* method, which follows later. The numerical technique is based on the trapping of the root between a positive and a negative value of the function, since the function must cross over at the root. The actual solution from the method is an interval in which the zero or the solution can be located. This interval is typically determined to a very small value, called the *tolerance*, typically of the order of 10E-08, but controlled by the user. Since this is a root-finding procedure, when applied to optimization it is applied to the gradient of the objective that must satisfy the unconstrained FOC. Establishing the minimum of the objective is thereby translated to locating the root of the gradient function. It is possible to develop a bisection technique based on the objective function itself.

The bisection method requires two limits which bracket the root to start the iterations, say α_1 and α_2. It is preferred that the value of ϕ at these points must be opposite in sign. These can be the side constraints of the 1D optimization problem. It is assumed that at least one solution exists in this initial interval. During each iteration, this interval is halved, with the midpoint of the interval replacing either α_1 or α_2, while keeping the root still trapped between function values of opposite signs. The iterative technique is expressed as the following algorithm.

Algorithm for Bisection Method (A 5.3):

Step 1: Choose α_1 and α_2 to start. Let $\alpha_1 < \alpha_2$
Step 2: Set $\alpha = \alpha_1 + (\alpha_2 - \alpha_1)/2$

Step 3:   **If** $\phi(\alpha) = 0.0$ – Converged Solution–**exit**
        Else If $(\alpha_2 - \alpha_1) < 10$ E-04 – tolerance met–**exit**
        Else If $\phi(\alpha)^*\phi(\alpha_1) > 0.0$;
        then $\alpha_1 \leftarrow \alpha$
        Else $\alpha_2 \leftarrow \alpha$
        Go to Step 2

The bisection method will identify that the root will be available between the final limits of α_1 and α_2

Example 5.1   The bisection technique is applied to FOC of Example 5.1.

$$\text{Minimize} \quad f(\alpha) = \sin(0.1 + 2\alpha)/(0.1 + \alpha) \tag{5.1a}$$

$$\text{Subject to:} \quad 0 \le \alpha \le 10 \tag{5.1b}$$

The FOC is

$$\phi(\alpha) = \frac{df}{d\alpha} = \frac{2\cos(0.1 + 2\alpha)}{(0.1 + \alpha)} - \frac{\sin(0.1 + 2\alpha)}{(0.1 + \alpha)^2} = 0 \tag{5.10}$$

MATLAB Code: The code for the bisection technique is available in **Bisection_Example5_1.m**. It will draw Figure 5.3 as well as generate the entries in Table 5.2. Symbolic computation is use for function evaluation in the code, though it could have easily been accomplished through numerical evaluation. Figure 5.3 is the final figure at the end of an animation. There are three points, using three different markers, placed on the figure as the iterations progress: The two limits and the midpoint. The markers illustrate the trapping of the root/minimum by the two limits. Also, during each iteration, only one limit marker moves to the midpoint. The code appears slow because of the pause statements for the animation effect. The code does not check for SOC so that the roots can identify maximum of the function, too. In fact, try the code with a different limit for the variables.

There are 16 iterations in Table 5.2 for the solution, compared to 5 iterations for the Newton–Raphson method. Both the methods found the solution. The lower number of iterations can be associated with a better quality of information in the Newton–Raphson procedure, since it uses the second derivative of the function for finding the root while the bisection method does not. However, gradient computations must be considered additional work and computationally expensive.

Both the methods presented so far handle the minimization indirectly as a matter of finding the root of the FOC for the unconstrained one-variable problem. The following two methods are used extensively for one-dimensional minimization. They handle the minimization problem directly.

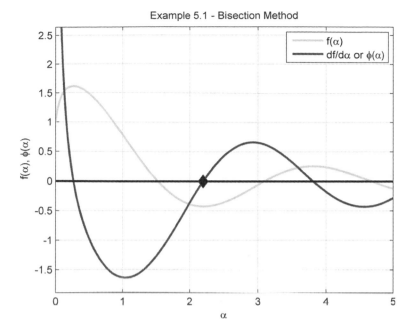

Figure 5.3 Bisection method: Example 5.1.

5.2.4 Polynomial Approximation

The method is simple in concept. Instead of minimizing a difficult function of one variable, minimize a polynomial that approximates the function. The optimal value of the variable that minimizes the polynomial is then considered to approximate the optimal value for the original function. It is rare for the degree of the approximating polynomial to exceed three. A quadratic approximation is standard unless the third degree is warranted. It is clear that serious errors in approximation are expected if the polynomial is to simulate the behavior of the original function over a large range of values of the variable. Mathematical theorems can justify a quadratic representation of the function, with a prescribed degree of error, within a small neighborhood of the minimum. We have seen this with the Taylor series representation in Chapter 4. This means that the polynomial approximation gets better in smaller neighborhoods around minimum.

In comparison to the earlier methods, there are no iterations involved. We are not applying the method dynamically in reduced neighborhoods. It is a single, non-iterative application of the method. Therefore, there is no *algorithm* for this method. A quadratic polynomial ($P(\alpha)$) is used for the approximation. This polynomial is expressed as follows:

$$P(\alpha) = b_0 + b_1\alpha + b_2\alpha^2 \tag{5.11}$$

Table 5.2   Bisection method: Example 5.1

ITERATIONS	α_A	α_B	MID α	$\phi(\alpha)$	F
1.0000	1.0000	4.0000	2.5000	0.4277	−0.3561
2.0000	1.0000	2.5000	1.7500	−0.8402	−0.2392
3.0000	1.7500	2.5000	2.1250	−0.1298	−0.4202
4.0000	2.1250	2.5000	2.3125	0.1823	−0.4145
5.0000	2.1250	2.3125	2.2188	0.0331	−0.4247
6.0000	2.1250	2.2188	2.1719	−0.0469	−0.4244
7.0000	2.1719	2.2188	2.1953	−0.0065	−0.4250
8.0000	2.1953	2.2188	2.2070	0.0134	−0.4250
9.0000	2.1953	2.2070	2.2012	0.0035	−0.4250
10.0000	2.1953	2.2012	2.1982	−0.0015	−0.4250
11.0000	2.1982	2.2012	2.1997	0.0010	−0.4250
12.0000	2.1982	2.1997	2.1990	−0.0003	−0.4250
13.0000	2.1990	2.1997	2.1993	0.0004	−0.4250
14.0000	2.1990	2.1993	2.1992	0.0001	−0.4250
15.0000	2.1990	2.1992	2.1991	−0.0001	−0.4250
16.0000	2.1991	2.1992	2.1991	−0.0000	−0.4250

The polynomial is completely defined if the coefficients b_0, b_1, and b_2 are known. To determine them, three data points $[(\alpha_1, f_1), (\alpha_2, f_2), (\alpha_3, f_3)]$ are generated from the function (5.1a). This sets up a linear system of three equations in three unknowns by requiring the values of the function and the polynomial to be the same at the three points. The polynomial approximates the original function, and its minimum can be obtained from the coefficients. The consideration of the inequality constraint in (5.2) presents a different problem 6 altogether. Here, two polynomials are necessary. The first polynomial will approximate the objective function that was handled. The second polynomial will approximate the constraint function, and its root is computed to see the change required to make it active. In the case of unconstrained minimum, for a good approximation, the points chosen to obtain the polynomial coefficients must bracket the minimum. A simple scanning procedure can be used find α_1, α_2, and α_3. Example 5.1, reproduced for convenience, is used to illustrate the scanning procedure

Example 5.1   Polynomial approximation technique is applied to the *objective function:*

$$\text{Minimize} \quad f(\alpha) = \sin(0.1 + 2\alpha)/(0.1 + \alpha) \qquad (5.1a)$$

$$\text{Subject to:} \quad 0 \le \alpha \le 10 \qquad (5.1b)$$

Scanning Procedure: The process is started from the *lower limit* for α. A value of zero for this value can be justified since it will refer to values at the current iteration in later discussion. A constant interval for α, $\Delta\alpha$, is also identified.

For a well-scaled problem, this value is usually 1. Starting at the lower limit, the interval is increased linearly, sometimes doubled, until the minimum is bracketed. This is the *upper limit* for the approximation. A third point can be chosen to lie midway between the lower and upper limit for a quadratic approximating polynomial. This gives us three points to evaluate the coefficients. This is a simple process. We can build in sophistication by allowing the lower limit to also move with more information so that the minimum is bracketed in a smaller range, yielding a better approximation. This is not attempted in the code, but you are encouraged to incorporate this change, since it is not difficult. The change in the lower limit is to avoid the positive slope of the function in the lower limit of 0, which will encourage a solution to the left. You can see this by running the code with the lower limit of α_0. For a minimum problem, the function is expected to decrease at the lower limit.

With respect to Example 5.1, the scanning procedure generates the following values:

$$\alpha = 0; \quad f(0) = 0.9983; \quad \textbf{(lower limit)}$$

$$\alpha = 1; \quad f(1) = 0.7847;$$

$$\alpha = 2; \quad f(2) = -0.3897; \quad \text{Minimum is not trapped}$$

$$\alpha = 3; \quad f(3) = -0.0588; \quad \textbf{(upper limit)}$$

The scanning process is not only simple but is indifferent to the example being solved. The important requirement of any such process is to ensure that the minimum lies between the limits established by the procedure. This procedure is developed as MATLAB m-file **UpperBound_1Var.m**.

The set of linear equations to establish the coefficients of the polynomial in (5.11) for Example 5.1, with $\alpha_1 = 0, \alpha_2 = 1, \alpha_3 = 3$, is

$$0.9983 = b_0$$

$$0.7847 = b_0 + b_1 + b_2$$

$$-0.0588 = b_0 + 3b_1 + 9b_2$$

Using MATLAB, the solution to the equations is

$$b_0 = 0.9983; \quad b_1 = -0.9441; \quad b_2 = 0.1972 \tag{5.12}$$

and the polynomial is

$$P(\alpha) = 0.9983 - 0.9441\alpha + 0.1972\alpha^2$$

The minimum for the polynomial is the solution to

$$\alpha_p^* = \alpha^* = 2.3932; \quad P(\alpha^*) = -0.1314$$

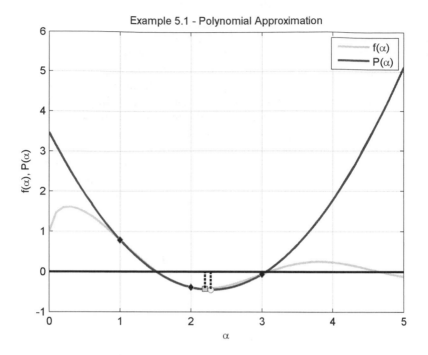

Figure 5.4   Polynomial approximation: Example 5.1.

Figure 5.4 describes polynomial approximation for Example 5.1. It is clear that the approximation leaves much to be desired. However, if the data for the polynomial coefficients were around a smaller region near $\alpha = 2.1$, then the results would be more impressive. Note, rather than use a higher-order polynomial, a better set of data is preferred.

MATLAB Code: The code for polynomial approximation for Example 5.1 can be found in **PolynomialApproximation_Example5_1.m**. The code is essentially used to create Figure 5.4 and will call **PolyApprox_1Var.m**, which in turn calls **UpperBound_1Var.m**, which in turn requires the code for the example that is being solved (**Example5_1.m**) So we can express the dependency of the code as follows:

```
PolynomialApproximation_Example5_1.m  (draws Figure 5.4)
  ▶ PolyApprox_1Var.m  (determines the polynomial)
    ▶ UpperBound_1Var.m (determines the upper bound)
      ▶ Example5_1.m  (the specific problem)
```

The strategy for the MATLAB code here is a little different from the previous numerical implementation. The two pieces of code that are called are written as *function m-files*. They are independent of any particular example. They can also be used by other forthcoming numerical techniques. Such generic code development is termed as a **block structured approach**. This way, large code development is possible using smaller pieces of existing code.

PolyApprox_1Var.m The code segment implements the polynomial approximation method for a function of one variable. This function uses **UpperBound_1Var.m** and the function file that contains the specific example whose minimum is to be found.

The input to the function is

functname—the function that is being approximated **name.m**

order—(2 or 3) of the approximation

lowbound—the start value of the scan passed to **UpperBound_1Var.m**

intvlstep—the scanning interval passed to **UpperBound_1Var.m**;

ntrials—the number of scanning steps passed to **UppperBound_1Var.m**

The output of the program is a vector of two values.

ReturnValue(1)—the *first element* is the location of the minimum.

ReturnValue(2)—the *second element* is the function value at this location.

The function referenced by the code must be a MATLAB m file, in the same directory (**Example5_1.m**). The input **for Example5_1.m** is the value at which the function needs to be computed, and its output is the value of the function.

Usage:

```
Value = PolyApprox_1Var('Example5_1',2,0,1,10)        (5.13)
```

The program uses the *switch/case* and the *feval* statements from MATLAB. The code also illustrates calling and returning from other functions.

UpperBound_1Var.m The code segment implements the determination of the upper bound of a function of single variable. The input to the function is

functname —name of the function m-file (e.g., **Example5_1.m**)

a0 —start of the scan, the scanning interval

da —scanning interval

ns —the number of scanning steps

The output of the program is a vector of two values.

ret(1)—the first element is the location of the upper bound.

ret(2)—the second element is the function value at this location.

The function referenced by the code must be a MATLAB m-file, in the same directory (**Example5_1.m**). The input for **Example5_1.m** is the value at which the function needs to be computed, and its output is the value of the function. The function name is passed by **PolyApprox_1Var.m**.

Usage:

$$\texttt{ret = UpperBound\_1Var('Example5\_1',0,1,10)} \qquad (5.14)$$

5.2.5   Golden Section Method

The *Golden Section* method is the best of interval-reducing methods. It reduces the interval by the same fraction with every iteration. The intervals are derived from the *golden ratio*, 1.61803. This ratio has significance in aesthetics as well as mathematics 3, 4. The method is simple to implement. It is indifferent to the shape and continuity properties of the function being minimized. Most important, the number of iterations to achieve a prescribed tolerance can be established before the iterations start. The Golden Section algorithm is similar to the Bisection Method, though the latter is used to obtain the root of the function. In the Golden Section method the minimum of the function is being found. A similarity between the two techniques is that for each iteration, only one of the limits is changed while two values in between are adjusted with previous known values. This can be seen in the animation involved with the code.

Algorithm for Golden Section Method (A 5.4):

Step 1: Choose $\alpha^{low}\alpha^{up}$
$\quad\quad\quad$ $\tau = 0.38197$ (from golden ratio),
$\quad\quad\quad$ $\varepsilon = \text{tolerance} = (\Delta\alpha)_{\text{final}}/(\alpha^{up} - \alpha^{low})$
$\quad\quad\quad$ $N = \text{number of iterations} = -2.078 \ln \varepsilon$
$\quad\quad\quad$ $i = 1$

Step 2: $\alpha_1 = (1 - \tau)\alpha^{low} + \tau\alpha^{up}; \quad f_1 = f(\alpha_1)$
$\quad\quad\quad$ $\alpha_2 = \tau\alpha^{low} + (1 - \tau)\alpha^{up}; \quad f_2 = f(\alpha_2)$
$\quad\quad\quad$ The points are equidistant from bounds.

Step 3: ***If*** $(i < N)$
$\quad\quad\quad$ ***If*** $f_1 > f_2$
$\quad\quad\quad$ $\alpha^{low} \leftarrow \alpha_1; \quad \alpha_1 \leftarrow \alpha_2; \quad f_1 \leftarrow f_2$
$\quad\quad\quad$ $\alpha_2 = \tau\alpha^{low} + (1 - \tau)\alpha^{up}; \quad f_2 = f(\alpha_2)$
$\quad\quad\quad$ $i \leftarrow i + 1$
$\quad\quad\quad$ ***Go*** to Step 3.
$\quad\quad\quad$ ***If*** $f_2 > f_1$
$\quad\quad\quad$ $\alpha^{up} \leftarrow \alpha_2; \quad \alpha_2 \leftarrow \alpha_1; \quad f_2 \leftarrow f_1$
$\quad\quad\quad$ $\alpha_1 = (1 - \tau)\alpha^{low} + \tau\alpha^{up}; \quad f_1 = f(\alpha_1)$

$i \leftarrow i + 1$

Go to Step 3.

The sign $\leftarrow$ implies replace the value of the left-hand side with the value from the right-hand side.

Example 5.1 The Golden Section method is applied to the *objective function:*

$$\text{Minimize} \quad f(\alpha) = \sin(0.1 + 2\alpha)/(0.1 + \alpha) \tag{5.1a}$$

$$\text{Subject to:} \quad 0 \le \alpha \le 10 \tag{5.1b}$$

MATLAB *Code:* The code for golden section for Example 5.1 can be found in **GoldenSection_Example5_1.m**. The code is used to create Figure 5.5 and will call **GoldSection_1Var.m**, which in turn calls **UpperBound_1Var.m**, which in turn requires the code for the example that is being solved (**Example5_1.m**) So we can express the dependency of the code as follows:

GoldenSection_Example5_1.m (draws Figure 5.4)
   ▶ GoldSection_1Var.m (determines the polynomial)
      ▶ UpperBound_1Var.m (determines the upper bound)
         ▶ Example5_1.m (the specific problem)

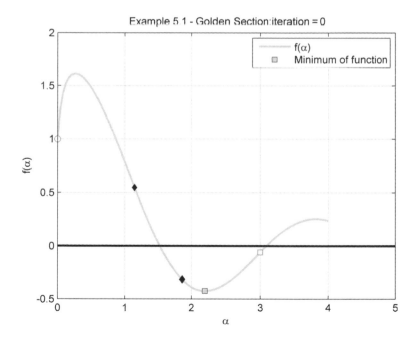

Figure 5.5 Golden Section: Example 5.1.

The code animates the location of α^{low}, α_1, α_2, and α^{up} with every iteration. You can see how only one of the outer limits are moved while the inside limits are adjusted. If the Command window is open, you will see the values of the variables and function printed synchronously for each iteration. Since we are reusing some of the code, the only new function m-file for this section is **GoldSection_1Var.m**.

GoldSection_1Var.m This code segment implements the golden section method for a function of one variable. This function uses **UpperBound_1Var.m** and the function file that contains the specific example whose minimum is to be found.

The input to the function is

functname —the function which is being approximated;

tol —(0.001) distance between outer limits at final iterarion;

lowbound —the start value of the scan passed to **UpperBound_1Var.m**;

intvl —the scanning interval passed to **UpperBound_1Var.m**;

ntrials —the number of scanning steps passed to **UppperBound_1Var.m**.

The output of the program is a vector **ReturnValue** of eight values. They are the four pairs of the variables and the function values in the final iteration. The function referenced by the code must be a MATLAB m-file, in the same directory (**Example5_1.m**). The input for **Example5_1.m** is the value at which the function needs to be computed, and its output is the value of the function.

Usage:

```
ret = GoldSection_1 Var('Example5_1',0.001,0,1,10)      (5.15)
```

Several Iterations of the Method: Several iterations of the method as seen in the Command window are reproduced here to verify the Golden Section algorithm. Iteration 0 is the start iteration.

```
iteration       0
             lower      left       right       upper
 alpha      0.0000     1.1459      1.8541      3.0000
    f       0.9983     0.5470     -0.3164     -0.0588

iteration       1
             lower      left       right       upper
 alpha      1.1459     1.8541      2.2918      3.0000
    f       0.5470    -0.3164     -0.4179     -0.0588

iteration       2
             lower      left       right       upper
```

```
alpha       1.8541        2.2918        2.5623        3.0000
   f       -0.3164       -0.4179       -0.3274       -0.0588


iteration       3
              lower         left         right         upper
alpha       1.8541        2.1246        2.2918        2.5623
   f       -0.3164       -0.4202       -0.4179       -0.3274


. . . . .
iteration      17
              lower         left         right         upper
alpha       2.1986        2.1989        2.1991        2.1995
   f       -0.4250       -0.4250       -0.4250       -0.4250
```

The tolerance $\alpha_2 - \alpha_1$ in the final iteration is about 0.001, as expected.

Comparison with Polynomial Approximation: There are two contenders for minimizing a function of one variable. The next section will highlight that this is a significant exercise of multi-variable minimization. The algorithms for Polynomial Approximation and the Golden Section have significant differences. The former is a one-shot approach and is accompanied by significant error in the estimation of the minimum, which will improve as the data for approximation get better. Implied in the polynomial approach is the continuity of the function.

The Golden Section method, by contrast, is an iterative technique. The number of iterations depends on the tolerance expected in the final result and is known prior to the start of the iterations—a significant improvement compared to the Newton-Raphson method where the number of iterations cannot be predicted *apriori*. The implementation is simple and the results are impressive, as it is able to home in on the minimum. It is indifferent to the nature of the function.

There is no reason why the two cannot be combined. The Golden Section method can be used to establish four data points with a reasonable tolerance (instead of a low value) and a cubic polynomial can be fit to identify the approximate location of minimum.

5.3   IMPORTANCE OF THE ONE-DIMENSIONAL PROBLEM

The one-dimensional problem in multi-variable optimization problem is employed for the determination of the step size after the search direction has been identified. It is best to understand these terms by recognizing the generic algorithm for unconstrained optimization. In order to focus the discussion, rather than the standard objective function representation, a specific one, Example 5.3, is introduced.

Example 5.3

$$\text{Minimize}\quad f(x_1, x_2, x_3) = (x_1 - x_2)^2 + 2(x_2 - x_3)^2 + 3(x_3 - 1)^2 \qquad (5.16)$$

A cursory glance at Example 5.3 indicates that the solution is at $x_1 = 1$; $x_2 = 1$; $x_3 = 1$; and the minimum value of f is 0. You can confirm the solution using the FOC and SOC.

Generic Algorithm (A 5.5): The generic algorithm is an iterative one and is also referred to as a *search algorithm*, as the iterations take place by moving along a *search direction*. These search directions can be determined in several ways. The algorithm without any convergence/stopping criteria can be expressed as

Step 1: Choose X_0 (initial guess)

Step 2: For each iteration i (starting $i = 0$)

Determine search direction S_{i+1} vector

Step 3: Calculate $\Delta X_{i+1} = \alpha_{i+1} S_{i+1}$

Note: ΔX_{i+1} is now a function of the scalar α_{i+1} as S_{i+1} is known from Step 2.

α_{i+1} is called the step size. as it establishes the length of ΔX_{i+1}

$X_{i+1} = X_i + \Delta X_{i+1}$

α_{i+1} is determined by minimizing $F(X_{i+1})$

This is referred to as one-dimensional step size computation.

$i \leftarrow i + 1$;

Go to Step 2.

Application of Generic Algorithm (A 5.5) to Example 5.3:

Step 1: $X_0 = \begin{bmatrix} 0.2 \\ 0.4 \\ 0.6 \end{bmatrix}$ (assume)

Step 2: $S_1 = \begin{bmatrix} 0.4 \\ 0.4 \\ 1.6 \end{bmatrix}$ (assume—this is negative of the gradient of f at X_0)

Step 3: $\Delta X_1 = \alpha_1 \begin{bmatrix} 0.4 \\ 0.4 \\ 1.6 \end{bmatrix}$; $\quad X_1 = \begin{bmatrix} 0.2 \\ 0.4 \\ 0.6 \end{bmatrix} + \alpha_1 \begin{bmatrix} 0.4 \\ 0.4 \\ 1.6 \end{bmatrix} = \begin{bmatrix} 0.2 + 0.4\alpha_1 \\ 0.4 + 0.4\alpha_1 \\ 0.6 + 1.6\alpha_1 \end{bmatrix}$

$$f(X_1) = f(\alpha_1) = (-0.2)^2 + 2(-0.2 - 1.2\alpha_1)^2 + 3(-0.4 + 1.6\alpha_1)^2$$
(5.17)

Minimizing $f(\alpha_1)$ the value of $\alpha_1 = 0.1364$. This is a one-dimensional optimization problem, even though there are three design variables.

$X_1 = \begin{bmatrix} 0.2525 \\ 0.4545 \\ 0.8182 \end{bmatrix}$; $\quad f = 0.4036$; $\quad i = 1$;

Go to Step 2.

Although the solution above is obtained analytically because it was a simple linear expression, the Polynomial Approximation or the Golden Section method could have been used instead.

5.4 ADDITIONAL EXAMPLES

In this section, additional examples of single-variable optimization problems are explored. The available code for the polynomial approximation and golden section should help solve any and all single-variable optimization problems. This is largely left to the reader to exploit. In this section only extensions, modifications, or creative application of the one-dimensional optimization problem are considered. First, Example 5.3 is revisited to modify the Golden Section method so that it can be used to calculate the step size in multi-dimensional problems. Example 5.4 is a solution to the Blassius problem. This is an example of the exact solution to the Navier–Stokes equation for flow over a flat plate. Example 5.5 is an examination of the inequality constraint in Example 5.2 by the Golden Section method so that equality and inequality constraints can be handled.

5.4.1 Example 5.3 – Golden Section Method for Many Variables

It was suggested in the previous section that the solution in (5.17) can be obtained numerically using the Golden Section or the Polynomial Approximation methods. In this subsection, we will apply the Golden Section method. In order to make it useful for the subsequent chapters, we will develop a **vector version** of the method rather than use the single-variable method developed earlier. An equivalent development with respect to the Polynomial Approximation method is left as an exercise for the student.

Example 5.3

$$\text{Minimize}: \quad f(x_1, x_2, x_3) = (x_1 - x_2)^2 + 2(x_2 - x_3)^2 + 3(x_3 - 1)^2 \quad (5.16)$$

MATLAB *Code:* The Golden Section method requires the establishment of the upper bound. Therefore, two functions will need to be changed to handle design vectors and the search direction vectors. We will call the files **UpperBound_nVar.m and GoldSection_nVar.m**. The modifications are not very challenging and can be inferred by comparing the usage of the two functions.

For the *single* variable:

Usage:

$$\texttt{UpperBound\_1Var('Example5\_1',0,1,10)} \quad (5.14)$$

For *many* variables

Usage:

$$\text{UpperBound\_nVar ('Example5\_3',x,s,0,1,10)} \qquad (5.18)$$

x: current position vector or design vector

s: prescribed search direction

Example5_3.m returns the value of the function corresponding to a vector of design variables. This implies that the function call to **Example5_3.m** can only take place after the design variables are evaluated from the value of step size applied along the prescribed search direction from the current location of the design.

A similar change to the code for applying the Golden Section method must take effect, since the method passes information to the upper bound calculation. For the *single* variable

Usage:

$$\text{Value = GoldSection\_1Var('Example5\_1',0.001,0,1,10)} \qquad (5.15)$$

For *many* variables

Usage:

$$\text{Value = GoldSection\_nVar ('Example5\_3',0.001,x,s,0,1,10)}$$
$$(5.19)$$

Please visit the code to see the details.

Solution of Example 5.3: The code **GoldSection_nVar.m** was run from the Command window using the following listing:

```
>> x = [0.2 0.4 0.6];
>> s = [0.4 0.4 1.6];
>> GoldSection_nVar('Example5_3',0.001,x,s,0,1,10)
```

The start iteration, with the first row being α and second row being f, is

```
start
alphal(low)    alpha(1)    alpha(2)    alpha{up)
          0      0.0764      0.1236      0.2000
     0.6000      0.4416      0.4054      0.4464
```

The final iteration is

```
iteration      11
alphal(low)    alpha(1)    alpha(2)    alpha{up)
```

| 0.1358 | 0.1361 | 0.1364 | 0.1368 |
| 0.4036 | 0.4036 | 0.4036 | 0.4036 |

The values returned from running the program is

| 0.1361 | 0.4036 | 0.2545 | 0.4545 | 0.8178 |

which translates to

$$\alpha_1 = .1361; \quad f(\alpha_1) = 0.4036; \quad x_1 = 0.2545; \quad x_2 = 0.4545; \quad x_3 = 0.8178$$

which is same as the previous calculation.

5.4.2 Example 5.4 — Two-Point Boundary Value Problem

A real example for one-variable optimization can be found in the numerical solution of the laminar flow over a flat plate.[5] The problem, usually attributed to Blassius, represents an example of the exact solution of the formidable Navier–Stokes equation. The mathematical formulation, allowing for similarity solutions, is expressed by a third-order nonlinear ordinary differential equations with boundary conditions specified at two points—a two-point boundary value problem (TPBVP). What follows is the essential mathematical description. Interested readers can follow the detail in most books on fluid mechanics, including the suggested reference.

Mathematical Formulation:

$$ff'' + 2f''' = 0 \tag{5.20}$$

$$x = 0; \quad f(0) = 0; \quad f'(0) = 0 \tag{5.21a}$$

$$x = \infty; \quad f'(\infty) = 1 \tag{5.21b}$$

The solution, nondimensional velocity in the boundary layer, is obtained as

$$\frac{u}{U_\infty} = f'(x) \tag{5.22}$$

Solution to the (5.20) to (5.21) is largely through special iterative techniques. These techniques use the numerical integration procedure such as the Runge–Kutta method to integrate the system by guessing and adjusting the missing boundary conditions at the initial point, such that the final point boundary conditions are realized. There are numerical techniques to solve boundary value problems, but we will find the solution using the procedure for an initial value problem because we can formulate it as a single variable optimization problem. In the Blasius problem, this would imply the following:

$$f''(0) = \alpha; \text{ (this is our design variable)} \tag{5.23}$$

The objective function therefore will be

$$\text{Minimize:} \quad h(\alpha) : (f'(\infty) - 1)^2 \tag{5.24}$$

Unlike previous examples, the objective function is implicit, and it is not a function but the difference in values at the final point. The squaring of the difference ensures the objective is being driven to zero when the boundary conditions will be met. However, we can define a certain tolerance in the satisfaction of the conditions at infinity. Implied in the formulation is that $f'(\infty)$ is obtained by integrating the differential Equation (5.20). (These can be considered as differential constraints.) There is one other consideration to take into account. The integration methods require the problem to be expressed in state space form, which requires an nth order differential equation to be expressed as a system of n *first-order* equations. The conversion is fairly standard and is done through transformation and introducing additional variables. For this example:

$$y_1 = f$$
$$y_1' = f' = y_2$$
$$y_2' = f'' = y_3$$
$$y_3' = f'''$$

The optimization can be formulated as: Find α.

$$\text{Minimize:} \quad [y_2(\infty) - 1]^2 \tag{5.25}$$

using Runge–Kutta method (or any other method) to integrate

$$\begin{bmatrix} y_1 \\ y_2 \\ y_2 \end{bmatrix}' = \begin{bmatrix} y_2 \\ y_3 \\ -0.5y_1y_3 \end{bmatrix} ; \quad \begin{bmatrix} y_1(0) \\ y_2(0) \\ y_3(0) \end{bmatrix} = \begin{bmatrix} 0 \\ 0 \\ \alpha \end{bmatrix} \tag{5.26}$$

From previous results the value of $\alpha^* = 0.3326$. The upper limit of infinity can be replaced without compromise by a value of 6.0 because of asymptotic nature of the solution. Using this value as another design variable, we should be able to define a corresponding two-variable optimization problem. In the implementation to follow the power and convenience of MATLAB is readily apparent.

MATLAB Code: The code is available in **GoldenSection_Example5_4.m**. This file is a wrapper that is used to create the plot and print the iteration information. If you are not interested in the plot, you can just call the **GoldSection_1Var.m** from the Command line to get the problem solved. Four m-files are used; two of them are part of our toolkit and have been used before. One new file is used to define the example, which will call the MATLAB's **ode45.m** function to integrate the state equations that are defined in another file. You are encouraged to go over the following code, as there are some new ideas and new programs that can be used elsewhere. In **GoldenSection_Example5_4.m**, the following dependency can be seen:

```
▶ GoldSection_1Var('Example5_4',tol,0,1,10)
    ▶UpperBound_1Var(functname,lowbound,intvl,ntrials)
        ▶ Example5_4(x)
            ▶ ode45('Ex5_4_state',tintval,bcinit);
                ▶ Ex5_4_State(t,y)
```

Figure 5.6 plots the state variables. Note that the boundary conditions are satisfied at $x = 6$.

The start iteration is

```
iteration      0
              lower      left      right      upper
alpha        0.0000    0.3820    0.6180    1.0000
    f        1.0000    0.0095    0.2632    1.1782
```

The final iteration is

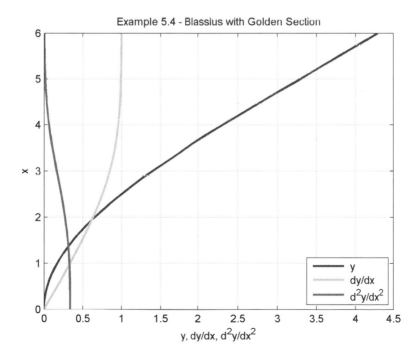

Figure 5.6 Graphical Solution: Example 5.4.

```
iteration      19
                lower      left      right      upper
    alpha      0.3325    0.3326     0.3326     0.3326
        f      0.0000    0.0000     0.0000     0.0000
```

The final solution for the missing boundary condition at the initial point corresponds to published value of 0.3326.

This is not a trivial problem, though the way we have approached it in a simple way. The Golden Section has produced an impressive solution to the TPBVP. We have also used one-dimensional minimization creatively. We are now pretty advanced in our use of MATLAB to solve problems. We are creating tools that can be reused. We can recover information and print them if necessary. We can create plots to display useful information.

5.4.3 Example 5.5 – Root Finding with Golden Section

The application of Golden Section method to the root-finding problem is important to accommodate equality and inequality constraints. Consider the following example.

Example 5.5

$$g(\alpha) : \alpha^2 - 7\alpha - 0.5 = 0 \tag{5.27}$$

This is the inequality constraint of Example 5.2 except that it is transformed to an equality constraint. In this form the α necessary to make it into an active constraint can be established. The simplest way to accomplish this is to convert this to a minimization problem for which the numerical techniques have already been established. Squaring (5.27) ensures that the minimum value would be zero (if it exists). Therefore solution to (5.27) is the same as

$$\text{Minimize:} \quad f(\alpha) : [\alpha^2 - 7\alpha - 0.5]^2 \tag{5.28}$$

which can be handled by **GoldSection_1Var.m**. Although this was quick and painless, there are some attendant problems using this approach. First, the nonlinearity has increased. This is usually not encouraged in numerical investigations. Second, the number of solutions has possibly been increased.

***Modification to* Upperbound_1Var.m:** Figure 5.7 plots both (5.27) and (5.28). Since we are dealing with the minimum of f, notice that f starts to increase at the start point of $x = 0$ and increases till around $x = 3.5$. This means that we have to change our scanning strategy since we are expecting f to decrease at the start of scanning. The next point that we accept should be such that the slope of f at the point must be zero, before we can start scanning. This is done by the piece of code in **UpperBound_1Var.m** (that was not necessary for earlier problems):

```
%%%%%%%%%%%%%%%%%%%%%%%%%%%%%%%%%%%%%%%%%%%%%%%%%%%%%%%%%%%%%%%%
%%%   for a minimum  and to start the scanning
%%%   we need the initial point to have a negative slope
%%%    this additional code should assist
% modification for a positive slope
% check if initial slope is positive
%%%%%%%%%%%%%%%%%%%%%%%%%%%%%%%%%%%%%%%%%%%%%%%%%%%%%%%%%%%%%%%
fstart = feval(functname,a0);
fnext = feval(functname,(a0 + 0.1*das));
slope =(fnext-fstart)/(0.1*das);
if (slope > 0)
    for i = 2:ntrials
        astart = a0 + i*das;
        anext = astart + 0.1*das;
        fstart = feval(functname,astart);
        fnext = feval(functname,anext);
        newslope = (fnext-fstart)/(anext-astart);
        if (newslope < 0)
            a0= astart;
            break;
        end
    end
end
```

MATLAB Code: The file **GoldenSection_Example5_5.m** is a wrapper that is used to create Figure 5.7, Graphical Description, Example 5.5, and print the iteration information to the Command window. You can directly call

```
GoldSection_1Var('Example5_5',0.001,0,1,20)
```

from the Command line. The values at the iterations are

```
start
                lower     left       right      upper
alpha    0.0000    3.0558     4.9442     8.0000
   f     0.2500    157.5690   113.7245   56.2500


iteration      23
                lower     left       right      upper
alpha    7.0707    7.0707     7.0707     7.0708
   f     0.0000    0.0000     0.0000     0.0000
```

The value of $x = 7.07$, where g is zero can be confirmed by the Figure 5.6. You can see the increase in the nonlinearity of $f = g^2$ in the figure, too.

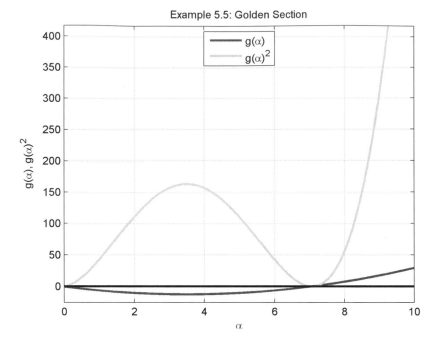

Figure 5.7 Graphical Solution: Example 5.5.

REFERENCES

1. Burden, R.L. and J.D. Faires. *Numerical Analysis*, 4th Ed. Boston: PWS_KENT Publishing Company.

2. Hosteller, G.H., M.S. Santina, and P.D. Montalou. *Analytical, Numerical, and Computational Methods for Science and Engineering*. Englewood Cliffs, NJ: Prentice Hall.

3. Vanderplaats, G.N. *Numerical Optimization Techniques for Engineering Design*. New York: McGraw-Hill Series in Mechanical Engineering, 1984.

4. Huntley, H.E. *The Divine Proportion: A Study in Mathematical Beauty*. New York: Dover Publications, 1970.

5. Schlichting, H. *Boundary-Layer Theory*. New York: McGraw-Hill. 1979.

PROBLEMS

5.1 Solve Example 5.1 using a vector version of *fsolve* using 10 starting points. Filter only the solutions that are a minimum and that meet side constraints. Identify the global minimum. For extra credit, place a marker at the minimum points.

5.2 Solve Example 5.2 using a vector version of *fsolve* using 10 starting points. Filter only the solutions that are a minimum, that are feasible, and that meets side constraints. Identify the global minimum. For extra credit, place a marker at the minimum points.

5.3 Reprogram the Newton–Raphson code so that the Newton–Raphson method is called by **Newton_Raphson_Example5_1.m** generically. The function and the derivatives are computed in different programs and are passed to the Newton–Raphson solver. Use numerical computation of derivatives instead of symbolic computation. *Hint*: See forward and central difference schemes for derivative computation. Verify using Example 5.1.

5.4 Second derivative computation is considered expensive. Verify if the Newton–Raphson technique will work for Example 5.1 if the second derivative during each iteration is unchanged from its starting value, while the first derivatives are evaluated numerically. Verify with Example 5.1.

5.5 Extend Example 5.1 to include two inequality constraints for which a solution exists. Display the functions graphically and identify the results.

5.6 Extend Example 5.1 to include two inequality constraints for which there is no solution. Display function graphically.

5.7 Identify the value of α where the constraint (5.2b) becomes active using the Newton–Raphson technique.

5.8 Implement a bisection procedure for the minimization of the function of one variable. Apply it to solve Example 5.1.

5.9 Verify that the bounds on the variable are decreasing by the same ratio in the Golden Section procedure. Why is this better than the bisection method?

5.10 How would you set up the Golden Section method to determine the *zero* of the function instead of the minimum directly? Apply it to Example 5.5.

5.11 As part of your own toolbox, develop code to combine the Golden Section and the Polynomial Approximation to find the minimum of the function. Apply it to Examples 5.1 and 5.4.

5.12 Obtain the solution to the Blasius equation using the Polynomial Approximation method.

5.13 Solve Problem 1.28 with your choice of parameters.

5.14 Solve Problem 1.29 with your choice of parameters.

5.15 5.15 Solve Problem 1.30 with your choice of parameters after including a suitable constraint. Explain your constraint.

5.16 5.16 Solve Problem 1.15 with your choice of parameters.

6

NUMERICAL TECHNIQUES FOR UNCONSTRAINED OPTIMIZATION

This chapter illustrates many numerical techniques for multivariable unconstrained optimization. Although unconstrained optimization is not a common occurrence in engineering design, nevertheless the numerical techniques included here demonstrate interesting ideas and also capture some of the early intensive work in the area of design optimization[1]. They also provide a vehicle to solve constrained problems after they have been transformed into an unconstrained one *(indirect methods–Chapter 7)*. Contemporary methods in global optimization (*Chapter 9*) still have difficulty in handling constraints directly.

We had used a brute force technique to find the minimum in Chapter 4 because of the considerable desktop power at everybody's disposal. It was called *scanning*, wherein we just picked points in order and checked if they satisfied the constraint and retained the smallest value that was feasible. While the procedure is unsophisticated, it is unmatched in the quality of the solution. The time increases exponentially with the number of variables, as well as the level of tolerance of the solution. It is a formidable approach at least to approximately locate the minimum. The use of nongradient techniques *(Section 6.2)* in this chapter continues this idea of harnessing the ubiquitous availability of incredible desktop computing power. The days of solving problems serviced by mainframe computers are long gone unless it can use parallel architecture, or the computations can be distributed among clusters of computers, or grid computation can be used. Today's desktop can run programs that require limited programming skills. The desire for global optimization has brought into focus numerical techniques that are largely heuristic, have limited need for sophisticated gradient-based methods, require limited programming resources, but take a lot of execution time. Personal desktops that are underused are an excellent vehicle for this implementation. While the techniques may compromise on elegance and sophistication,

290

it is important to recognize the opportunity to engage idle desktop resources to greater usefulness.

6.1 PROBLEM DEFINITION

The unconstrained optimization problem requires only the objective function. The book has chosen to emphasize an accompanying set of side constraints to restrict the solution to an acceptable design space/region. The numerical techniques in this chapter ignore the side constraints in laying out the method because it assumes that a single minimum exists in the interior of the design space. It is the responsibility of the designer to include the side constraints as part of the exploration of the optimum. The problem for this chapter can be defined as follows:

$$\text{Minimize} \quad f(X); \quad [X]_n \tag{6.1}$$

$$\text{Subject to:} \quad x_i^l \le x_i \le x_i^u; \quad i = 1, 2 \ldots n \tag{6.2}$$

6.1.1 Example 6.1

A two-variable problem is used to illustrate all of the algorithms in this chapter. This provides a geometric support to the numerical techniques. The two variables allow us to express the techniques in vector notation, which then represents the approach for any number of variables without any extra effort. Most books, including the previous edition of this one, use simple polynomial problems because they are well behaved and things happen as expected without surprises. This time we will follow a nontrivial problem with both minimum and maximum present within the design space, typical of real problems, so that we can get feedback if there is any problem with the technique. Real problems are not usually well behaved and will require intervention, even with a good code.

$$\text{Minimize} \quad f(X) = f(x_1, x_2) = 3 \cdot \sin(0.5 + 0.25x_1x_2) \cdot \cos(x_1) \tag{6.3}$$

$$\text{Subject to:} \quad 0 \le x_1 \le 5; \quad 0 \le x_2 \le 8; \tag{6.4}$$

6.1.2 Graphical Solution

Figure 6.1 displays the graphical solution to the problem represented by (6.3). It is generated by running **GraphicalSolution_Example6_1.m**.* The solution is in the interior of the feasible region at

$$x_1^* = 3.1416; \quad x_2^* = 1.3634; \quad f^* = -3.0 \tag{6.5}$$

Figure 6.1 also illustrates the presence of a minimum and a maximum in the interior of the design region. When seen in color there also appears to be additional local minimums on the boundary at $x_2 = 8$.

*Files to be downloaded from the web site are indicated by boldface courier type.

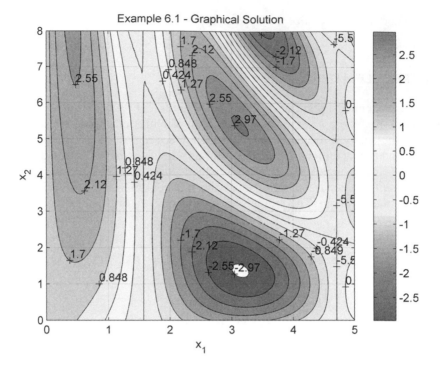

Figure 6.1 Graphical solution: Example 6.1.

6.1.3 Necessary and Sufficient Conditions

The necessary and sufficient conditions for unconstrained problem (also called the Kuhn-Tucker conditions) in Example 6.1 require the gradients must vanish at the solution:

$$\frac{\partial f}{\partial x_1} = 0.75 \cos(0.5 + 0.25 x_1 x_2) x_2 \cos x_1 - 3 \sin(0.5 + 0.25 x_1 x_2) \sin x_1 = 0$$

$$(6.6a)$$

$$\frac{\partial f}{\partial x_2} = 0.75 \cos(0.5 + 0.25 x_1 x_2) x_1 \cos x_1 = 0 \tag{6.6b}$$

Equations (6.6) look difficult to solve. A little persistence can be rewarding. Three possible solutions to the problem can be deduced from (6.6b).

$$(i) \quad x_1 = 0 \tag{6.7a}$$

$$(ii) \quad \cos(x_1) = 0 \tag{6.7b}$$

$$(iii) \quad \cos(0.5 + 0.25 x_1 x_2) = 0 \tag{6.7c}$$

First we will consider what the solution x is (6.7c)

$$0.5 + 0.25x_1x_2 = \pm n\frac{\pi}{2}$$

Substituting in (6.6a) will require

$$x_1 = \pm m\pi$$

where m are integers. For a solution in the positive quadrant:

$$x_1 = \pi = 3.1416; \quad x_2 = \frac{0.5\pi - 0.5}{0.25\pi} = 1.3633; \tag{6.8}$$

This solution will be verified through MATLAB using symbolic and numerical calculations.

MATLAB *Code:* The code in **GraphicalSolution_Example6_1.m** will also attempt to solve Example 6.1. The symbolic approach yields three different analytical solutions for the FOC, but none of them satisfy the side constraints. This information is printed to the Command window and is not recorded here. The numerical solution, by contrast, depends on the starting value, and will find the solution that matches (6.8)

```
Numerical solution

Initial Guess:  x1 =   2.0000    x2 =   2.0000
Final Solution: x1 =   3.1416    x2 =   1.3634
f = -3.0000
eigenvalues:   1.5014,   3.6977
```

The SOC is also applied to the numerical solution to verify it is a minimum. The eigenvalues are both positive, suggesting that the Hessian matrix is positive definite. The numerical solution requires the file **FOC_Example6_1.m**.

6.1.4 Elements of a Numerical Technique

The elements of a typical numerical technique can be associated with the *generic algorithm* introduced during the discussion of the relevance of the one-dimensional optimization for multi-variable problems *(Chapter 5)*. This algorithm is iterative and is also referred to as a *Search Algorithm*, as the iterations take place by moving along a *search direction* or a *search vector* during an iteration. These search directions can be determined in many ways. The different techniques that are presented in this chapter primarily differ in how the search direction is established. The remaining procedures, except for the convergence/stopping criteria, are the same for almost all of the methods in this chapter. The general algorithm can be expressed as:

General Algorithm (A 6.0):

Step 1: Choose X_i (initial design).

Step 2: For each iteration i:
Determine search direction vector S_i
(It would be nice if the objective decreased along this direction).

Step 3: Calculate $\Delta X_i = \alpha_i S_i$ (change in design).
Note: ΔX_i is now a function of the scalar α_i as S_i is known from Step 2
α_i is called the step size as it establishes the length of ΔX_i
α_i is determined by Minimizing $f(X_{i+1})$, where
$$X_{i+1} = X_i + \Delta X_i$$
Here the convergence criteria (FOC/SOC), and the stopping criteria (Is design changing? Exceeding iteration count? etc.) can be checked.
$i \leftarrow i + 1$;
Go to Step 2

In this chapter, this algorithmic structure will be basic for all the methods. The one-dimensional optimization will be implemented using the golden section method. Example 6.1 will be used to develop the numerical methods.

6.2 NUMERICAL TECHNIQUE – NONGRADIENT METHODS

These methods are also called *zero-order* methods. The order, refers to the order of the derivative of the objective function needed to establish the search direction during any iteration. Zero order signifies no derivatives are used, which in turn implies that only function values are used to establish the search vector[2]. There is another important significance that accompanies the missing derivative computation–*the FOC cannot be applied* in these methods since they will require computation of the derivatives. Only the changes in the objective function or the design variables can provide convergence and/or stopping criteria.

Four techniques are included in this section. The first is a scanning technique that is implemented in a nesting fashion. It keeps zooming in on the minimum by establishing the solution more precisely. We will call it *Scan and Zoom*. The second is a heuristic one based on random search directions and is called the *Random Walk*. Most current methods of global optimization are heuristic. The third one, the *Pattern Search*, cycles through a search direction based on each variable. The last is the Powell's method which has the property of *Quadratic Convergence*.

6.2.1 Scan and Zoom

Unlike the remaining methods in this section the Scan and Zoom method is not built around the general algorithm. It can be considered a brute-force technique since the method will go through a disciplined search for the minimum in the design region. To establish the solution with significant accuracy, it would require the solution be searched at extremely large number of points in the domain with significant increase in time. In this method, the solution is searched over a coarse grid but over several levels. Each level implies a finer grid than the previous level. The best value, at the end of scanning in a particular level is used to anchor the design space at the next level. This becomes the center of the search for the next level. The search domain is reduced to half the extent of the previous level. This allows a brief violation of the side constraints to track solutions at the boundary. This also implies that you are basically zooming around the best solution at the current level. Each level will give a better estimate of the solution in terms of significant figures. A prescribed number of levels can be searched. It is possible to stop after a prescribed tolerance in the value of f. The algorithm is iterative, with a diminishing design region and an improved estimate for the minimum.

Algorithm: Scan and Zoom (A 6.1):

Step 1: Choose X_{ci} (**center of design region**), $\pm \Delta X_i$: extant of the of the design region. nL: number of levels, nP: number of points along each variable.

Step 2: For each Level i
Scan and store the best objective and location
$f_{i,\min}, X_{i,\min}$

Step 3: Calculate
$\Delta f = abs(f_{i,\min} - f_{i-1,\min})$
If Δf < tolerance: exit
Else: $X_{ci} = X_{i,\min}; \Delta X_i = 0.5 * \Delta X_{i-1}$
$i \leftarrow i + 1$
Go to Step 2

The algorithm is straightforward and unsophisticated. At each level it will define the design space around the centric point, use nP points linearly distributed along each variable, scan for the minimum, store the best value, and move on to next level until it meets the stopping criteria or the number of levels. The shrinking design space is both flexible and improves in accuracy with each level.

Example 6.1:

Minimize $f(X) = f(x_1, x_2) = 3 \cdot \sin(0.5 + 0.25x_1x_2) \cdot \cos(x_1)$ (6.3)

Subject to: $0 \leq x_1 \leq 5; \quad 0 \leq x_2 \leq 8;$ (6.4)

MATLAB **Code:** The algorithm is implemented in scan_Zoom_Example6_1.m. The number of points per level was 11 for each variable. The number of levels was set at 20. The tolerance in the change in the objective was 1.0e008. The results after the first and the last level are copied from Command window.

Example 6.1

```
Zoom Level ( 1 ):
x1* =   3.0000000
x2* =   1.6000000
 f* =  −2.9452222
Zoom Level (17):  (final)
x1* =   3.1415939
x2* =   1.3633781
 f* =  −3.0000000
```

The solution matches the analytical solution (6.8). The more amazing experience with this algorithm is that it took about 0.5 seconds *cpu* time on the desktop and about 0.012 seconds on a laptop in spite of the printing to the screen. These are contemporary machines and based on using MATLAB in interpreted mode. I am sure even with a lot more variables, but with executed code, the time required for establishing the solution will be nominal at current computational resources. This solution will also be the best solution in the design region. Now, that is certainly a welcome qualification. You are strongly encouraged to go over the code to see the simple manner in which the algorithm is implemented.

6.2.2 Random Walk

The search direction is a random direction. Random numbers and sets of random numbers are usually available through software. MATLAB includes the ability to generate random numbers and matrices. Most computer generated random numbers are called *pseudo random numbers*, as they cycle after a sufficient number of them has been created. The one-dimensional step size computation in this methos is done using the Golden Section method.

Algorithm: Random Walk (A 6.2):

Step 1: Choose X_0, N (number of iterations)
        Set $i = 1$
Step 2: For each iteration i
        S_i = Random vector
Step 3: $X_{i+1} = X_i + \alpha_i S_i$
        α_i is determined by Minimizing $f(X_{i+1})$
        $i \leftarrow i + 1$
        If $i < N$ *Go* to Step 2
        else Stop (iterations exceeded)

If the function increases along the search direction then a new direction is chosen. Prior to developing the code please see the help on random numbers in

MATLAB *(>> help rand)*. Notice that the random search direction vector created by MATLAB always points in the positive quadrant. Even though the design space is in the positive quadrant, the search for the next step can be in any direction. To allow for all directions, the elements of the search vector will switch sign based on a test of random number.

MATLAB Code: The code `RandomWalk_Example6_1.m` is a wrapper code that will call the function m-file `RandomWalk.m` to solve Example 6.1 using the Random Walk algorithm. Figure 6.2 tracks the iterations on the contour plot. It will also print the iteration history to the Command window. You can also call `RandomWalk.m` from the Command line directly and recover values for the final iteration if the iteration details are not necessary. This book strongly encourages translation of all algorithms into functional code components. You are strongly encouraged to examine the code as there are many more details then are included in this section. The code is well commented at most places. Many features of MATLAB are exploited to graphically describe the movement of the iterations in the design space. For its *relative simplicity* as well as total *lack of sophistication* the Random Walk method functions incredibly well, sometime approaching close to the solution in 30 iterations. Many current methods for global optimization use heuristic techniques that use random direction. The algorithm implemented here also includes these following additional features:

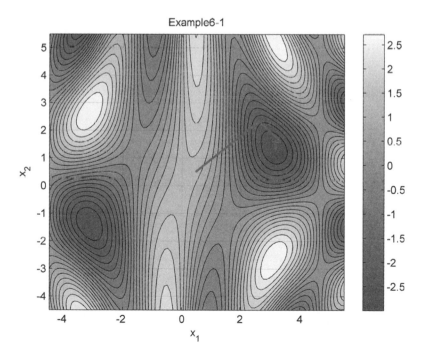

Figure 6.2 Random walk: Example 6.1.

- For two-variable problems the contours of the objective function are drawn.
- For two-variable problems the successful iterations are tracked on the contour plot. Iteration number can be printed on the tracks if needed.
- The multi-variable Golden Section method (developed in Chapter 5) is used for the one-dimensional step size computation.
- The multi-variable upper bound calculation (developed in Chapter 5) is used for scanning the upper bound.
- The search directions that cause the function to increase are ignored.
- The design variable and function value for the iterations is saved and available in a global statement for additional processing if needed.
- The function will return the values of the design at the last iteration.

Usage:

$$\text{RandomWalk('Example6\_1'[0.5\ 0.5],30,0.001,0,1,20)} \qquad (6.9)$$

Input:

'Example6_1' :	The file **Example6_1.m** where the objective function is described
[0.5 0.5]	starting design vector (dvar0)
0.001:	tolerance fro the golden section procedure (tol)
0	initial stepsize for scanning upper bound (lowbound)
1	step interval for scanning in upper bound (intvl)
20	number of scanning steps for upper bound (ntrials)

Output:

[xs, fs]:	Values of design variables, value of function
xs:	A vector of $n + 1$ values of design variables at the last iteration.
fs:	Value of objective function

The iterations are printed to the Command window. The truncated history is reproduced:

```
iterations    x(1)          x(2)          f

    1        0.50000       0.50000       1.40405
    2        0.50000       0.50000       1.40405
    3        2.63009       2.32132      -2.34928

   30        3.14155       1.36386      -3.00000
Total cpu time (s) =  1.6563
```

These values compare well to the actual solution (6.8). This is remarkable for a method that has no sense of preferred direction. You can rerun the method on the same problem with identical information and often arrive at a different solution, because this example has multiple solutions and the side constraints are not checked in this implementation. Once more we have reused code that we had developed earlier.

6.2.3 Pattern Search

The Pattern Search method is a minor modification to the *Univariate* method with a major impact. In the Univariate method, not presented, also known as *Cyclic Coordinate Descent* method, each design variable (considered a coordinate) provides a search direction. This is also referred to as a coordinate direction, and is easily expressed through the unit vector for that coordinate. The search direction is cycled through the number of variables in an orderly manner. Therefore, each cycle will have used *n (number of design variables)* iterations for each of the *n* search directions. It can be shown by application that for problems with considerable nonlinearity, the Univariate method tends to get locked into a zigzag pattern of smaller and smaller moves as it approaches the solution.

The Pattern Search procedure attempts to disrupt this zigzag behavior by executing *one additional* iteration for each cycle. In each cycle, at the end of *n* *Univariate* directions, the *n* + 1 search direction is assembled as the sum of the product of the previous *n* search directions and the optimum value of the stepsize for that direction. A one-dimensional optimal stepsize is computed, and the next cycle of iteration begins.

Algorithm: Pattern Search (A 6.3):

Step 1: Choose X_1, N_c (number of cycles)
$f_c(1) = f(X_1); X_c(1) = X_1$
$\varepsilon_1, \varepsilon_2$: tolerance for Stopping criteria
Set $j = 1$ (initialize cycle count)

Step 2: For each cycle j
For $i = 1, n$
$S_i = \hat{e}_i$ (univariate step)
$X_{i+1} = X_i + \alpha_i S_i$
α_i is determined by Minimizing $f(X_{i+1})$
end of For loop
$S_j = \sum_{i=1}^{n} \alpha_i^* S_i \equiv X_{n+1} - X_1$ (Pattern step)
$X_j = X_{n+1} + \alpha_j S_j$ (best stepsize)
$X_c(j+1) \leftarrow X_j; f_c(j+1) = f(X_j)$ (store cycle values)

Step 3: $\Delta f = f_c(j+1) - f_c(j); \Delta X = X_c(j+1) - X_c(j)$
If $|\Delta f| \leq \varepsilon_1; stop$
If $\Delta X^T \Delta X \leq \varepsilon_2; stop$
If $j = N_c; stop$

$$X_1 \leftarrow X_j; f(X_1) \leftarrow f(X_j)$$
$$j \leftarrow j + 1$$
Go to Step 2

The Pattern Search implemented here method has a few additional stopping criteria. These are based how much the function is decreasing for each cycle and how much change is taking place in the design variable itself. For the latter this information is based on the length of the change in the design vector for the cycle. Once again the approach is simple and the implementation is direct. Programming such a technique will not be too forbidding. The file for the objective function is the same as that used in the previous examples.

Example 6.1:

Minimize: $f(X) = f(x_1, x_2) = 3 \cdot \sin(0.5 + 0.25x_1x_2) \cdot \cos(x_1)$ (6.3)

Subject to: $0 \le x_1 \le 5$; $0 \le x_2 \le 8$;   (6.4)

MATLAB **Code:** `PatternSearch.m` is implemented very differently from the other algorithms that have been translated into code so far. It is a stand alone program. There are prompts for user inputs with appropriate default values in case the user decides not to enter a value. The Command window must be visible when you run this program so you can read the prompts. The one- dimensional step-size computation is implemented by the `GoldSection_nVar.m.` The interesting features are listed here:

- It is an interactive program.
- The file containing the objective function is selected through a list box.
- Several program control parameters are entered from the keyboard after a suitable prompt, indicating default values.
- Default values are used if user chooses not to enter the values.
- Initial design vector *must be entered* by the user.
- The iterations are printed after program completes.
- Since the upper bound calculations are based on positive values of stepsize, the search direction is reversed if the returned stepsize is the same as the lower bound.
- Iteration counts are recorded.
- For two-variable problems, the objective function contours are plotted and the initial iterations are tracked on the contour plot.
- Only the first nine or less iterations are traced on the contour plot for clarity. The univariate and the pattern steps are color coded on the plot (not seen in the book).

Usage:

PatternSearch   (6.10)

Output (in MATLAB Command window):   With default program control values, the starting vector of [0.5, 0.5], and applied to Example 6.1, the following
are written to the Command window.

```
» PatternSearch
The function for which the minimum is sought must be
 a MATLAB function M-File.  Given a vector dependent
variable it must return a scalar value.  This is the
function to be MINIMIZED. Please select function name
in the dialog box and hit return:

The function you have chosen is::Example6_1
 maximum number of cycles [1000]:

 convergence tolerance for difference in f[1e-8]:

 convergence tolerance on change in design x [1e-8]:
Input the starting design vector. This is mandatory
as there is no default vector setup. The length of your
vector indicates the number of unknowns.
Please enter it now and hit return
[0.5 0.5]

 The initial design vector [   0.50     0.50 ]
Convergence in f:      2.822E-012  reached in       6 cycles
Number of useful calls to the Golden Section Search
Method:    19
Total number of calls to the Golden Section Search
Method:    23

The values for x and f and stepsize are
    0.5000     0.5000     1.4041          0
    3.2389     0.5000    -2.3479     2.7393
    3.2389     1.3222    -2.9858     0.8222
    3.1662     1.3004    -2.9965     0.0268
    3.1564     1.3004    -2.9967     0.0102
    3.1564     1.3561    -2.9997     0.0567
    3.1554     1.3616    -2.9997     0.0988
    3.1420     1.3616    -3.0000     0.0134
    3.1420     1.3629    -3.0000     0.0013
    3.1417     1.3629    -3.0000     0.0255
    3.1417     1.3629    -3.0000     0.0000
    3.1417     1.3633    -3.0000     0.0004
    3.1417     1.3633    -3.0000     0.1341
    3.1416     1.3633    -3.0000     0.0001
```

3.1416	1.3634	−3.0000	0.0000
3.1416	1.3634	−3.0000	0.2208
3.1416	1.3634	−3.0000	0
3.1416	1.3634	−3.0000	0.0000
3.1416	1.3634	−3.0000	0.2076

As you scan the output, notice that for some iterations, only one design variable changes. Those are the univariate directions. After two univariate searches then we have the combined search, which is the pattern direction, after which the cycle repeats itself. The number of cycles for convergence for this example, this starting point, and these program control values is 6. The useful calls signify those iterations for which the step size is not zero, or the design vector moved closer to the solution. It can be noted that the difference between two iterations change lincarly. Such methods are expected to run for a large number of iterations. The solution is more accurate than the *Random Walk* method and compares excellently with (6.8). Figure 6.3 illustrates the algorithm graphically for the first few cycles. The univariate and the pattern directions are quite distinguishable. Readers are encouraged to go over the code and change the color to their preferences.

6.2.4 Powell's Method

If there were only one zero-order method that must be programmed, the overwhelming choice would be the *Powell's method*[3]. The principal reason for the

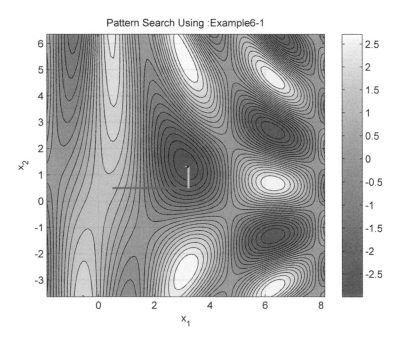

Figure 6.3 Pattern search: Example 6.1.

decision would be that it has the property of *quadratic convergence*. This property can be simply stated, "For a quadratic problem with n-variables convergence will be achieved in less than or equal to n Powell cycles." A *quadratic problem* is an unconstrained minimization of a function that is expressed as a *quadratic polynomial* – a polynomial with no term having a degree greater than two. Engineering design optimization problems are rarely described by a quadratic polynomial. This does not imply that you cannot use the Powell's method. What this means is that the solution should not be expected to converge quadratically. In practice, as the solution is approached iteratively, the objective can be approximated very well by a quadratic function, so that quadratic convergence property is realized in the computations in the neighborhood of the minimum.

The actual algorithm (**A 6.4**) is a simple (the word is not being used lightly) modification to the pattern search algorithm. In each cycle, instead of using univariate directions in the first n iterations, the search directions are obtained from the previous cycle. The new search directions are obtained by *left shifting* the directions of the previous cycle. In this way, a history of the previous search directions is used to establish the next. With this change, the Powell's method should converge in three iterations for Example 6.1, rather than the several listed in the previous section.

Algorithm: Powell Method (A 6.4):

Step 1: Choose $\mathbf{X}_1, \mathbf{N}_c$ (number of cycles)
$f_c(1) = f(\mathbf{X}_1); X_c(1) = X_1$
$\varepsilon_1, \varepsilon_2$: tolerance for Stopping criteria
Set $j = 1$ (initialize Powell cycle count)
For $i = 1, n$
$\mathbf{S}_i = \hat{\mathbf{e}}_i$ (univariate step)

Step 2: For each cycle j
For $i = 1, n$
If $(j >= 2)$ $S_i \leftarrow S_{i+1}$ (Powell shift)
$X_{i+1} = X_i + \alpha_i S_i$
α_i is determined by Minimizing $f(X_{i+1})$
end of For loop
$S_j^p = S_{i+1} = \sum_{i=1}^{n} \alpha_i^* S_i = X_{n+1} - X_1$ (Pattern step)
$X_j^p = X_{n+1} + \alpha_j S_j^p$
$X_c(j+1) \leftarrow X_j^p; f_c(j+1) = f(X_j^p)$ (store cycle values)

Step 3: $\Delta f = f_c(j+1) - f_c(j); \Delta X = X_c(j+1) - X_c(j)$
If $|\Delta f| \leq \varepsilon_1$; *stop*
If $\Delta X^T \Delta X \leq \varepsilon_2$; *stop*
If $j = N_c$; *stop*
$X_1 \leftarrow X_j$
$j \leftarrow j + 1$
Go to Step 2

Application of the Powell's Method: The translation of the algorithm into
MATLAB code is left as an exercise for the student. Step by step application to
Example 6.1 is shown here. Example 6.1 is reintroduced

Example 6.1:

$$\text{Minimize:}\quad f(X) = f(x_1, x_2) = 3 \cdot \sin(0.5 + 0.25x_1x_2) \cdot \cos(x_1) \qquad (6.3)$$

$$\text{Subject to:}\quad 0 \le x_1 \le 5; \quad 0 \le x_2 \le 8; \qquad (6.4)$$

Step 1: $X_1 = \begin{bmatrix} 0.5 \\ 0.5 \end{bmatrix}$; $f(X_1) = 2.7395$; $\varepsilon_1 = 1.0E - 08$; $\varepsilon_2 = 1.0E - 08$

$S_1 = \begin{bmatrix} 1 \\ 0 \end{bmatrix}$; $S_2 = \begin{bmatrix} 0 \\ 1 \end{bmatrix}$

$j = 1$

Step 2: $j = 1$

$i = 1$; $X_2 = \begin{bmatrix} 0.5 \\ 0.5 \end{bmatrix} + \alpha_1 \begin{bmatrix} 1 \\ 0 \end{bmatrix} = \begin{bmatrix} 0.5 + \alpha_1 \\ 0.5 \end{bmatrix}$

$f(\alpha_1) = 3 \sin(9/16 + 1/8\alpha) \cos(1/2 + \alpha)$; $\alpha_1^* = 2.7395$;

$X_2 = \begin{bmatrix} 3.2395 \\ 0.5 \end{bmatrix}$; $f(X_2) = -2.3479$

$i = 2$: $X_3 = \begin{bmatrix} 3.2395 \\ 0.5 \end{bmatrix} + \alpha_2 \begin{bmatrix} 0 \\ 1 \end{bmatrix} = \begin{bmatrix} 3.2395 \\ 0.5 + \alpha_2 \end{bmatrix}$

$\alpha_2^* = 0.8222$; $X_3 = \begin{bmatrix} 3.2395 \\ 1.3222 \end{bmatrix}$; $f(X_3 = -2.9856)$

$i = 3$: $S_3 = 2.7395 \begin{bmatrix} 1 \\ 0 \end{bmatrix} + 0.8222 \begin{bmatrix} 0 \\ 1 \end{bmatrix} = \begin{bmatrix} 2.7395 \\ 0.8222 \end{bmatrix}$

$X_4 = \begin{bmatrix} 3.2395 \\ 1.3222 \end{bmatrix} + \alpha_3 \begin{bmatrix} 2.7395 \\ 0.8222 \end{bmatrix}$;

$\alpha_3^* = -0.0271$; $X_4 = \begin{bmatrix} 3.1653 \\ 1.2999 \end{bmatrix}$; $f(X_4) = -2.9965$

Step 3: $|\Delta f| = 5.736 > \varepsilon_1$;   *continue*

$|\Delta X^T \Delta X| = 7.7437 > \varepsilon_2$;   *continue*

$j = 2$: One cycle over

Go to Step 2

Step 2: $j = 2$

$i = 1$;

$X_1 = \begin{bmatrix} 3.1653 \\ 1.2999 \end{bmatrix}$; $S_1 = \begin{bmatrix} 0 \\ 1 \end{bmatrix}$; $\alpha_1^* = 0.0533$;

$X_2 = \begin{bmatrix} 3.1653 \\ 1.3532 \end{bmatrix}$; $f(X_2) = -2.9992$

$i = 2$

$$X_2 = \begin{bmatrix} 3.1653 \\ 1.3532 \end{bmatrix}; \; S_2 = \begin{bmatrix} 2.7395 \\ 0.8222 \end{bmatrix}; \; \alpha_2^* = -0.0065;$$

$$X_3 = \begin{bmatrix} 3.1475 \\ 1.3478 \end{bmatrix}; \; f(X_3) = -2.9998$$

$i = 3$

$$X_3 = \begin{bmatrix} 3.1475 \\ 1.3478 \end{bmatrix}; \; S_3 = \begin{bmatrix} -0.0178 \\ 0.0479 \end{bmatrix}; \; \alpha_3^* = 0.3259;$$

$$X_4 = \begin{bmatrix} 3.1417 \\ 1.3634 \end{bmatrix}; \; f(X_4) = -3.0000$$

Step 3: $|\Delta f| = 0.0006 > \varepsilon_1$;   *continue*

        $|\Delta X^T \Delta X| = 0.0026 > \varepsilon_2$;   *continue*

Step 2: $j = 3$

        $i = 1$

$$X_1 = \begin{bmatrix} 3.1417 \\ 1.3634 \end{bmatrix}; \; S_1 = \begin{bmatrix} 2.7395 \\ 0.8222 \end{bmatrix}; \; \alpha_1^* = 3.76e - 05;$$

$$X_2 = \begin{bmatrix} 3.1417 \\ 1.3634 \end{bmatrix}; \; f(X_2) = -3.0000$$

The stepsize is very small and the values of the variables are the same as (6.8). We will stop further iterations here, even though we should have checked for the values at the end of the cycle.

The following observations can be made about the Powell's method:

- The method only took a little more than two cycles to converge.
- This was not a quadratic problem, and hence the two cycles are impressive but may be coincidence.
- For an algorithm that is very similar to the pattern search, it took four cycles *less* to converge. The extent of computation per cycle is the same as the pattern search but the number of iterations is significantly less.
- The negative stepsize is the search in the reverse direction.
- While not illustrated here, the maximum number of cycles to convergence is not dependent on the starting point for a quadratic problem

Conjugacy: The formal reason for the *quadratic convergence* property for Powell's method is that the search directions at the $n + 1$ iterations are *conjugate with respect to the Hessian matrix*. For a quadratic problem, the Hessian matrix is a constant matrix. If search directions S_i and S_j are conjugate directions then the conditions for conjugacy are

$$S_i^T[H]S_j = 0 \tag{6.11}$$

Verify if it is true for the example.

6.3 NUMERICAL TECHNIQUE–GRADIENT-BASED METHODS

In light of our definition of the zero-order method, the gradient- based methods can be referred to as *first-order* methods. The search directions will be constructed using the gradient of the objective function. Since gradients are being computed, the Kuhn–Tucker conditions (FOC) for unconstrained problems, $\nabla f = 0$, can be used to check for convergence. The SOC are usually never applied. One of the reasons is that it would involve the computation of an $n \times n$ second derivative matrix which is considered computationally expensive, particularly if the evaluation of the objective function requires a call to a finite element method for generating required information. This may not be an excuse in these days of practically inexhaustive computational power, but another significant reason for not calculating the *Hessian*, is that the existence of the second derivative in a real design problem is not certain, even though it is computationally possible or feasible.

These methods therefore require user vigilance to ensure that the solution obtained is a minimum rather than a maximum or a saddle point, particularly if the SOC is not implemented. A simple way to verify this is to perturb the objective function through small changes in the design variables around the solution and verify it is a local minimum. This brings up an important property of these methods–*they only find local optimums*. Usually, this will be close to the design where the iterations are begun. Before concluding the design exercise, it is necessary to execute the method from several starting points to explore if other minimums exist and select the best one by comparison. Needless to say, the bulk of existing unconstrained and constrained optimization methods belongs to this category.

Four methods are presented. The first is the *Steepest Descent* method. Although this method is not used in practice it provides an excellent example for understanding the algorithmic principles for the gradient based techniques. The second is the *Conjugate Gradient* technique which is a classical workhorse for industrial usage. It is used in numerical methods in several disciplines. The third and fourth belong to the category of *Variable Metric* methods, or *Quasi-Newton* methods. This category is the best among the first-order methods. They have been popular for some time, and will possibly stay that way for a long time to come. They require very little more effort than the basic steepest descent method.

The general problem is reproduced for convenience

$$\text{Minimize} \quad f(X); \quad [X]_n \tag{6.1}$$

$$\text{Subject to:} \quad x_i^l \leq x_i \leq x_i^u; \quad i = 1, 2 \ldots n \tag{6.2}$$

6.3.1 Steepest Descent Method

This method provides a natural evolution for the gradient-based techniques[4]. The *gradient* of a function at a point is the *direction* of the *most rapid increase in the value of the function* at that point. The *descent* direction can be obtained by reversing the gradient (or multiplying it by − 1). The next step would be to regard the descent vector as a *search direction*; after all, we are attempting to decrease the function through successive iterations. This series of steps give rise to the Steepest Descent algorithm.

Algorithm: Steepest Descent (A 6.5):

Step 1: Choose X_1, N (number of iterations)
$f_s(1) = f(X_1);$ $X_s(1) = X_1$ (store values)
$\varepsilon_1, \varepsilon_2, \varepsilon_3$: (tolerance for stopping criteria)
Set $i = 1$ (initialize iteration counter)

Step 2: $S_i = -\nabla f(X_i)$ (this is computed in Step 3 for subsequent iterations)
$X_{i+1} = X_i + \alpha_i S_i$
α_i is determined by Minimizing $f(X_{i+1})$
$X_s(i + 1) \leftarrow X_{i+1};$ $f_s(i + 1) = f(X_{i+1})$ (store values)

Step 3: $\Delta f = f_s(i + 1) - f_s(i);$ $\Delta X = X_s(i + 1) - X_s(i)$
If $|\Delta f| \leq \varepsilon_1;$ *stop* (function not decreasing)
If $\Delta X^T \Delta X \leq \varepsilon_2;$ *stop* (design not changing)
If $i + 1 = N;$ *stop*
If $\nabla f(X_{i+1})^T \nabla f(X_{i+1}) \leq \varepsilon_3;$ *convergerd*
$i \leftarrow i + 1$
Go to Step 2

Example 6.1:

Minimize: $f(X) = f(x_1, x_2) = 3 \cdot \sin(0.5 + 0.25 x_1 x_2) \cdot \cos(x_1)$ (6.3)

Subject to: $0 \leq x_1 \leq 5;$ $0 \leq x_2 \leq 8;$ (6.4)

MATLAB **Code:**   Once again we employ a wrapper script m-file that will initialize variables and call the Steepest Descent algorithm and print the iterations to the screen. This is **SteepestDescent_Example6_1.m**. The actual algorithm is in **SteepestDescent.m**. This m-file executes algorithm A.65. It uses the golden section and the upper bound scanning process. The features of this program are as follows:

- For two variables, it will draw the contour plot.
- For two variables, the design vector changes are tracked graphically in alternate color. You can see that the search directions are the gradients to the function at that point.
- The design variables, the function value, and the square of the gradient vector (convergence criteria), stepsize α at each iteration, and change in the design vector are displayed in the Command window at completion of the number of iterations. This is done through the wrapper file.

- The gradient of the function is numerically computed using first forward finite difference. The gradient computation is therefore automatic. No additional function m-file is needed.

The algorithm can be called without the wrapper function. It will only return the value of the final design. The usage is given here.

Usage:

```
SteepestDescent('Example6_1',[0.5 0.5], 20,1e-08,0,1,20)
```
(6.12)

Input:

'Example 6_1':	The file **Example6_1.m** defines objective function
[0.5 0.5]	Starting design vector (dvar0)
20	Number of iterations of the steepest descent method (niter)
1e−08:	Tolerance for the golden section procedure and this method (tol)
0	Initial stepsize for scanning upper bound (lowbound)
1	Step interval for scanning in upper bound (intvl)
20	Number of scanning steps for upper bound (ntrials)

Output:

[designvar, fsd]	Values of design variables, value of function
designvar:	A vector of $n + 1$ values of design variables at the last iteration.
fsd:	Value of objective function

The output in the Command window is partially reproduced to fit the page. The values of α and the change in design are not reproduced. First, notice the condition for the exit of the program. The design is not changing. It is due to the value of zero stepsize for the last search direction. The value of the design vector in the last two iterations match to five places after the decimal. If this accuracy is insufficient, then the tolerance must be reset in **GoldenSection.m.**

```
Exit: Design not changing

Example 6.1 (Steepest Descent)
*******************************

-The design vector, function value, KT condition etc.
    during the iterations

   iter  x(1)       x(2)        f         KT Cond
   ─────────────────────────────────────────────────

     1  0.50000   0.50000   1.40405    3.171e−001
```

```
2    2.14931   −0.43793   −0.42913   9.572e−001
3    3.16033    1.34155   −2.99930   2.820e−003
4    3.14088    1.35182   −2.99987   5.403e−004
5    3.14474    1.36002   −2.99998   9.840e−005
6    3.14180    1.36087   −2.99999   1.269e−005
7    3.14154    1.36339   −3.00000   3.100e−006
8    3.14154    1.36339   −3.00000   3.100e−006
```

```
Total cpu time (s)=  0.4063
```

Figure 6.4 represents the graphical solution with the tracking of the design variables. To see the third and higher iterations, you will have to zoom around the solution point. The zoom is active for the figure. You should be able to move the cursor into the region of the figure and click to keep zooming to the solution.

You can start from different points and may finally discover a point when the method stalls and zigzags in very small steps toward the solution. Figure 6.4 does not illustrate any stalling for the chosen starting point. In such instances, the Steepest Descent method is woefully inadequate compared to the Powell's method for the same starting point, even if the latter is a zero-order method. This can be

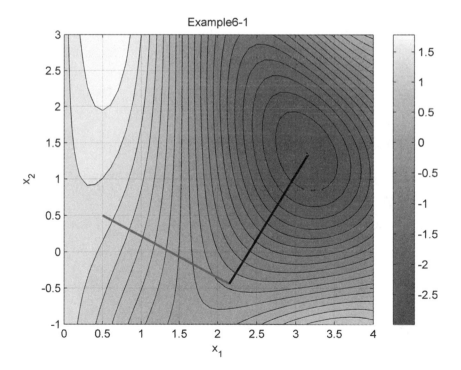

Figure 6.4 Steepest descent: Example 6.1.

experienced by solving one of the problems at the end of this chapter. This stalling property of the algorithm is the reason it is not used much for optimization. Stable methods are expected to overcome this behavior and proceed to the solution with acceleration as they get close to the minimum. The improvement of the Pattern Search over the Univariate method can be attributed to the former being able to break out of the zigzag steps by executing an extra iteration during the cycle. An iteration breaking out of the zigzag ladder (or preventing getting locked into one) is necessary to improve the method. Such a feature is implemented in the remaining methods of this section.

6.3.2 Conjugate Gradient (Fletcher–Reeves) Method

The conjugate gradient, originally due to Fletcher-Reeves, is a small modification to the Steepest Descent method with an enormous effect on performance[5]. This improvement mirrors the achievement in the transformation of the Pattern Search to the Powell's method. The most impressive gain is the property of *quadratic convergence* because the search directions are *conjugate* with respect to the Hessian matrix at the solution. Therefore, a quadratic problem in $n-$variables will converge in no more than n iterations.

Algorithm: Conjugate Gradient (A 6.6):

Step 1: Choose X_1, N (number of iterations)
$f_s(1) = f(X_1);$ $X_s(1) = X_1$ (store values)
$\varepsilon_1, \varepsilon_2, \varepsilon_3$: (tolerance for stopping criteria)
Set $i = 1$ (initialize iteration counter)

Step 2: *If* $i = 1, \mathbf{S_i} = -\nabla \mathbf{f(X_i)}$
Else $\beta = \dfrac{\nabla f(X_i)^T \nabla f(X_i)}{\nabla f(X_{i-1})^T \nabla f(X_{i-1})}$
$S_i = -\nabla f(X_i) + \beta \, S_{i-1}$
$X_{i+1} = X_i + \alpha_i \mathbf{S_i}$
α_i is determined by Minimizing $f(X_{i+1})$
$X_s(i+1) \leftarrow X_{i+1};$ $f_s(i+1) = f(X_{i+1})$ (store values)

Step 3: $\Delta f = f_s(i+1) - f_s(i);$ $\Delta X = X_s(i+1) - X_s(i)$
If $|\Delta f| \leq \varepsilon_1;$ *stop* (function not decreasing)
If $\Delta X^T \Delta X \leq \varepsilon_2;$ *stop* (design not changing)
If $i + 1 = N;$ *stop*
If $\nabla f(X_{i+1})^T \nabla f(X_{i+1}) \leq \varepsilon_3;$ *convergerd*
$i \leftarrow i + 1$
Go to Step 2

Example 6.1:

Minimize $f(X) = f(x_1, x_2) = 3 \cdot \sin(0.5 + 0.25 x_1 x_2) \cdot \cos(x_1)$ (6.3)

Subject to: $0 \leq x_1 \leq 5;$ $0 \leq x_2 \leq 8;$ (6.4)

MATLAB Code: The wrapper script m-file that will initialize variables and call the Conjugate Gradient algorithm and print the iterations to the screen is `ConjugateGradient_Example6_1.m`. The values are transferred through the global statement which can be ignored if you do not wish to use the wrapper script file. The method itself can be run from the command line with the usage described below. The algorithm (A. 6.6) is programmed in `ConjugateGradient.m`. You will notice that it is quite similar in style to the Steepest Descent method and shares a lot of the code layout and the statements. The method uses the Golden Section and the upper bound scanning process. The features of this program are the same as the Steepest Descent method and are not listed here.

The algorithm can be called without the wrapper function. It will only return the value of the final design. The usage is given here.

Usage:

```
ConjugateGradient('Example6_1', [0.5 0.5], 20,1e-08,0,1,10)
```
$$(6.13)$$

Input:

'Example6_1' :	The file **Example6_1.m** defines objective function
[0.5 0.5]	starting design vector (dvar0)
20	no. of iterations of the Conjugate Gradient method (niter)
1e−08:	tolerance for the golden section procedure and this method (tol)
0	initial stepsize for scanning upper bound (lowbound)
1	step interval for scanning in upper bound (intvl)
10	number of scanning steps for upper bound (ntrials)

Output:

[designvar, ffp]	values of design variables, value of function
designvar:	A vector of $n + 1$ values of design variables at the last iteration.
ffp:	value of objective function

The output in the Command window is partially reproduced to fit the page. The values of α and the change in design are not reported. First, notice the condition for the exit of the program. The design is not changing. It is due to the value of zero stepsize for the last search direction. The value of the design vector in the last two iterations match to five places after the decimal. If this accuracy is insufficient, then the tolerance must be reset in `GoldenSection.m`. The method has found another solution with the same objective value.

```
Exit: Design not changing
Example 6.1 (Conjugate Gradient)
```

```
********************************
```

-The design vector, function value, KT condition etc.
   during the iterations

iter	x(1)	x(2)	f	KT Cond
1	0.50000	0.50000	1.40405	2.396e−001
2	2.14931	−0.43793	−0.42913	9.572e−001
3	2.68002	−0.43511	−0.55592	3.100e+000
4	5.78797	3.06439	−2.57496	1.601e+000
5	6.14284	2.97857	−2.77816	2.773e+000
6	6.35664	2.68067	−2.98852	1.537e−001
7	6.32658	2.65096	−2.99661	1.662e−002
8	6.28185	2.67247	−2.99964	5.790e−003
9	6.28122	2.68248	−2.99999	2.615e−005
10	6.28269	2.68133	−3.00000	1.510e−006
11	6.28337	2.68150	−3.00000	1.386e−005
12	6.28337	2.68150	−3.00000	1.386e−005

Total cpu time (s) = 1.1094

Figure 6.5 captures the application of the Conjugate Gradient method to Example 6.1, starting at the initial design of $x_1 = 0.5, x_2 = 0.5$. For the same starting point, the Conjugate Gradient method identifies a different solution with more iterations than the Steepest Descent method. This is highly unexpected since the Conjugate Gradient algorithm is more stable of the two. It is also regarded as more efficient. If we had enforced the side constraints (remember this is a technique for unconstrained minimum) then the method would report that no solution was found. In effect this is a counterexample for general observations about the various optimization techniques, particularly the Conjugate Gradient method. The properties of these methods are usually compared with respect to quadratic problems. Example 6.1 is not a quadratic objective function. In fact, with a quadratic objective function the properties of the techniques are easily established. A problem at the end of the chapter should verify that the Conjugate Gradient method is better than the Steepest Descent method.

A comparison between the algorithms for the Steepest Descent and the Conjugate Gradient method reveals that the latter carries one additional computation— β and the successive adjustment of the search direction incorporating this value. β represents the ratio of the square of the current gradient vector to the square of the previous gradient vector. The first thing to notice is that a degree of robustness is built into the method by carrying information from the previous iteration. This is like maintaining a history of the method, albeit for just one iteration. Since the FOC is based on the length of the gradient approaching zero at the solution,

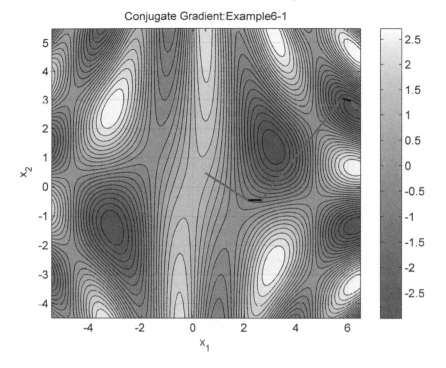

Figure 6.5 Conjugate gradient method: Example 6.1.

this particular form of incorporation of β is ingenious. If the previous iteration is close to the solution, then β is large and the previous iteration plays a significant role in the current iteration. On the other hand, if β is small, suggesting the solution is still far away, then the current value of the gradient determines the new search direction.

6.3.3 Davidon-Fletcher-Powell Method

The Davidon-Fletcher-Powell (DFP) method belongs to the family of *Variable Metric Methods (VMM)*[6]. The method was first introduced by Davidon and several years later was developed in its current form by Fletcher and Powell. The VMM were designed to replace the Conjugate Gradient technique and improve its robustness. At a minimum they must deliver *quadratic convergence*. Generally VMM go beyond that by behaving like a second-order method (Newton's method—Section 6.4) as they approach the solution. The second order method should *locate the solution to a quadratic problem in one iteration*. The Newton-like behavior of the VMM has earned them the label of *Quasi-Newton or Newton-like Methods*. These methods, at the solution will also determine some form of the Hessian, so that the second-order conditions (SOC) can be established.

The DFP method is presented here for historical interest and also because it is a little easier to understand than the others. In fact Huang documents a generic procedure from which most of the popular methods can be obtained, and from which you could even derive your own[7].

To understand why these methods do so well it is necessary to go back to the Conjugate Gradient method. The improvement of Conjugate Gradient technique with respect to the Steepest Descent method was possible because of the inclusion of the history from the previous iteration. In the quasi-Newton methods the history from all previous iterations is available. This information is collected in a $n \times n$ *matrix* called the *metric*. The metric is updated with each iteration and is used to establish the search direction. An initial choice for the metric is also required. It must be a *symmetric positive definite* matrix. For the method to converge, the metric must hold on to its positive definite property through the iterations. In DFP method, the metric approaches the inverse of the Hessian at the solution.

Algorithm: Davidon-Fletcher-Powell (A 6.7):

Step 1:   Choose X_1, $[A_1]$ (initial metric), N
$\varepsilon_1, \varepsilon_2, \varepsilon_3$: (tolerance for stopping criteria)
Set $i = 1$ (initialize iteration counter)

Step 2: $S_i = -[A_i] \nabla f(X_i)$
$X_{i+1} = X_i + \alpha_i S_i$; $\Delta X = \alpha_i S_i$
α_i is determined by Minimizing $f(X_{i+1})$

Step 3:   *If* $\nabla f(X_{i+1})^T \nabla f(X_{i+1}) \leq \varepsilon_3$;   *convergerd*
If $|f(X_{i+1}) - f(X_i)| \leq \varepsilon_1$;   *stop* (function not decreasing)
If $\Delta X^T \Delta X \leq \varepsilon_2$;   *stop* (design not changing)
If $i + 1 = N$;   *stop* (iteration limit)
Else
$Y = \nabla f(X_{i+1}) - \nabla f(X_i)$
$Z = [A_i] Y$
$[B] = \dfrac{\Delta X \, \Delta X^T}{\Delta X^T Y}$
$[C] = -\dfrac{Z \, Z^T}{Y^T Z}$
$[A_{i+1}] = [A_i] + [B] + [C]$
$i \leftarrow i + 1$
Go to Step 2

Here, the matrices are in square parenthesis. The initial choice of the metric is a positive definite matrix. The *identity* matrix is a safe choice.

Example 6.1:

Minimize $f(X) = f(x_1, x_2) = 3 \cdot \sin(0.5 + 0.25 x_1 x_2) \cdot \cos(x_1)$ (6.3)

Subject to: $0 \leq x_1 \leq 5$; $0 \leq x_2 \leq 8$;   (6.4)

Matlab Code: The wrapper script m-file that will initialize variables and call the DFP algorithm and print the iterations to the screen is **DFP_Example6_1.m.** The values are transferred through the global statement which can be ignored if you do not wish to use the wrapper script file. The method itself can be run from the Command line with the usage described next. The algorithm (A 6.7) is programmed in **DFP.m.** The method uses the golden section and the upper bound scanning process. The features of this program are the same as the Steepest Descent method and are not listed here.

The algorithm can be called without the wrapper function. It will only return the value of the final design. The usage is given here.

Usage:

$$\text{DFP('Example6\_1',[0.5 0.5],20,1e-08,0,1,10)} \qquad (6.14)$$

Input:

'Example6_1' :   The file **Example6_1.m** defines objective function
[0.5 0.5]   starting design vector (dvar0)
20   Number of iterations of the Conjugate Gradient method (niter)
1e−08:   tolerance for the golden section procedure and this method (tol)
0   Initial stepsize for scanning upper bound (lowbound)
1   Step interval for scanning in upper bound (intvl)
10   Number of scanning steps for upper bound (ntrials)

Output:

[designvar, fcg]   values of design variables, value of function
designvar: A vector of $n + 1$ values of design variables at the last iteration.
fcg: Value of objective function

A partial output is included here. The values of α and the change in design are not reported. The program is exited because the design is not changing.

```
Exit: Design not changing

Example 6.1 (DFP)
***********************

-The design vector,function value, KT condition etc.
    during the iterations
```

iter	x(1)	x(2)	f	KT Cond
1	0.50000	0.50000	1.40405	3.171e-001

2	2.14931	-0.43793	-0.42913	9.572e-001
3	2.68488	-0.43426	-0.55737	3.126e+000
4	3.77169	0.15432	-1.45825	4.310e+000
5	3.81436	1.08319	-2.34462	3.412e+000
6	3.14263	1.40890	-2.99804	9.248e-003
7	3.13808	1.36548	-2.99998	7.435e-005
8	3.14146	1.36282	-3.00000	6.576e-007
9	3.14145	1.36284	-3.00000	6.000e-007
10	3.14145	1.36284	-3.00000	6.000e-007

```
The Final Metric
    0.3332    -0.1451
   -0.1451     0.6015
Total cpu time (s)=  1.0313
```

The inverse of the Hessian at the solution is

$$A_{act} = \begin{bmatrix} 0.3333 & -0.1447 \\ -0.1447 & 0.6032 \end{bmatrix}$$

This is very close to the values of the final metric of the DFP method. Figure 6.6 tracks the iterations of the DFP method to Example 6.1, starting at the initial design of $x_1 = 0.5, x_2 = 0.5$. For the same starting point and the

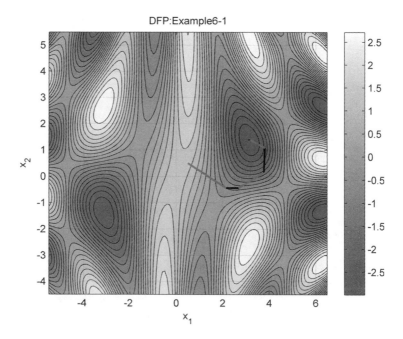

Figure 6.6 DFP method: Example 6.1.

same stopping criteria, the Steepest Descent takes fewer iterations to find a better-quality solution. This is a surprising and unexpected result; after all the VMM methods are consider quite superior to the standard gradient-based techniques. It did better than the Conjugate Gradient method since it found the solution. There are other interesting observations, too. Unlike the Steepest Descent and the Conjugate Gradient technique, the KT conditions are not uniformly decreasing. The quality of the solution is also lacking compared to (6.8). In fact the solution provided by the Steepest Descent method is far better than the DFP method. The easy explanation is that this is not a *quadratic problem* and hence there should be no expectations. However, the advantage of VMM methods is precisely for these kinds of problem. Another explanation is that the particular starting point makes it traverse a difficult region, and is just a matter of chance. Once more, this example provides a counterexample for general statement about the efficacy of the various methods. For a quadratic example, you will find that the method will certainly do very well.

6.3.4 Broydon-Fletcher-Goldfarb-Shanno (BFGS) Method

If you were to program only one gradient-based method, then BFGS method would be the one[8]. It is a quasi-Newton method and currently is the most popular of the Variable Metric methods. It enjoys the property of quadratic convergence and has robustness by carrying forward information from the previous iterations. The difference between the DFP and BFGS is the way the metric are updated. The former converges to the inverse of the Hessian, while the latter converges to the Hessian itself. In a sense the BFGS is more direct. The BFGS has replaced the Conjugate Gradient techniques as a workhorse in solving nonlinear equations.

For convergence, the metric must be a positive definite. An initial choice of positive definite matrix for the metric is usually sufficient to ensure this property. The symmetric positive definite identity matrix is usually a default choice.

Algorithm: Broyden-Fletcher-Goldfarb-Shanno (BFGS) Method (A 6.8):

Step 1: Choose $X_1, [A_1]$ (initial metric), N
$\varepsilon_1, \varepsilon_2, \varepsilon_3$: (tolerance for stopping criteria)
Set $i = 1$ (initialize iteration counter)

Step 2: The search direction is obtained as a solution to
$[A_i] S_i = -\nabla f(X_i)$
$X_{i+1} = X_i + \alpha_i S_i; \quad \Delta X = \alpha_i S_i$
α_i is determined by Minimizing $f(X_{i+1})$

Step 3: *If* $\nabla f(X_{i+1})^T \nabla f(X_{i+1}) \leq \varepsilon_3; \quad convergerd$
If $|f(X_{i+1}) - f(X_i)| \leq \varepsilon_1; \quad stop$ (function not decreasing)
If $\Delta X^T \Delta X \leq \varepsilon_2; \quad stop$ (design not changing)
If $i + 1 = N; \quad stop$ (iteration limit)
Else

$$Y = \nabla f(X_{i+1}) - \nabla f(X_i)$$

$$[B] = \frac{Y\,Y^T}{Y^T\,\Delta X}$$

$$[C] = \frac{\nabla f(X_i)\,\nabla f(X_i)^T}{\nabla f(X_i)^T\,S_i}$$

$$[A_{i+1}] = [A_i] + [B] + [C]$$

$$i \leftarrow i + 1$$

Go to Step 2

Example 6.1:

Minimize: $\quad f(X) = f(x_1, x_2) = 3 \cdot \sin(0.5 + 0.25x_1x_2) \cdot \cos(x_1)$ $\quad$ (6.3)

Subject to: $\quad 0 \le x_1 \le 5; \quad 0 \le x_2 \le 8;$ $\qquad\qquad\qquad$ (6.4)

Example 6.1 BFGS Method: The method is applied to Example 6.1. The first iteration is shown in detail. The next two are shown with final values. The last iteration is also included. The symbolic computation was difficult to implement since it would not determine the value for the optimum step size. The computations were performed numerically using the Golden Section method implemented earlier. The Hessian matrix, computed analytically, is then compared to the final metric of the BFGS method applied numerically.

Step 1: $\quad X_1 = \begin{bmatrix} 0.5 \\ 0.5 \end{bmatrix}; \quad f(X_1) = 1.4041; \quad [A_1] = \begin{bmatrix} 1 & 0 \\ 0 & 1 \end{bmatrix};$

$\qquad\qquad \nabla f(X_1) = \begin{bmatrix} -0.4895 \\ 0.2784 \end{bmatrix}$

Step 2: $\quad S_1 = \begin{bmatrix} 0.4895 \\ -0.2784 \end{bmatrix}; \quad \alpha_1 = 3.3693; \quad X_2 = \begin{bmatrix} 2.1493 \\ -0.4379 \end{bmatrix};$

$\qquad\qquad \nabla f(X_2) = \begin{bmatrix} -0.4833 \\ -0.8506 \end{bmatrix}$

Step 3: $\quad Y = \begin{bmatrix} 0.0062 \\ -1.1280 \end{bmatrix}; \quad \Delta X = \begin{bmatrix} 1.6493 \\ -0.9379 \end{bmatrix}$

$$[B] = \frac{\begin{bmatrix} 0.0062 \\ -1.1290 \end{bmatrix} [0.0062 - 1.1290]}{[0.0062 - 1.1290] \begin{bmatrix} 1.6493 \\ -0.9379 \end{bmatrix}} = \begin{bmatrix} 0.0000 & -0.0066 \\ -0.0066 & 1.1922 \end{bmatrix}$$

$$[C] = \frac{\begin{bmatrix} -0.4833 \\ -0.8506 \end{bmatrix} [-0.4833 - 0.8506]}{[-0.4833 - 0.8506] \begin{bmatrix} 0.4895 \\ -0.2784 \end{bmatrix}} = \begin{bmatrix} -0.7556 & 0.4297 \\ 0.4297 & -0.2444 \end{bmatrix}$$

$$[A_2] = [A_1] + [B] + [C] = \begin{bmatrix} 0.2444 & 0.4231 \\ 0.4231 & 1.9478 \end{bmatrix}$$

$i = 2$;

Go to Step 2

Step 2: $S_2 = \begin{bmatrix} 1.9576 \\ 0.0114 \end{bmatrix}$; $\alpha_2 = 0.2720$; $X_3 = \begin{bmatrix} 2.6817 \\ -0.4348 \end{bmatrix}$;

$f(X_3) = -0.5564$; $\nabla f(X_3) = \begin{bmatrix} 0.0107 \\ -1.7631 \end{bmatrix}$

$Y = \begin{bmatrix} 0.4940 \\ -0.9125 \end{bmatrix}$; $\Delta X = \begin{bmatrix} 0.5324 \\ 0.0031 \end{bmatrix}$

$[B] = \begin{bmatrix} 0.9381 & -1.7327 \\ -1.7327 & 3.2004 \end{bmatrix}$; $[C] = \begin{bmatrix} -0.2444 & -0.4301 \\ -0.4301 & -0.7570 \end{bmatrix}$;

$[A_3] = \begin{bmatrix} 0.9381 & -1.7396 \\ -1.7396 & 4.3912 \end{bmatrix}$

$i = 3$;

Go to Step 2

Step 2: $S_2 = \begin{bmatrix} 2.7630 \\ 1.4961 \end{bmatrix}$; $\alpha_2 = 0.3943$; $X_3 = \begin{bmatrix} 3.7710 \\ 0.1550 \end{bmatrix}$;

$f(X_3) = -1.4602$; $\nabla f(X_3) = \begin{bmatrix} 0.9892 \\ -1.8247 \end{bmatrix}$

$Y = \begin{bmatrix} 0.9784 \\ -0.0615 \end{bmatrix}$; $\Delta X = \begin{bmatrix} 1.0894 \\ 0.5899 \end{bmatrix}$

$[B] = \begin{bmatrix} 0.9299 & -0.0585 \\ -0.0585 & 0.0037 \end{bmatrix}$; $[C] = \begin{bmatrix} -0.0000 & 0.0072 \\ 0.0072 & -1.1919 \end{bmatrix}$;

$[A_4] = \begin{bmatrix} 1.8679 & -1.7909 \\ -1.7909 & 3.2030 \end{bmatrix}$

$i = 7$;

Go to Step 2

Step 2 (last iteration): $S_2 = \begin{bmatrix} 0.0040 \\ -0.0031 \end{bmatrix}$; $\alpha_2 = 1.0140$; $X_3 = \begin{bmatrix} 3.1415 \\ 1.3628 \end{bmatrix}$;

$f(X_3) = -3.0000$; $\nabla f(X_3) = 1.0e - 03 \begin{bmatrix} 0.7657 \\ -0.2267 \end{bmatrix}$

$[A_7] = \begin{bmatrix} 3.2314 & 0.8508 \\ 0.8508 & 1.8467 \end{bmatrix}$;

$[X_{act}] = \begin{bmatrix} 3.1415939 \\ 1.3633781 \end{bmatrix}$ $[A_{act}] = \begin{bmatrix} 3.3485 & 0.8031 \\ 0.8031 & 1.8506 \end{bmatrix}$

The translation of BFGS algorithm into code is left as an exercise for the student. Figure 6.7 illustrates the track of the BFGS method for Example 6.1. There are seven iterations, three iterations fewer than the DFP method. Table 6.1 is the iteration history for Example 6.1. There are slight differences between the actual values and the numerically computed values in the design as well as the Hessian. It is recommended that the calculation of the search direction in Step 2

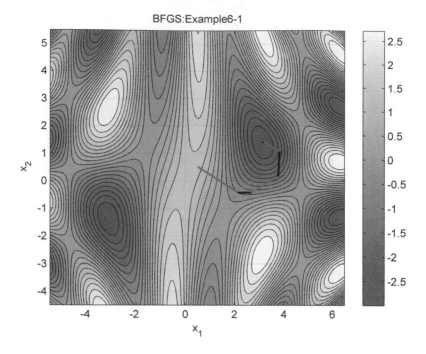

Figure 6.7 BFGS: Example 6.1.

Table 6.1 BFGS iterations for Example 6.1

x_1	x_2	f	α	$\nabla f^T \nabla f$
0.5000	0.5000	1.4041	0	0.3171
2.1493	−0.4379	−0.4291	3.3693	0.9572
2.6817	−0.4348	−0.5564	0.2720	3.1088
3.7710	0.1550	−1.4602	0.3943	4.3079
3.8264	1.0661	−2.3206	1.5450	3.4901
3.1436	1.4150	−2.9974	0.3503	0.0123
3.1374	1.3660	−3.0000	0.9675	0.0001
3.1415	1.3628	−3.0000	1.0140	0.0000

be accomplished as a solution to a set of linear equations (as listed), rather than inverting the matrix A.

6.4 NUMERICAL TECHNIQUE – SECOND ORDER

Second-order methods for unconstrained optimization are not used because computation of the Hessian matrix is considered computationally expensive. A second-order method with the property of quadratic convergence is very impressive. *An **n**-variable problem can converge in one iteration*. As mentioned before, for real design problems, where decisions are required to be made on the discrete nature of some of the variables, the existence of the first derivative, leave alone second derivatives, is questionable. Moreover, the quasi-Newton methods of the last two sections are able to effectively function as a second order method as they approach the solution–*and they do not need the estimation of second derivatives*. Second-order method is presented here for the sake of completeness.

There is only one second-order technique for unconstrained optimization. It is based on the extension of the Newton-Raphson technique to multi-variable problems. Many different extensions are available, but in this text, a direct extension respecting the general algorithmic structure (A 6.1) is presented. Once again, the general problem for this section is

$$\text{Minimize} \quad f(X); \quad [X]_n \tag{6.1}$$

$$\text{Subject to:} \quad x_i^l \le x_i \le x_i^u; \quad i = 1, 2 \ldots n \tag{6.2}$$

6.4.1 Modified Newton's Method

The Newton-Raphson method, used for a single variable, solves the problem ($\phi(x) = 0$). This equation can represent the FOC for an unconstrained. in one variable (Chapter 5). The iterative change in the variable is computed through

$$\Delta x = -\frac{\phi(x)}{\phi'(x)} = -\frac{f'(x)}{f''(x)} \tag{6.15}$$

where f is the single-variable objective function. The multi-variable extension to computing a similar change in the variable vector is

$$\Delta X = -[H]^{-1} \nabla f(X) \tag{6.16}$$

The original Newton-Raphson is not known for its robustness or stability, and (6.16) shares the same disadvantage. To control the design variable changes and to bring it under the scheme of algorithm (A 6.1) the right-hand side is used to determine the search direction vector S followed with a standard one-dimensional stepsize computation. This is termed as the Modified Newton method. The complete algorithm:

Algorithm: Modified Newton Method (A 6.9):

Step 1: Choose X_1, N
        $\varepsilon_1, \varepsilon_2, \varepsilon_3$: (tolerance for stopping criteria)
        Set $i = 1$ (initialize iteration counter)

Step 2:   The search direction is obtained as a solution to
$$[H(X_i)] S_i = -\nabla f(X_i); \quad [H] \text{ is the Hessian}$$
$$X_{i+1} = X_i + \alpha_i S_i; \quad \Delta X = \alpha_i S_i$$
α_i is determined by Minimizing $f(X_{i+1})$

Step 3:   **If** $\nabla f(X_{i+1})^T \nabla f(X_{i+1}) \le \varepsilon_3;$ *convergerd*
If $|f(X_{i+1}) - f(X_i)| \le \varepsilon_1;$ *stop* (function not decreasing)
If $\Delta X^T \Delta X \le \varepsilon_2;$ *stop* (design not changing)
If $i + 1 = N;$ *stop* (iteration limit)
Else $i \leftarrow i + 1$
Go to Step 2

Example 6.1:

Minimize: $f(X) = f(x_1, x_2) = 3 \cdot \sin(0.5 + 0.25 x_1 x_2) \cdot \cos(x_1)$ (6.3)

Subject to: $0 \le x_1 \le 5; \quad 0 \le x_2 \le 8;$ (6.4)

Solution to Example 6.1:   Algorithm (A 6.9) is used to solve Example 6.1. Only essential computations are included:

Step 1: $X_1 = \begin{bmatrix} 0.5 \\ 0.5 \end{bmatrix}; \quad f(X_1) = 1.4041$

Step 2: $\begin{bmatrix} -1.7302 & 0.3828 \\ 0.3828 & -0.0219 \end{bmatrix} S_1 = -\begin{bmatrix} -0.4886 \\ 0.2784 \end{bmatrix}; \quad S_1 = \begin{bmatrix} -0.8829 \\ -2.7143 \end{bmatrix}$

$\alpha_1 = 0.2722; \quad X_2 = \begin{bmatrix} 0.2596 \\ -0.2390 \end{bmatrix}; \quad f(X_2) = 1.3504;$

$i = 2;$

Step 2: $\begin{bmatrix} -1.2738 & 0.6024 \\ 0.6024 & -0.0057 \end{bmatrix} S_1 = -\begin{bmatrix} -0.5120 \\ 0.1665 \end{bmatrix}; S_1 = \begin{bmatrix} -0.2739 \\ 0.2708 \end{bmatrix}$

$\alpha_1 = 0;$ *Stop : cannot decrease function*

The method cannot find the optimum from the starting design of $x_1 = 0.5$ and $x_2 = 0.5$. The method has also failed to live up to expectation since it was expected to converge faster than the quasi-Newton methods, as we are providing better information to establish the search directions–namely the Hessian. However, this is not a quadratic problem. In the exercises you will apply the technique to a quadratic problem and see that it does converge in one iteration.

Figure 6.8 shows the single move of the method. It appears that it become stuck in an area where the function is increasing in all directions. Starting at $x_1 = 2.5$ and $x_2 = 1.0$, the same method determines the solution in three iterations. Therefore one can conclude that the method's performance is based on the starting value of the design. The Newton's method has been used to generate examples of *Chaos* through nonlinear examples that are sensitive to initial conditions[10]. Translation of the algorithm into MATLAB code is once again left as an exercise.

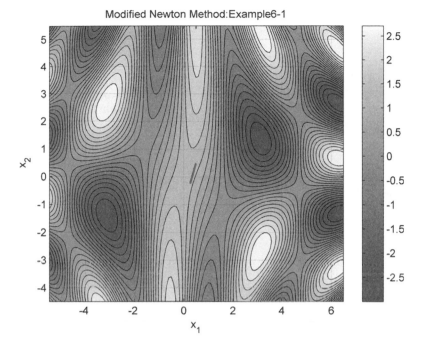

Figure 6.8 Modified Newton's Method: Example 6.1.

6.5 ADDITIONAL EXAMPLES

Three additional examples are presented in this section. The examples will be solved with the numerical techniques already programmed. Obtaining the solution is a matter of judicious application of the numerical procedures. The examples also illustrate how optimization can be used creatively to solve examples that are not formulated in the standard mold. The first example is the Rosenbrock problem [9]. This example was one of those created to challenge the numerical techniques for unconstrained minimization. It is also sometimes referred to as the *banana function*. If you have a copy of the *Optimization Toolbox* from MATLAB you will see this as part of the toolbox demo. You will also note that the toolbox contains many of the techniques that was developed in this section. The second example is a solution to a nonlinear two-point boundary value problem that is due to the Navier-Stokes equations describing flow due to a spinning disc. The last is an example from electrical engineering.

6.5.1 Example 6.2–Rosenbrock Problem

The Rosenbrock problem is:

$$\text{Minimize} \quad f(x_1, x_2): \quad 100(x_2 - x_1^2)^2 + (1 - x_1)^2 \tag{6.17}$$

The side constraints are used for drawing contours. The solution to this problem is

$$x_1^* = 1.0; \quad x_2^* = 1.0; \quad f^* = 0.0 \tag{6.18}$$

The problem is notorious for large number of iterations for convergence as well as very small changes in design as the solution is being approached.

MATLAB Code: Three methods are used to solve the unconstrained example 6.2. The file **Scan_Zoom_Example6_2.m** will solve it using the simplest of the zero-order method. The solution is reached in 20 levels of zoom, and the values are

```
Example 6.2
_____

 Zoom Level ( 1):
x1* =   0.0000000
x2* =   0.0000000
 f* =   1.0000000


Example 6.2
_____

 Zoom Level (20):
x1* =   0.9999996
x2* =   0.9999990
 f* =   0.0000000
Total cpu time (s)=   0.2500
```

The simple and direct Scan and Zoom technique has found the solution to a simple problem that is numerically challenging because of strong variation and a narrow valley where the solution is located.

The Random Walk method, executed with **RandomWalk_Example6_2.m**, after 300 iterations starting from $x_1 = 0.0$, $x_2 = 0.0$, yields the following:

$$x_1 = 1.00290 \ x_2 = 1.00582 \ f = 0.00001 \tag{6.19}$$

Figure 6.9 illustrates the iterations of the method.

The Conjugate Gradient method, through **ConjugateGradient_Example6_2.m**, starting from $x_1 = 0.0$, $x_2 = 0.0$, yields the following:

$$x_1 = 1.01740; x_2 = 1.03543; f = 0.00031 \tag{6.20}$$

It exits after 10 iterations, with the information that the design is not changing. Figure 6.10 tracks the algorithm on the contour plot of Example 6.2. The Random Walk method appears to have done better than the Conjugate Gradient technique.

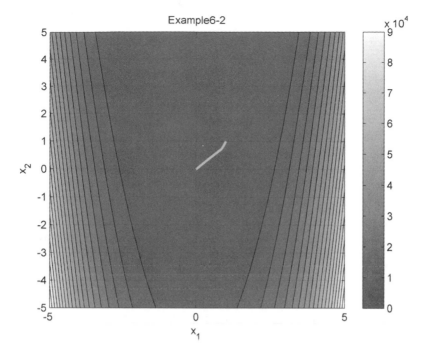

Figure 6.9 Random walk: Example 6.2.

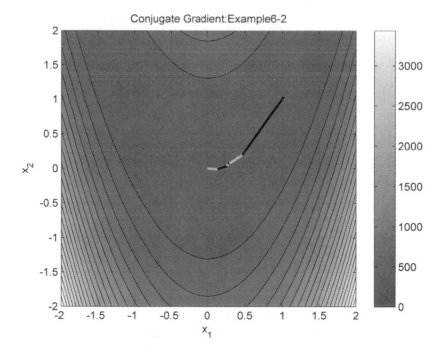

Figure 6.10 Conjugate Gradient: Example 6.2.

The variable metric methods, or at least the implementation in this chapter, are no better in solving this problem. It appears that the zero-order methods (Scan and Zoom and the Random Walk) have distinguished themselves in this problem compared to the more sophisticated gradient-based techniques. The Scan and Zoom has the best solution.

6.5.2 Example 6.3–Three-Dimensional Flow near a Rotating Disk

The example represents the viscous flow of a fluid around a flat disk that rotates about an axis perpendicular to its plane with a constant angular velocity. Figure 6.11 illustrates the dependent variables–the three velocity components. An example is the air flow created by an open CD player. In the model, the flow field is infinite in extent. Hock and Schittkowski start with the Navier-Stokes equation in cylindrical coordinates, and through several persuasive assumption reduce the problem from a coupled set of partial nonlinear equations to a coupled set of nonlinear ordinary differential equations in scaled variables[11]. The final mathematical model, a nonlinear two-point boundary value problem, and the boundary conditions at two points are shown next. F, G, H are scaled velocities V_r, V_0, V_z in r, θ, and z coordinate directions. The derivatives are with respect to the scaled z variable. See the reference for the complete derivation[11].

$$2F + H' = 0 \tag{6.21a}$$

$$F^2 + F'H - G^2 - F'' = 0 \tag{6.21b}$$

$$2FG + HG' - G'' = 0 \tag{6.21c}$$

$$F(0) = 0; \quad G(0) = 1; \quad H(0) = 0 \tag{6.21a}$$

$$F(\infty) = 0; \quad G(\infty) = 0 \tag{6.21b}$$

Equations (6.21) are converted to state space form (see Chapter 5, Example 5.4). The design variables are the missing boundary conditions on $F'(0)$, $G'(0)$ [to make it into an initial value problem that can be numerically integrated directly], and the final value of the independent variable (in lieu of ∞). The objective function is the error in the integrated value at the final condition and the boundary conditions (6.21b).

The Optimization Problem:

$$\text{Minimize} \quad f(x_1, x_2, x_3): \quad y_1(x_3)^2 + y_3(x_3)^2 \tag{6.22}$$

where the state variables are obtained from the solution of the following initial value problem:

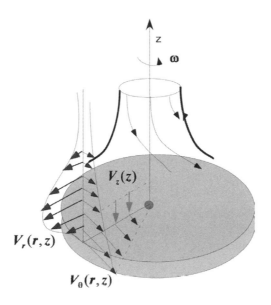

Figure 6.11   Problem variables: Example 6.3.

$$
\begin{bmatrix} y_1' \\ y_2' \\ y_3' \\ y_4' \\ y_5' \end{bmatrix} = \begin{bmatrix} y_2 \\ y_1^2 + y_2 y_5 - y_3^2 \\ y_4 \\ 2 y_1 y_3 + y_4 y_5 \\ -2 y_1 \end{bmatrix} ; \quad \begin{bmatrix} y_1(0) \\ y_2(0) \\ y_3(0) \\ y_4(0) \\ y_5(0) \end{bmatrix} = \begin{bmatrix} 0 \\ x_1 \\ 1 \\ x_2 \\ 0 \end{bmatrix} \tag{6.23}
$$

The state equations (6.23) are integrated using the MATLAB function MATLAB Command **ode45**.

MATLAB Code: The three variable problem is solved first using the scan and zoom method through **Scan_Zoom_Example6_3.m.** For each combination of x_1, x_2, x_3 the differential system is integrated form 0 to x_3 using the MATLAB function MATLAB Command **ode45**, and the best value of the objective function is returned. The dependency of the function m-files is

```
Scan_Zoom_Example6_3.m
   ▶ Example6_3.m (MATLAB code ode45 is called here)
      ▶ Ex6_3_state.m
```

The starting level and the last level are shown here. The solution takes 366 CPU seconds.

```
Example 6.3
```

```
Zoom Level ( 1):
x1* =   0.8000000
x2* =  -1.2000000
x3* =  10.0000000
 f* =   0.3318969
Example 6.3
```
———————————————

```
Zoom Level (18):
x1* =   0.5101990
x2* =  -0.6159180
x3* =  10.1889549
 f* =   0.0000000
Total cpu time (s)= 365.4219
```

The published values are:

$$x_1^* = 0.5101; \quad x_2^* = -0.6159; \quad x_3^* = \text{usually not specified;} \quad f^* = 0 \quad (6.24)$$

The zero-order method has performed remarkably well.

The problem was also solved by the DFP method. The call to DFP method with the initial design vector of $[1 -1\ 8]$, 20 iterations of DFP method, a tolerance for the golden section of 1.0e-08 was made directly to the algorithm without a wrapper function. The usage is included below.

Usage:

$$\text{DFP('Example6\_3',[1 -1 8],20,1.0e-08,0,1,20)} \quad (6.25)$$

The final value for the design after 20 iterations with the message that the design is not changing is

$$x_1^*(DFP) = 0.5101; \quad x_2^*(DFP) = -0.6157;$$

$$x_3^*(DFP) = 10.7627; \quad f^*(DFP) = 0 \quad (6.26)$$

which is remarkably close to the published values. The value of x_3^* is usually not a design variable in the published information, but is suspected to be 7. Since it is a substitute for infinity, the larger number should not be a problem.

Example 6.3 is not a trivial problem. The integration is highly sensitive to the initial values. A reasonable starting point is essential to prevent the integration from generating NaNs (not a number). The second design variable must have a negative starting value.

The example illustrates the application of standard optimization technique to solve a nonlinear differential system. A similar application can be made to problems in system dynamics and optimal control. It is essential to understand that these algorithms are numerical tools that transcend any particular discipline. It is the responsibility of the user to use it appropriately and creatively.

There are numerical techniques that address two-point nonlinear boundary value problem more efficiently. However, computing resources, especially on a PC, is not a concern. Therefore, the procedure adopted in this example is an acceptable approach to this class of problems. Even the Random Walk method turns in an impressive performance for this example.

6.5.3   Example 6.4 — An Electrical Engineering Problem

Figure 6.12 illustrates the electric circuit, which represents an interstage coupling of a tuned amplifier[12]. The two design variables are the variable capacitances $(C_{1,2})$ that can be adjusted to provide the maximum output power through the resistance R_2 for a given carrier frequency. The parameters are the resistance R_1, R_2, and the indicator, L. From circuit calculations, the power output is

$$P_o = \frac{|V_o(j\omega)|^2}{R_2} \tag{6.27}$$

$$V_o(j\omega) = \frac{-R_1 R_2 I(j\omega)}{(j\omega)^3(LR_1C_1R_2C_2) + (j\omega)^2 L(R_1C_1 + R_2C_2)} \tag{6.28a}$$
$$+ (j\omega)(L + R_1R_2C_2 + R_1R_2C_1) + R_1 + R_2$$

$$|V_o(j\omega)| = \frac{IR_1}{\sqrt{\left(1 + \dfrac{R_1}{R_2} - \dfrac{\omega^2 LR_1C_1}{R_2} - \dfrac{\omega^2 LR_2C_2}{R_2}\right)^2 + \left(\dfrac{\omega L}{R_2} + \dfrac{\omega R_1R_2C_2}{R_2} + \omega R_1C_1 - \dfrac{\omega^3 LR_1C_1R_2C_2}{R_2}\right)^2}} \tag{6.28b}$$

Define new parameters and new design variables to eliminate the dependence on the service frequency:

$$a = \frac{R_1}{R_2}; \quad b = \frac{\omega L}{R_2}; \quad x_1 = \omega R_1 C_1; \quad x_2 = \omega R_2 C_2 \tag{6.29}$$

To convert this to the standard format that defines a minimization problem, the objective function can be set up to minimize the inverse of the output power and should be optimized for a chosen value of a and b.

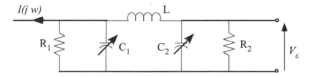

Figure 6.12   Circuit for Example 6.4.

Example 6.4:

$$\text{Min } f(x_1, x_2) = (1 + a - bx_1 - bx_2)^2 + (b + x_1 + ax_2 - bx_1x_2)^2 \quad (6.30)$$

for $a = 10$, and $b = 1$.

MATLAB Code: The first solution is obtained by the Scan and Zoom zero-order method. It has turned in incredible performance so far. The solution is obtained by running `Scan_Zoom_Example6_4.m`. Only a portion of the output in the Command window is copied here. The results:

```
Example 6.4
_____

 Zoom Level ( 1):
x1* = 12.0000000
x2* =  4.0000000
 f* = 50.0000000


Example 6.4
_____

 Zoom Level (20):
x1* = 13.0000000
x2* =  4.0000000
 f* = 40.0000000


Hessv =      ( % Hessian at solution is positive definite)
     20     16
     16     20
Total time (s)=  0.2813  (on a contemporary laptop)
```

This final values in Zoom level 20 is the actual solution to the problem. The Scan and Zoom method has been successful over all of the examples in this chapter. It has also found the best solution for all of the examples.

`Conjugate_GradientExample6_4.m` will execute the conjugate gradient method for this example. Starting from $x_1 = 10$, $x_2 = 2$, the edited iterations are shown:

```
Example 6.4 (Conjugate Gradient)
*******************************
-The design vector,function value, KT condition etc.
     during the iterations
  iter    x(1)         x(2)           f          KT Cond
_____

     1   10.00000    2.00000    122.00000    3.999e+002
     2   13.08296    1.69152     92.88944    2.243e+003
```

```
 3   15.55552    2.46081    57.54888    3.549e+002
 4   15.50754    2.78241    54.54361    1.099e+002
 5   13.21480    3.62593    40.65830    2.566e+001
 6   13.20397    3.84714    40.14225    2.018e+000
 7   13.09519    3.86880    40.06578    1.465e+000
 8   13.03195    3.98524    40.00479    2.182e-001
 9   13.01596    3.98170    40.00123    1.172e-002
10   13.00650    3.99748    40.00022    1.396e-002
11   13.00070    3.99897    40.00000    5.805e-005
12   13.00070    3.99897    40.00000    5.805e-005
Total cpu time (s)=  1.2656
```

Since side constraints are not enforced it is possible to discover the other solution at $x_1 = 7$ and $x_2 = -2$ starting from a different point. Figure 6.13 illustrates the two solutions for the problem.

The DFP method called from the command line

Usage:

```
DFP('Example6_4',[10 2],20,1.0e-08,0,1,20)
```

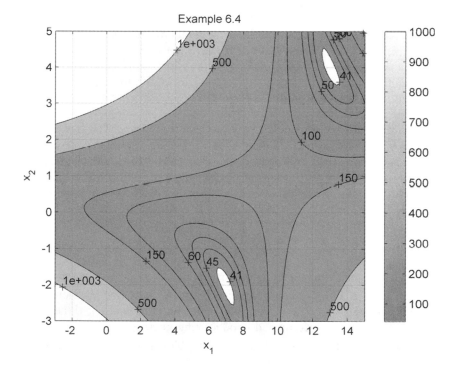

Figure 6.13 Contour plot for Example 6.4.

The solution is

$$x_1 = 12.9994; \quad x_2 = 4.0002; \quad f = 40.0000$$

6.6 SUMMARY

Many algorithms were introduced in this chapter to solve the unconstrained optimization problem in several variables. They were classified into zero-order methods, first-order methods, and second-order methods. The zero-order methods used only the value of the objective function to determine the optimum values. The first-order methods used the objective function and the gradient of the objective function to construct search directions during each iteration. The second-order methods used second-derivative or the Hessian to construct the search directions. These methods also represent increasing sophistication in the numerical techniques for establishing the minimum.

The popular techniques are the first-order or the gradient-based techniques. Many important ideas about the performance of these techniques are based on a quadratic objective function. An important aspect of the algorithms in this chapter is that they identify a local minimum only. This means a solution close to the point at which you start. If the objective function is more nonlinear, then many of the properties of the various algorithms cannot be guaranteed. In this chapter we have focused on a nonquadratic example to illustrate that even the most sophisticated gradient methods may be unable to determine the solution, under normal circumstances. In order to ensure a successful resolution of the solution many additional considerations have to be incorporated into the code. Most importantly, it is necessary to implement the side constraints.

Through several examples, it is shown that the zero-order methods deliver great performance at significantly more computation time. These methods are quite robust and can be easily implemented without any optimization software. It is not surprising that contemporary methods are mostly based on the zero-order methods. For real problem existence of derivatives is also of concern. In summary, based on our experience in this chapter, we can make the following recommendations:

- Do not discount zero-order methods.
- Use side constraints to define acceptable design region.
- Have more than one method in our arsenal to solve the optimization problem.
- Have a good initial guess for the solution to solve the problem. This is a challenging recommendation. For more than two variables, the graphics cannot be used to recommend good starting guesses.
- For nonlinear problems with several variables, it is difficult to know how many solutions are there. Start from several starting designs throughout the design region.

REFERENCES

1. Fletcher, R. *Practical Methods for Optimization*, *Vol. 1*, John Wiley & Sons, Inc. Hoboken, NJ: 1980.
2. Brent, R. P. *Algorithms for Minimization without Derivative*. Prentice-Hall, Englewood Cliffs, NJ: 1973.
3. Powell, M.J.D. *Nonlinear Optimization*. Academic Press, New York, 1981.
4. Fox, R. L. *Optimization Methods for Engineering Design*. Addison-Wesley, Reading MA: 1971.
5. Fletcher, R., and R.M. Reeves. "Function Minimization by Conjugate Gradients." *The Computer Journal*, Vol. 7, pp 149–180, 1964.
6. Davidon, W.C. "Variable Metric Methods for Minimization," *US Atomic Energy Commission Research and Development Report* No. ANL–5990, Argonne National Laboratory: 1959.
7. Huang, H. Y. "Unified Approach to Quadratically Convergent Algorithms for Function Minimization." *Journal of Optimization Theory and Applications*, Vol. 5, pp. 405–423, 1970.
8. Vanderplaats, G.N. *Numerical Optimization Techniques for Engineering Design*. McGraw-Hill, New York: 1984.
9. Strogatz, S. H. *Nonlinear Dynamics and Chaos–with Applications to Physics, Biology, Chemistry, and Engineering*. De capo Press, 2001.
10. Hock, W., and K. Schittkowski. *Test Examples for Non Linear Programming Codes, Lecture notes in Economic and Mathematical Systems*, 187, Springer-Verlag, Berlin: 1980.
11. Schlichting, H., *Boundary Layer Theory*. McGraw-Hill Book Company, New York: 1979.
12. Pierre, A. A. *Optimization Theory with Applications*, Dover Publications, 1896.

PROBLEMS

You can also solve many of the problems using the Optimization Toolbox (see Chapter 10).

6.1 Apply the Random walk method to Example 5.3

6.2 Solve using Random Walk method

$$\text{Minimize} \quad f(x_1, x_2) = x_1^4 - 2x_1^2 x_2 + x_1^2 + x_2^2 - 2x_1 + 4$$

6.3 Apply the Pattern Search method to Example 5.2

6.4 Solve using Pattern Search method.

$$\text{Minimize} \quad f(x_1, x_2) = x_1^4 - 2x_1^2 x_2 + x_1^2 + x_2^2 - 2x_1 + 4$$

6.5 Modify Pattern Search so that program control parameters for contour plotting, golden section, and upper bound calculation can be set by the user. Include the prompt and allow for default values in case user decides not to take advantage of it.

6.6 Translate the Powell's method into working MATLAB code. Verify solution to Example 6.1 and Example 5.2. Start from several points and verify that the number of cycles to converge is the same.

6.7 Verify that the search directions are *conjugate* with respect to the Hessian matrix for Example 6.1 and Example 5.3

6.8 Solve using Powell's method and verify any two search directions are *conjugate* with respect to the Hessian matrix at the solution.

$$\text{Minimize} \quad f(x_1, x_2) = x_1^4 - 2x_1^2 x_2 + x_1^2 + x_2^2 - 2x_1 + 4$$

6.9 Solve Example 5.3 using the Steepest Descent, Conjugate Gradient, DFP, and BFGS methods

6.10 Solve using the Steepest Descent, Conjugate Gradient, DFP, and BFGS method the problem

$$\text{Minimize} \quad f(x_1, x_2) = x_1^4 - 2x_1^2 x_2 + x_1^2 + x_2^2 - 2x_1 + 4$$

6.11 Verify if the matrix **A** at the solution is the inverse of the Hessian for Example 5.3 when DFB method is used.

6.12 Verify if the matrix **A** at the solution is the inverse of the Hessian for

$$\text{Minimize} \quad f(x_1, x_2) = x_1^4 - 2x_1^2 x_2 + x_1^2 + x_2^2 - 2x_1 + 4$$

6.13 Develop the BFGS method into MATLAB code and verify the calculations in Section 6.3.4.

6.14 Develop the Modified Newton method into MATLAB code and apply it to Example 5.3.

6.15 How will you incorporate the side constraints into the code for all the various methods in this section? Implement them in the numerical procedures.

6.16 Solve Examples 6.2, 6.3, and 6.4 by one other method of the section.

6.17 Identify and solve a system dynamics problem using any method of this section.

6.18 Identify and solve your own curve fit problem.

6.19 Start the Steepest Descent method from different points for Example 6.1 and identify where the solution requires more than 15 iterations.

$$\text{Minimize} \quad f(X) = f(x_1, x_2) = 3 + (x_1 - 1.5x_2)^2 + (x_2 - 2)^2 \quad (6.3)$$

$$\text{Subject to:} \quad 0 \le x_1 \le 5; \quad 0 \le x_2 \le 5; \quad (6.4)$$

6.20 Start the Conjugate Gradient method from different points for Example 6.1 and identify where the solution requires more than 15 iterations.

6.21 Example 6.2 was explicitly created to challenge the numerical techniques for unconstrained optimization. There are many other similar test problems. A problem with shallow minimum is

$$\text{Minimize} \quad f(x_1, x_2) : (x_1^2 - x_2)^2 + (1 - x_1)^2$$

Obtain minimum (i) analytically; (ii) graphically; (iii) using Scan and Zoom; (iv) using BFGS; (v) using Newton's method.

6.22 Example 6.2 was explicitly created to challenge the numerical techniques for unconstrained optimization. There are many other similar test problems. A problem with steep minimum is

$$\text{Minimize} \quad f(x_1, x_2) : (x_1^2 - x_2)^2 + 100(1 - x_1)^2$$

Obtain minimum (i) analytically; (ii) graphically; (iii) using random walk; (iv) using conjugate gradient; (v) using Modified Newton's method.

6.23 Example 6.2 was explicitly created to challenge the numerical techniques for unconstrained optimization. There are many other similar test problems. A problem with steep minimum is

$$\text{Minimize} \quad f(x_1, x_2) : 100(x_1^3 - x_2)^2 + (1 - x_1)^2$$

Obtain minimum (i) analytically; (ii) graphically; (iii) using Scan and Zoom; (iv) using Steepest Descent; (v) using Modified Newton's method.

6.24 Example 6.2 was explicitly created to challenge the numerical techniques for unconstrained optimization. There are many other similar test problems. A problem with steep minimum (Beale, *Survey of Integer Programming*) is

$$\text{Minimize} \quad f(x_1, x_2, x_3) : \sum_{i=1}^{3} \left[c_i - x_1(1 - x_2^i) \right]^2; \quad c_1 = 1.5;$$

$$c_2 = 2.25; \quad c_3 = 2.625$$

Obtain minimum (i) analytically; (ii) using Scan and Zoom; (iii) using DFP; (iv) using Modified Newton's method.

6.25 Example 6.2 was explicitly created to challenge the numerical techniques for unconstrained optimization. There are many other similar test problems. A problem with steep minimum (Powell) is

$$\text{Minimize} \quad f(X) : (x_1 + 10x_2)^2 + 5(x_3 - x_4)^2 + (x_2 - 2x_3)^4$$
$$+ (10x_1 - x_4)^4$$

Obtain minimum (i) analytically; (ii) using Pattern Method; (iii) using DFP method; (iv) using Modified Newton's method.

6.26 Example 6.2 was explicitly created to challenge the numerical techniques for unconstrained optimization. There are many other similar test problems. A problem with steep minimum (Fletcher and Powell) is

Minimize $\quad f(x_1, x_2, x_3) : 100[(x_3 - 10\theta)^2 + (r - 1)^2] + x_3^2;$

where $\quad x_1 = r \cos 2\pi\theta; \quad x_2 = r \sin 2\pi\theta; \quad -\pi/2 < 2\pi\theta < 3\pi/2$

Obtain minimum (i) analytically; (ii) using Random Walk; (iii) using DFP method (iv) using Modified Newton's method.

6.27 Solve Problem 1.38.

7

NUMERICAL TECHNIQUES FOR CONSTRAINED OPTIMIZATION

The numerical techniques in Chapters 5 and 6 solved unconstrained optimization problems. We saw some useful examples where this category was sufficient for problem definition. Most practical optimization problems will include constraints, even as basic as the side constraints. This chapter introduces algorithms/methods that handle the general constrained optimization problem. Both equality and inequality constraints are included. For an engineering problem, this will involve bringing several additional nonlinear relations into the design space. This is certain to increase the degree of difficulty in obtaining the solution. For the designer, there is an additional requirement of being more attentive to the design changes and the corresponding strategy to coax the solution if the mathematical definition of the problem is particularly severe. In all of these problems, there are two principal outcomes that the algorithms, or parts of it, seek to accomplish. The first is to ensure that the design is feasible *(satisfies all constraints)* and the second that it is optimal *(satisfies the Kuhn-Tucker (KT) conditions)*. While the goal is to determine the latter, in times of difficulty, it is essential to remember that *feasibility* is more important that *optimality*, and that the optimal solution must be feasible. In this chapter, there are two distinct approaches for handling the constrained optimization problem. The first approach is termed as the *indirect* approach and solves the problem by transforming it into an *unconstrained problem*. The second approach it to handle the constraints without transformation–the *direct* approach.

The *indirect* approach is an expression of incremental development of the subject to take advantage of prevailing techniques. For example, it leverages the *DFP method* to handle constrained optimal problems. Two indirect methods are presented, the *Exterior Penalty Function (EPF) method*, and the *Augmented Lagrange Multiplier (ALM) method*. The first method is more historic and is

337

illustrative of the indirect approach. The second is very popular and is quite competitive with the direct methods. Many of the contemporary global optimization techniques are still not very efficient at handling the constraints directly. They prefer to use the indirect methods. The *direct* approach handles the constraints and the objective together without any transformation. Five methods are presented. The first method is our champion from the previous chapter, the *Scan and Zoom method* which is extended to handle constraints. The remaining are gradient-based techniques. The first of these is included to illustrate some of the ideas of the direct approach. These methods are *Sequential Linear Programming (SLP), Sequential Quadratic Programming (SQP), Generalized Reduced Gradient Method (GRG)*, and *Sequential Gradient Restoration Algorithm (SGRA)*. Unlike the code in previous chapters, the code here is largely limited to the examples being discussed. The algorithms are also coded for symbolic solution since the example is simple. A code that will solve the general problem will require very significant effort and can easily be exploited commercially. For example, the SQP is used in the MATLAB's optimization toolbox.

7.1 PROBLEM DEFINITION

The standard format of the Nonlinear Programming Problem (NLP) reproduced here for convenience

$$\text{Minimize} \quad f(x_1, x_2, \ldots x_n) \tag{7.1}$$

$$\text{Subject to:} \quad h_k(x_1, x_2, \ldots, x_n) = 0, \quad k = 1, 2, \ldots, l \tag{7.2}$$

$$g_j(x_1, x_2, \ldots, x_n) \leq 0, \quad j = 1, 2, \ldots, m \tag{7.3}$$

$$x_i^l \leq x_i \leq x_i^u \quad i = 1, 2, \ldots, n \tag{7.4}$$

In vector notation

$$\text{Minimize} \quad f(X), \quad [X]_n \tag{7.5}$$

$$\text{Subject to:} \quad [h(X)]_l = \mathbf{0} \tag{7.6}$$

$$[g(X)]_m \leq \mathbf{0} \tag{7.7}$$

$$X^{\text{low}} \leq X \leq X^{\text{up}} \tag{7.8}$$

For this chapter the following indices are reserved: *n* is number of variables; *l* is number of equality constraints; *m* is number of inequality constraints. Many of the algorithm and methods present a reasonable level of difficulty. A two-variable problem (for graphical support), with a single equality and inequality constraint is used for illustrating the various algorithms. All of the functions are nonlinear though the constraint functions have simple geometric shapes.

7.1.1 Problem Formulation — Example 7.1

Example 7.1 is a simple mathematical formulation. It does not represent any engineering problem. It is constructed to have a solution of $x_1^* = 1$; and $x_2^* = 1$ for the constrained problem and $x_1^* = 0.8520$, $x_2^* = 0.8520$ for the unconstrained problem. The two constraint functions have simple nonlinearity. The equality constraint is a circle while the inequality constraint is an ellipse.

$$\text{Minimize} \quad f(x_1, x_2): x_1^4 - 2x_1^2 x_2 + x_1^2 + x_1 x_2^2 - 2x_1 + 4 \qquad (7.9)$$

$$\text{Subject to:} \quad h(x_1, x_2): x_1^2 + x_2^2 - 2 = 0 \qquad (7.10\text{a})$$

$$g(x_1, x_2): 0.25 x_1^2 + 0.75 x_2^2 - 1 \le 0 \qquad (7.10\text{b})$$

$$0 \le x_1 \le 4; \ 0 \le x_2 \le 4 \qquad (7.10\text{c})$$

Figure 7.1 displays the graphical solution to the problem. It is clear from the figure that the solution is at $x_1^* = 1$, $x_2^* = 1$ (why?). The value of the function is **3**. A simple approach to the solution of the problem is to use (7.10a) to eliminate one of the variables from the problem, thereby reducing it to a single-variable constrained problem. This is not done here but it suggested for the reader to

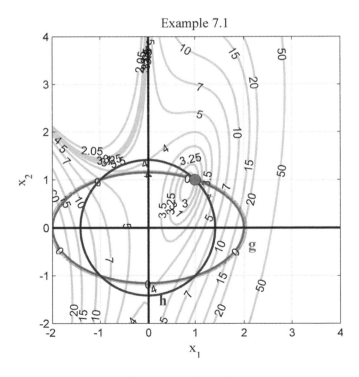

Figure 7.1 Example 7.1: Graphical Optimum.

explore. It is something an engineer should consider when dealing with difficult optimization problems.

7.1.2 Necessary Conditions

The necessary conditions, or the Kuhn-Tucker (KT) conditions, for the problem were developed in Chapter 4. They are based on the minimization of the Lagrangian function (F). The conditions are recollected here as:

$$\text{Minimize} \quad F(x_1 \ldots x_n, \lambda_1 \ldots \lambda_l, \beta_1 \ldots \beta_m) = f(x_1, \ldots, x_n)$$

$$+ \lambda_1 h_1 + \ldots + \lambda_l h_l$$

$$+ \beta_1 g_1 + \ldots + \beta_m g_m \quad (7.11)$$

subject to the constraints. There are $n + l + m$ *unknowns* in the Lagrangian. The same number of equations are required to solve the problem. These are provided by the **FOC** or the **Kuhn-Tucker** conditions. Here n equations are obtained as

$$\frac{\partial F}{\partial x_i} = \frac{\partial f}{\partial x_i} + \lambda_1 \frac{\partial h_1}{\partial x_i} + \ldots + \lambda_l \frac{\partial h_l}{\partial x_i}$$

$$+ \beta_1 \frac{\partial g_1}{\partial x_i} + \ldots + \beta_m \frac{\partial g_m}{\partial x_i} = 0; \quad i = 1, 2, \ldots, n \quad (7.12)$$

l equations are obtained directly through the equality constraints

$$h_k(x_1, x_2, \ldots, x_n) = 0; \quad k = 1, 2, \ldots, l \quad (7.13)$$

m equations are applied through the 2^m cases.

$$\beta_j g_j = 0 \rightarrow if \ \beta_j = 0 \quad then \quad g_j < 0$$

$$if \ g_j = 0 \quad then \quad \beta_j > 0 \quad (7.14)$$

The KT conditions apply only if the points *regular*.

Application of the KT Conditions to Example 7.1:   The Lagrangian is

$$F(x_1, x_2, \lambda, \beta) : \quad x_1^4 - 2x_1^2 x_2 + x_1^2 + x_1 x_2^2 - 2x_1 + 4$$

$$+ \lambda(x_1^2 + x_2^2 - 2) + \beta \, (0.25 \, x_1^2 + 0.75 \, x_2^2 - 1) \quad (7.15)$$

Three equations are set up:

$$\frac{\partial F}{\partial x_1} = 4x_1^3 - 4x_1x_2 + 2x_1 + x_2^2 - 2 + 2\lambda x_1 + 0.5\,\beta\,x_1 = 0 \qquad (7.16a)$$

$$\frac{\partial F}{\partial x_2} = -2x_1^2 + 2x_1x_2 + 2\lambda x_2 + 1.5\,\beta\,x_2 = 0 \qquad (7.16b)$$

$$h = x_1^2 + x_2^2 - 2 = 0 \qquad (7.16c)$$

The fourth equation is developed through the two cases:
Case (a): $\beta = 0$ With this information, (7.16) becomes

$$4x_1^3 - 4x_1x_2 + 2x_1 + x_2^2 - 2 + 2\lambda x_1 = 0 \qquad (7.17a)$$

$$-2x_1^2 + 2x_1x_2 + 2\lambda x_2 = 0 \qquad (7.17b)$$

$$x_1^2 + x_2^2 - 2 = 0 \qquad (7.17c)$$

and (7.17) can be solved for the three unknowns x_1, x_2, and λ.

Case (b): $g = 0$ This establishes four equations as

$$4x_1^3 - 4x_1x_2 + 2x_1 - 2 + x_2^2 + 2\lambda x_1 + 0.5\,\beta\,x_1 = 0 \qquad (7.18a)$$

$$-2x_1^2 + 2x_1x_2 + 2\lambda x_2 + 1.5\,\beta\,x_2 = 0 \qquad (7.18b)$$

$$x_1^2 + x_2^2 - 2 = 0 \qquad (7.18c)$$

$$0.25\,x_1^2 + 0.75\,x_2^2 - 1 = 0 \qquad (7.18d)$$

Equations (7.18) are solved for the four unknowns x_1, x_2, β, and λ. Here (7.18c) and (7.18d) are first solved for the design variables x_1, and x_2. This is then used in (17.18a) and (7.18b) to solve for λ and β.

MATLAB Code: `Graphical_Solution_Example7_1.m`* will plot Figure 7.1 and report the solution to the problem in the Command Window. While there are eight possibilities for Case (a), none of them yield the solution. Case (b) examines four possibilities and outputs the only solution as:

```
Solution Example 7.1 Case (b):
      x1* =   1.0000
      x2* =   1.0000


  Functions
  f*    =   3.0000
  g*    =   0.0000
```

*Files to be downloaded from the web site are indicated by boldface courier type.

```
h*   =   0.0000

Multipliers
beta*   =   1.0000
lamda*  =  -0.7500
```

The calculations are performed symbolically and the solution is also plotted on the figure. The reader is encouraged to walk through the code as it filters out the imaginary candidate solutions and only stores feasible solutions. The positive sign of beta suggests that the **KT-conditions** have been met.

7.1.3 Elements of a Numerical Technique

There will be a difference in the numerical techniques from chapter 6 since constraints are present. The idea of the search direction is still relevant in this chapter. One of the significant differences among the various gradient-based algorithms of the previous chapter was the property of the search direction. To understand the necessary changes, the general approach for unconstrained optimization, as seen in the previous chapter was as follows:

General Algorithm (A 6.0):

Step 1: Choose X_0

Step 2: For each iteration i
Determine search direction vector S_i
(It would be nice if the objective decreased along this direction)

Step 3: Calculate $\Delta X_i = \alpha_i S_i$
Note: ΔX_i is now a function of the scalar α_i as S_i is known from

Step 2: α_i is called the step size as it establishes the length of ΔX_i
α_i is determined by Minimizing $f(X_{i+1})$, where
$$X_{i+1} = X_i + \Delta X_i$$

As theses steps are reviewed, in Step 2 an important dilemma needs to be resolved. *Should the search direction S_i decrease the function or should it attempt to satisfy the constraints?* (It is assumed that it cannot do both). Since feasibility is more important than optimality it is easier to identify the latter as a significant influence in determining the direction. This is more difficult than it sounds as each constraint can have its own favorite direction at the current point. Such occasions call for a trade-off and see if it is possible to negotiate a decrease in the objective function as well. It is clear that Step 2 will involve investment of effort and some sophistication.

Step 3 is not far behind in the requirement of special handling. Once the direction is established (Step 2), what kind of step size will be acceptable? Several results are possible. First, the objective function can be decreased along the direction. Second, current active constraints (including the equality constraints) can become inactive, violated, or can still preserve its active state. Third, current inactive constraints can undergo a similar change of state. Fourth, probably most important, current violated constraints become active or inactive. It is difficult to encapsulate this discussion in a generic algorithm. Some algorithms combine Step 2 and Step 3 into a single step. Some divide up the iteration into a feasibility component and an optimality component. Therefore, the generic algorithm for this chapter, which is not particularly useful, is this:

Generic Algorithm: (A 7.0):

Step 1:   Choose X_0

Step 2:   For each iteration i
          Determine search direction vector S_i
          (It would be nice if the objective decreased and the constraints were still feasible)

Step 3:   Calculate $\Delta X_i = \alpha_i S_i$
          $X_{i+1} = X_i + \Delta X_i$
          α_i must try to decrease the objective and improve feasibility

7.2   INDIRECT METHODS FOR CONSTRAINED OPTIMIZATION

These methods were developed to take advantage of codes that solve unconstrained optimization problems. They are also referred to as *Sequential Unconstrained Minimization Techniques (SUMT)*.[1] The idea behind the approach is to repeatedly call the *unconstrained optimization algorithm* using the results of the current iteration to start the next one. The unconstrained algorithm itself will execute many iterations. A robust unconstrained minimizer, like the variable metric method, should be used for this approach to be successful. The BFGS method is robust and impressive over a large class of problems.

The principal component of the indirect method is the transformation of the constrained problem into an unconstrained problem. This is the difference among the various algorithms. The constraint functions must be folded into the objective function. All algorithms accomplish this, by adding to the original objective function, additional functions that will reflect the violation of the constraints in a significant way. These functions are referred to as the *penalty functions*. There was significant activity in this area at one time which led to entire families of different types of *penalty function* methods[2]. The first of these was the *Exterior Penalty Function (EPF)* method which is easy to appreciate and implement. The first version of *ANYSY*[3] that incorporated optimization in its finite element program relied on EPF. The EPF is presented in the next section. The EPF had several shortcomings. To address those, the *Interior Penalty Function* (IPF) methods were developed finally leading to the *Variable Penalty Function* (VPF) methods. In this text, only the EPF is addressed largely due to academic interest. In view of the excellent performance of the ***direct methods*** these methods will probably not be used today for continuous problems. They are once again important in *global optimization* techniques for constrained problems. The second method in this section, the *Augmented Lagrange Method* (ALM), is the best of the SUMT. Its exceeding simple implementation, its quality of solution, and its ability to generate information on the Lagrange multipliers allows it to seriously challenge the direct techniques.

7.2.1 Exterior Penalty Function (EPF) Method

The transformation of the optimization problem (7.1) to (7.4) to an unconstrained problem is made possible through a penalty function formulation. The transformed unconstrained problem is:

$$\text{Minimize} \quad F(X, r_h, r_g): \quad f(X) + P(X, r_h, r_g) \tag{7.19}$$

$$x_i^l \leq x_i \leq x_i^u \quad i = 1, 2, \ldots, n \tag{7.4}$$

where $P(X, r_h, r_g)$ is the penalty function, r_h, r_g are penalty constants (also called penalty multipliers), $F(X, r_h, r_g)$ is referred to as the Augmented function. The penalty function is developed as

$$P(X, r_h, r_g) = r_h \left[\sum_{k=1}^{l} h_k(X)^2 \right] + r_g \left[\sum_{j=1}^{m} \left(\max\left\{0, \ g_j(X)\right\} \right)^2 \right] \tag{7.20}$$

In (7.20) if the equality constraints are not zero then their value gets squared and multiplied by the penalty multiplier and gets added to the objective function. If the inequality constraint is in violation, it too gets squared and added to the objective function after being amplified by the penalty multipliers. In a sense, if the constraints are not satisfied then they are penalized, hence, the reason for the function's name. Also, the squaring of the function increases its importance in comparison to the original objective function, emphasizing the priority of a feasible solution. It can be shown that the transformed unconstrained problem solves the original constrained problem as the multipliers r_h, r_g approach ∞. In order for P to remain finite at these values of the multipliers (needed for a valid solution), the constraints must be satisfied. In computer implementation, this limit is replaced by a large value instead of ∞. Another facet of computer implementation of this method is that a large value at the first iteration is bound to create numerical difficulties. These multipliers are started with small values and are updated geometrically with each iteration. In the examples in the book, the unconstrained technique DFP, will be used to solve (7.19) for a known value of the multipliers. The solution returned from the DFP can be considered as a function of the multiplier and can be thought of as

$$X^* = X^*(r_h, r_g) \tag{7.21}$$

The SUMT iteration involves updating the multipliers and the initial design vector and calling the unconstrained minimizer again. In the algorithm, the DFP method is used, since we have the code. Any method from Chapter 6 can be used while BFGS is recommended.

Algorithm: Exterior Penalty Function (EPF) Method (A 7.1):

Step 1: Choose X^1, N_s (no. of SUMT iterations),
   N_u (no. of DFP iterations)
   ε_i's (for convergence and stopping)
   r_h^1, r_g^1 (initial penalty multipliers)
   c_h, c_g (scaling value for multipliers)
   $q = 1$ (SUMT iteration counter)

Step 2: Call DFP to minimize $F(X^q, r_h^q, r_g^q)$
   Output: X^{q*}

Step 3: Convergence for EPF
   If $h_k = 0$, for $k = 1, 2, \ldots l$;
   If $g_j \leq 0$, for $j = 1, 2, \ldots m$;
   If all side constraints are satisfied
   Then Converged, Stop
   Stopping Criteria:
   $\Delta F = F_q - F_{q-1}$, $\Delta X = X^{q*} - X^{(q-1)*}$
   If $\Delta F^T \Delta F \leq \varepsilon_1$: Stop (function not changing)
   Else If $\Delta X^T \Delta X \leq \varepsilon_1$: Stop (design not changing)
   Else If $q = N_s$: Stop (maximum iterations reached)
   Continue
   $q \leftarrow q + 1$
   $r_h^q \leftarrow r_h^q * C_h$; $r_g^q \leftarrow r_g^q * C_g$
   $X^q \leftarrow X^{q*}$
   Go to Step 2

The EPF is sensitive to the starting value of the multipliers, and to the scaling factors, as well. Different problems respond favorably to different values of the multipliers. It is recommended that the initial values of the multipliers be chosen as the ratio of the objective function to the corresponding term in the penalty function at the initial design. This insures that both the objective function and the constraints are equally important to determine the changes in the design for the succeeding iteration.

One reason for the term *exterior penalty function* is that at the end of each SUMT iteration the solution will be outside the feasible region or *infeasible* (until the solution is obtained). This implies that the method determines design values that are approaching the feasible region from the outside. This is a serious drawback if the method fails prematurely, as it will often do. The information generated so far is not useful as the designs were never feasible. As seen in the following example below the EPF severely increases the nonlinearity of the problem, creating conditions for the method to fail. It is expected that the increase in nonlinearity is balanced by a closer starting value for the design, as each SUMT iteration starts closer to the solution than the previous one.

In the following, the EPF is applied to Example 7.1 through a series of calculations rather than through the translation of the algorithm into MATLAB code. There are a couple of changes with respect to algorithm (A 7.1). To resolve the penalty function with respect to the inequality constraint, the constraint is assumed to be always in violation, so that the return from the *max* function is the constraint function itself. This will drive the inequality constraint to be active, which we know to be true for this example. Numerical implementation as outlined in the algorithm should allow determination of inactive constraints. Instead of numerical implementation of the unconstrained problem, an analytical solution is determined using MATLAB symbolic computation. Example 7.1 is reproduced for convenience.

Example 7.1:

$$\text{Minimize} \quad f(x_1, x_2): \quad x_1^4 - 2x_1^2 x_2 + x_1^2 + x_1 x_2^2 - 2x_1 + 4 \quad (7.9)$$

$$\text{Subject to:} \quad h(x_1, x_2): \quad x_1^2 + x_2^2 - 2 = 0 \quad (7.10a)$$

$$g(x_1, x_2): \quad 0.25\, x_1^2 + 0.75\, x_2^2 - 1 \le 0 \quad (7.10b)$$

$$0 \le x_1 \le 4; \quad 0 \le x_2 \le 4 \quad (7.10c)$$

The augmented function for the EPF method for Example 7.1 is

$$F(x_1, x_2, r_h, r_g) = (x_1^4 - 2x_1^2 x_2 + x_1^2 + x_1 x_2^2 - 2x_1 + 4) + r_h\,(x_1^2 + x_2^2 - 2)^2$$
$$+ r_g\,(0.25\, x_1^2 + 0.75\, x_2^2 - 1)^2 \quad (7.22)$$

For known values of the penalty multipliers the FOC is

$$\frac{\partial F}{\partial x_1} = 4x_1^3 - 4x_1 x_2 + 2x_1 + x_2^2 - 2 + 4r_h(x_1^2 + x_2^2 - 2)x_2$$

$$+ r_g(0.25\, x_1^2 + 0.75\, x_2^2 - 1)x_1 = 0$$

$$\frac{\partial F}{\partial x_2} = 2x_1^2 + 2x_1 x_2 + x_2^2 - 2 + 4r_h(x_1^2 + x_2^2 - 2)x_1$$

$$+ 3r_g(0.25\, x_1^2 + 0.75\, x_2^2 - 1)x_2 = 0 \quad (7.23)$$

The solution of (7.23), for a given value of r_g and r_h, can be expressed as

$$x_1 = x_1(r_h, r_g); \quad x_2 = x_2(r_h, r_g);$$

MATLAB Code: Figure 7.2 compares the augmented function and the original objective function and constraints for the penalty multipliers $r_g = 1$, and $r_h = 1$. Figure 7.3 does the same for the penalty multipliers $r_g = 10$, and $r_h = 10$. The increasing nonlinearity can be seen through the contour values of the augmented function suggesting steep valleys with increasing penalty constants. This creates difficulties for the numerical methods. These plots are created through **GraphicalPlot_EPF_Example7_1.m**.

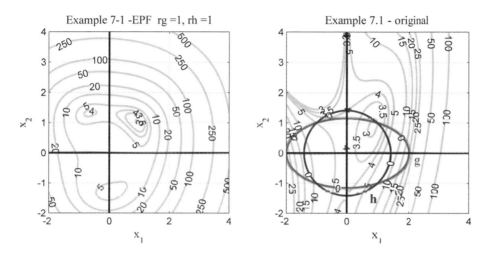

Figure 7.2 Exterior penalty function method: Example 7.1 ($r_h = 1, r_g = 1$).

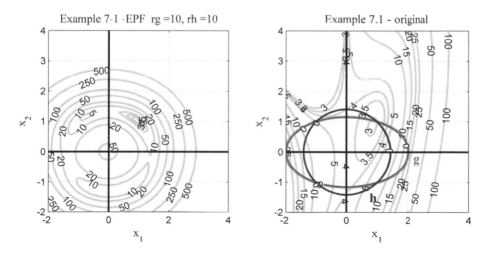

Figure 7.3 Exterior penalty function method: Example 7.1 ($r_h = 10, r_g = 10$).

MATLAB Code: `AnalyticalSolution_EPF_Example7_1.m` will determine the solution symbolically for a given value of r_h and r_g. The output in the Command window for the first is as follows:

```
Solution Example 7.1 EPF:
      x1* =   0.9228
      x2* =   1.0391
       f*  =   2.9578
       g*  =   0.0227
       h*  = -0.0687
Penalty multiplers
rg  =   1.0000
rh  =   1.0000
```

The multipliers are increased by a factor of 10 for each iteration. The iteration history is relayed in the Command window. The stopping criteria was 1.0e-06 for both the function value and the design vector. The solution was reached when the multipliers reached a value of 100,000.

```
Solution Example 7.1 EPF:
      x1* =   1.0000
      x2* =   1.0000
       f*  =   3.0000
       g*  =   0.0000
       h*  = -0.0000
Penalty multiplers
rg  = 100000.0000
rh  = 100000.0000
```

```
Program exited-Design not changing
```

Table 7.1, also printed to the command window, illustrates the values for several entities with each iteration. The solution for the first iteration.

The solution for the first iteration, for $r_h = 1$ and $r_g = 1$, is infeasible as both g and h constraints are in violation. The solution for $r_h = 100,000$ and

Table 7.1   Exterior Penalty Function Method: Example 7.1

Iteration	r_h	r_g	x_1	x_2	f	h	g
1	1	1	0.9228	1.0391	2.9578	−0.0687	0.0227
2	5	5	0.9464	1.0364	2.9651	−0.0302	0.0295
3	25	25	0.9775	1.0165	2.9810	−0.0112	0.0138
4	125	125	0.9942	1.0044	2.9944	−0.0027	0.0037
5	625	625	0.9988	1.0009	2.9988	−5.9775e−004	7.5097e−004

$r_g = 100,000$ is optimal, since both g and h are satisfied. The change in design was measured to the sixth decimal place. There are other issues that you may experience in running the code. If you attempt to tighten the tolerance on the design even more, the solutions start to diverge again and are no longer feasible, due to round-off errors for large values of the penalty multipliers. It was only possible to obtain symbolic solutions for numeric values of the penalty multipliers. Even then, there were nine possible solutions to (7.23). Side constraints are checked and the best feasible solution is chosen. The following characteristics of the exterior penalty function method can be noted with respect to Example 7.1 as the multipliers increase:

- The design approaches the optimal value
- The constraint violations decrease
- The solution is being approached from outside the feasible region.

7.2.2   Augmented Lagrange Multiplier Method

This is the most robust of the penalty function methods. More importantly, it provides information on the Lagrange multipliers at the solution. This is achieved by not solving for the multipliers, but merely updating them during successive SUMT iterations.[2,4] It overcomes many of the difficulties associated with the penalty function formulation without any significant overhead. The general optimization problem in equations (7.1) to (7.3):

$$\text{Minimize} \quad f(x_1, x_2, \ldots x_n) \tag{7.1}$$

$$\text{Subject to:} \quad h_k(x_1, x_2, \ldots, x_n) = 0, \quad k = 1, 2, \ldots, l \tag{7.2}$$

$$g_j(x_1, x_2, \ldots, x_n) \le 0, \quad j = 1, 2, \ldots, m \tag{7.3}$$

is transformed as an unconstrained optimization problem in the augmented function:

Minimize:

$$F(X, \lambda, \beta, r_h, r_g): \quad f(X) + r_h \sum_{k=1}^{l} h_k(X)^2 + r_g \sum_{j=1}^{m} \left(\max \left[g_j(X), -\frac{\beta_j}{2r_g} \right] \right)^2$$

$$+ \sum_{k=1}^{l} \lambda_k h_k(X) + \sum_{j=1}^{m} \beta_j \left(\max \left[g_j(X), -\frac{\beta_j}{2r_g} \right] \right) \tag{7.24}$$

$$x_i^l \le x_i \le x_i^u \quad i = 1, 2, \ldots, n \tag{7.4}$$

Here λ is the multiplier vector tied to the equality constraints, β is the multiplier vector associated with the inequality constraints, r_h, r_g are the penalty multipliers used similar to the EPF method. F is solved as an unconstrained function for defined values of λ, β, r_h, r_g. Therefore the solution at the end of each SUMT iteration is

$$X^* = X^*(\lambda, \beta, r_h, r_g)$$

Before the start of the next SUMT iteration, the values of the multipliers and penalty constants are updated. The latter is usually geometrically scaled, but unlike EPF do not have to be driven to ∞ for convergence.

Algorithm: Augmented Lagrange Multiplier (ALM) Method (A 7.2):

Step 1: Choose X^1, N_s (no. of SUMT iteraions),
   N_u (no. of DFP iterations)
   ε_i's (for convergence and stopping)
   r_h^1, r_g^1 (initial penalty multipliers)
   c_h, c_g (scaling value for multipliers)
   r_h^{max}, r_g^{max} (maximum value for multipliers)
   λ^1, β^1 (initial multiplier vectors)
   $q = 1$ (SUMT iteration counter)

Step 2: Call DFP to minimize $F(X^q, \lambda^q, \beta^q, r_h^q, r_g^q)$
   Output: X^{q*}

Step 3: Convergence for ALM
   If $h_k^q = 0$, for $k = 1, 2, \ldots l$;
   If $g_j^q \leq 0$, for $j = 1, 2, \ldots m$;
   (*If* $\beta_j^q > 0$ for $g_j = 0$)
   (*If* $\nabla f + \sum \lambda_k^q \nabla h_k^q + \sum \beta_j^q \nabla g_j^q = 0$)
   If all side constraints are satisfied
   Then Converged, Stop
   Stopping Criteria:
   $\Delta F = F^q - F^{q-1}$, $\Delta X = X^{q*} - X^{(q-1)*}$
   If $\Delta F^T \Delta F \leq \varepsilon_1$: Stop (function not changing)
   Else If $\Delta X^T \Delta X \leq \varepsilon_1$: Stop (design not changing)
   Else If $q = N_s$: Stop (maximum iterations reached)
   Continue
   $q \leftarrow q + 1$
   $\lambda^q \leftarrow \lambda^q + 2r_h^q h(X^{q*})$
   $\beta^q \leftarrow \beta^q + 2r_g^q (\max[g(X^{q*}), -\beta^q/2r_g^q])$
   $r_h^q \leftarrow r_h^{q*} C_h$; $r_g^q \leftarrow r_g^{q*} C_g$
   $X^q \leftarrow X^{q*}$
   Go to Step 2

To apply the ALM method to Example 7.1 we will have to first transform the problem to the unconstrained form.

Example 7.1:

$$\text{Minimize} \quad f(x_1, x_2): \quad x_1^4 - 2x_1^2 x_2 + x_1^2 + x_1 x_2^2 - 2x_1 + 4 \qquad (7.9)$$

$$\text{Subject to:} \quad h(x_1, x_2): \quad x_1^2 + x_2^2 - 2 = 0 \qquad (7.10a)$$

$$g(x_1, x_2): \quad 0.25\,x_1^2 + 0.75\,x_2^2 - 1 \le 0 \qquad (7.10b)$$

$$0 \le x_1 \le 4; \quad 0 \le x_2 \le 4 \qquad (7.10c)$$

Minimize

$$
\begin{aligned}
F(x_1, x_2, r_g, r_h, \lambda, \beta): \quad & x_1^4 - 2x_1^2 x_2 + x_1^2 + x_1 x_2^2 - 2x_1 + 4 \\
& + \lambda(x_1^2 + x_2^2 - 2) + r_h(x_1^2 + x_2^2 - 2)^2 \\
& + \beta \max\left[0.25\,x_1^2 + 0.75\,x_2^2 - 1)^2, \; -\frac{\beta}{2r_g} \right] \\
& + r_g \left[\max(0.25\,x_1^2 + 0.75\,x_2^2 - 1)^2, \; -\frac{\beta}{2r_g} \right]^2
\end{aligned}
$$
$$(7.25)$$

For a known value of the multipliers and constants we can find the solution X^*

MATLAB **Code:** Example 7.1 is solved using the wrapper function **ALM_Example7_1.m**. All initialization and program control is performed here and the ALM algorithm, available in **ALM.m** is called with the initial guess. The number of iteration of the DFP method is counted in the Command window. The total number of the ALM iterations is also shown including the summary printout. The start iteration and the final iteration, in this case the fourth, are copied from the Command window below.

```
**********************
*ALM iteration number:      0
**********************
Design Vector (X):       3    2
Objective function:      64
Sum of Squared Error in constraints(h, g):    1.2100e+002
1.8063e+001
Lagrange Multipliers (lamda beta):      1    1
Penalty Multipliers (rh rg):    5.2893e-001  3.5433e+000

**********************
*ALM iteration number:      4
**********************
Design Vector (X):                9.9940e-001  1.0002e+000
```

```
Objective function:                      2.9994e+000
Sum of Squared Error in constraints(h, g):   6.1709e-007
   3.5508e-011
Lagrange Multipliers (lamda beta):           9.0778e-002
   9.5157e-001
Penalty Multipliers (rh rg):                 5.2893e+002
   3.5433e+003
```

The track of other quantities like the penalty multipliers are also reported to the Command window. The graphical plot of the functions and design changes in the ALM iterations are plotted in Figure 7.4. The solution from ALM is close to the analytical one for the design and the objective. The multipliers do differ substantially. The ALM calls the DFP method to solve the sequence of unconstrained optimization problems. The implementation of the ALM algorithm is a serious piece of code. These algorithms are not trivial. To translate algorithm (A 7.2) into working code is a tremendous accomplishment, particularly if your exposure to MATLAB happened through this book. Please walk through the code line by line. The organization and structure are kept simple and straightforward. Comments have been liberally interspersed in the code.

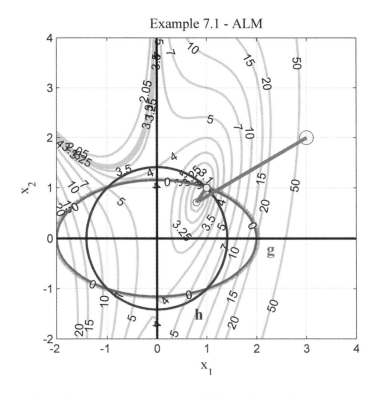

Figure 7.4 Augmented Lagrange Multiplier Method: Example 7.1.

The significant features of the code are as follows:

- The initial design information is written in the wrapper file.
- The program control can be changed for different problems and therefore placed in the wrapper file.
- The initial guess for the penalty and Lagrange multipliers are defined in the wrapper file.
- The side constraints are only checked in the wrapper file and are not part of the algorithm implemented.
- The function names that define objective, constraints, and the augmented function are defined in the wrapper file. The ALM algorithm is independent of specific file names.
- Separate files are used for each type of function. Multiple constraints of the same type should be returned as a vector.
- The **DFP.m** is used for unconstrained optimization. This program uses **GoldSection_nVar.m**, **UpperBound_nVar.m**, and **gradfunction.m**.
- The program uses global statements to communicate values between the wrapper file and the ALM algorithm.
- The initial penalty and Lagrange multipliers are adjusted based on the initial guess.
- The track of the design changes is plotted through the wrapper file.

The ALM solution for Example 7.1 satisfies the KT conditions. The values of the multipliers, obtained using symbolic computation are quite different. The tolerance has to sustainably decrease to obtain a modest improvement. You can experiment with different initial guesses and conclude that the side constraints must be part of the problem definition. This is left as an exercise for the student.

The ALM method is an effective and a useful method in spite of belonging to the class of indirect methods. It has several significant advantages over the other penalty function formulations.[2] Some of them can be listed as:

- The method is not sensitive to the initial penalty multipliers or their scaling strategy.
- The method does not require the penalty multipliers to be driven to extreme values to establish convergence.
- The equality and active constraint can be satisfied precisely.
- The starting design can be feasible or infeasible.
- The initial choices for the multipliers can be relaxed.
- There is only a modest increase in the calculations with respect to other penalty formulations but the robustness has increased significantly.
- At convergence the Lagrange multipliers will be driven to their optimum values. This allows verification of KT conditions through the sign of the multipliers associated with the active inequality constraints.

7.3   DIRECT METHODS FOR CONSTRAINED OPTIMIZATION

In this edition we introduce a constrained version of the Scan and Zoom algorithm that was used in Chapter 6 for unconstrained minimization. This will be a *zero-order* method where only function values are used to drive the search for the optimum. The remaining methods are all *first-order* methods. Both the zero-order and first-order direct methods must involve both the objective and the constraints in the search for the optimal solution. While the first-order methods do not involve conversion to a different class of problems (like Section 7.2), many of them are based on ***linearization*** of the functions (objective and constraints, too) about the current design point. Linearization is based on expansion of the function about the current variable values using the Taylor series (Chapter 4).

7.3.1   Constrained Scan and Zoom

It is difficult to characterize the Constrained Scan and Zoom method effectively through an algorithm. Unlike most of the methods in this chapter, it does not deal with search directions. It executes a disciplined search for the minimum in the design region. Its primary input is the extant of the design region through the prescription of the lower and upper bounds for the design variables. Using a coarse grid, the maximum values of the functions are noted. These values are the limits of the function. For multiple constraints of the same type, the maximum constraint dissatisfaction in the set is used to drive the search. This is the first zoom level. Understand that these maximum values could be at different values for the design.

During the first pass, it tries to identify the best design, that at the least is less than the maximum values of the function. It is expected that there will be several of them. The best design at any zoom level is that which decreases the dissatisfaction in the equality constraint while ensuring that the objective does not increase. If there are no equality constraints then the search will be driven by the maximum violated inequality constraints. Once the design is identified, the search region is shrunk to half the size of the previous zoom level centered at the current best design. The limits are also decreased by half, and the process is repeated.

If the coarse grid at the next level does not identify a better solution, the number of grid points is increased and the design region searched again. This is done three times, and if no better solution is found, then the search is stopped and the program exits. The algorithm was developed using Example 7.1 and is not extensively tested. Each level will give a better estimate of the solution in terms of significant figures. A prescribed number of levels can be searched. It is possible to stop after a prescribed tolerance in the value of f. The algorithm is iterative with a diminishing design region and an improved estimate for the minimum.

Algorithm: Scan and Zoom (A 7.3):

Step 1:   Choose X_{ci} (center of design region) with $\pm \Delta X_i$ being the extant of the of the design region.

nL: number of levels,

nP: number of points for each variable.

Initialize iLevel $= 1$,

Grid design region using nP points and identify f_limit $=$ max(f)

h_limit $=$ max($h_1, h_2, \ldots h_k$)

g_limit $=$ max($g_1, g_2, \ldots g_m$)

Initialize Best fsol, hsol, gsol, xsol

Step 2:   For each Level iLevel

Grid design region using nP points

Scan all the points: For each point ij

If $h_{ij} <=$ h_limit & $g_{ij} <=$ g_limit & $f_{ij} <$ f_limit

If $h_{ij} <$ hsol & $f_{ij} <=$ fsol

hsol $- h_{ij}$

fsol $= f_{ij}$

gsol $= g_{ij}$

xsol $- [ij]$

If finished scanning, store the best values for each level

Xbest(iLevel) $=$ xsol

Fbest(iLevel) $=$ fsol

Hbest(iLevel) $-$ hsol

Gbest(iLevel) $=$ gsol

Step 3:   If design is not changing

$nP = nP + 10$;

Go to Step 2

Calculate $\Delta f =$ change in best f between current and previous level

If $\Delta f <$ tolerance: exit

Else: $X_{ci} = X_{i,\min}$; $\Delta X_i = 0.5 * \Delta X_{i-1}$

f_limit $= 0.5*$f_limit

h_limit $= 0.5*$h_limit

g_limit $= 0.5*$g_limit

ilevel $\leftarrow$ ilevel $+ 1$

Go to Step 2

The algorithm is straightforward and unsophisticated. This algorithm assumes that both h and g exist. At each level it will define the design space around the centric point, use nP points linearly distributed along each variable, scan for the minimum, store the best value, and move on to next level until it meets the stopping criteria or the number of levels. The shrinking design space is both flexible and improves in accuracy with each level.

Example 7.1:

$$\text{Minimize} \quad f(x_1, x_2): \quad x_1^4 - 2x_1^2 x_2 + x_1^2 + x_1 x_2^2 - 2x_1 + 4 \quad (7.9)$$

$$\text{Subject to:} \quad h(x_1, x_2): \quad x_1^2 + x_2^2 - 2 = 0 \quad (7.10a)$$

$$g(x_1, x_2): \quad 0.25\, x_1^2 + 0.75\, x_2^2 - 1 \le 0 \quad (7.10b)$$

$$0 \le x_1 \le 4; \quad 0 \le x_2 \le 4 \quad (7.10c)$$

MATLAB **Code:** Example 7.1 is solved using the wrapper function **Constrained_Scan_Zoom_Example7_1.m**. Algorithm A 7.3 is programmed in **Constrained_Scan_Zoom.m**. The initialization, program control, identifying the function names is performed in the wrapper file. For two variables the functions are plotted and the iterations are traced on the plot. The function files are the ones used in the earlier methods. Using the design range in (7.10c), with 21 points in each direction, the solution is dramatically found in a single iteration. This is a coincidence because one of the grid points lies on the minimum. Only a portion of the Command window is reproduced to illustrate the solution. The selected output is:

```
* * * * * * * * * * * * * * * *
*STARTING VALUE:
* * * * * * * * * * * * * * * *
Midpoint Design Vector (Xmid)      :        2      2
Largest Objective function  (flimit)  :     268
Largest Equality Constraint Value :      30
Largest Inequality Constraint value:      15
Lower Design Space Limit (X)     :       0      0
Upper Design Space Limit (X)     :       4      4


* * * * * * * * * * * * * *
*ZOOM LEVEL:       1
* * * * * * * * * * * * * * * *
Design Vector (X)     :        1      1
Objective function   :      3
f_limit   :     268
Equality Constraints :      0
h_limit:      30
Inequality Constraints:      0
g_limit:      15

Lower Limit Design (X)     :        0      0
Upper Limit Design (X)     :        2      2
Number of scan steps:      21
```

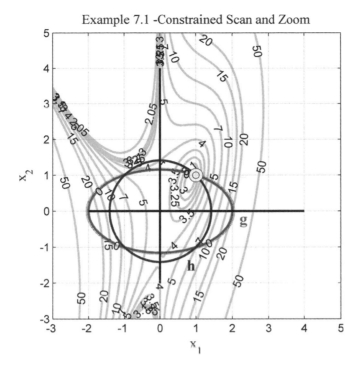

Figure 7.5 Constrained scan and zoom method: Example 7.1.

The algorithm will execute three times and add more grid points and try and improve the current solution. When the design does not change then the program stops and returns the control to the wrapper function. Figure 7.5 is the plot of the functions and the trace of the midpoint of the design region for the constrained scan and zoom method. As you can see, the solution was obtained at the first zoom level. If we rerun the wrapper file with a very large design region as reported in the starting iteration, then the solution is still obtained after five zoom levels:

```
Example 7.1 (CONSTRAINED SCAN ZOOM)
*******************************
*****************
*STARTING VALUE:
*****************
Midpoint Design Vector (Xmid) :   1.0000e+001  5.5000e+000
Largest Objective function  (flimit)  :      402804
Largest Equality Constraint Value :    984
Largest Inequality Constraint value:   426
Lower Design Space Limit (X)   :     -5    -8
Upper Design Space Limit (X)   :     25    19
```

```
* * * * * * * * * * * * * * * * * * * * * * * * * * * * * * * * * * * * * * * * * * * * * * * * * * * * * *
*CANNOT IMPROVE SOLUTION-ZOOM LEVEL ITERATIONS:      7
* * * * * * * * * * * * * * * * * * * * * * * * * * * * * * * * * * * * * * * * * * * * * * * * * * * * * * *
Design Vector (X)    :     1.0000e+000  1.0000e+000
Objective function   :      3
f_limit   :   1.2588e+004
Equality Constraints :   1.7764e-015
h_limit:   3.0750e+001
Inequality Constraints:  -1.3323e-015
g_limit:   1.3313e+001
```

Running the wrapper file with yet another design region, the possible solution, after eight zoom levels is close and not so great because the final limits do not include the solution. The values in the Command window are as follows:

```
Example 7.1 (CONSTRAINED SCAN ZOOM)
* * * * * * * * * * * * * * * * * * * * * * * * * * * * * * * * *
* * * * * * * * * * * * * * * * * *
*STARTING VALUE:
* * * * * * * * * * * * * * * * * *
Midpoint Design Vector (Xmid):   -5.5000e+000  1.0000e+001
Largest Objective function  (flimit)  :        50884
Largest Equality Constraint Value :    623
Largest Inequality Constraint value:   3.5525e+002
Lower Design Space Limit (X)   :      -15     0
Upper Design Space Limit (X)   :       4     20


* * * * * * * * * * * * * * * * * * * * * * * * * * * * * * * * * * * * * * * * * * * * * * * * * * * * *
*CANNOT IMPROVE SOLUTION-ZOOM LEVEL ITERATIONS:     12
* * * * * * * * * * * * * * * * * * * * * * * * * * * * * * * * * * * * * * * * * * * * * * * * * * * * * *
Design Vector (X)    :     8.9780e-001  9.9107e-001
Objective function   :   2.9443e+000
f_limit   :   1.9877e+002
Equality Constraints :   2.1174e-001
h_limit:   2.4336e+000
Inequality Constraints:  -6.1823e-002
g_limit:   1.3877e+000

The limits for the design variables are
Lower Limit Design (X)    :     8.6069e-001  9.5201e-001
Upper Limit Design (X)    :     9.3491e-001  1.0301e+000
```

Note, that the range for x_1 does not include the solution. At the same time the constraints are not satisfied. These design values do not provide a solution.

The Constrained Scan and Zoom method is definitely able to move the solution towards the optimum. It still needs some work in ensuring that the adjustments of the limits do not prevent a possible solution. It should be possible to perturb the limits if the increase in grid points is not effective when the design does not change for three successive increments in the number of grid points. This is suggested as an exercise among the problem sets. The method is definitely simple, easy to apply, and very natural.

7.3.2 Expansion of Functions

The Taylor (Section 4.2.3) series for a *two-variable* problem expanded quadratically about the point (x_{1p}, x_{2p}) is expressed as:

$$
f(x_{1p} + \Delta x_1, x_{2p} + \Delta x_2) = f(x_{1p}, x_{2p}) + \left[\frac{\partial f}{\partial x_1} \bigg|_{(x_{1p}, x_{2p})} \Delta x_1 \right.
$$

$$
\left. + \frac{\partial f}{\partial x_2} \bigg|_{(x_{1p}, x_{2p})} \Delta x_2 \right] + \frac{1}{2} \left[\frac{\partial^2 f}{\partial x_1^2} \bigg|_{(x_{1p}, x_{2p})} (\Delta x_1)^2 \right.
$$

$$
\left. + 2 \frac{\partial^2 f}{\partial x_1 \partial x_2} \bigg|_{(x_{1p}, x_{2p})} \Delta x_1 \Delta x_2 + \frac{\partial^2 f}{\partial x_2^2} \bigg|_{(x_{1p}, x_{2p})} (\Delta x_2)^2 \right] \tag{7.26}
$$

If the displacements are organized as a column vector $[\Delta x_1 \Delta x_2]^T$, the expansion in (7.26) can be expressed in a condensed manner as

$$
f(x_{1p} + \Delta x_1, x_{2p} + \Delta x_2) = f(x_{1p}, x_{2p}) + \nabla f|_{(x_{1p}, x_{2p})}^T \begin{bmatrix} \Delta x_1 \\ \Delta x_2 \end{bmatrix}
$$

$$
+ \frac{1}{2} \begin{bmatrix} \Delta x_1 & \Delta x_2 \end{bmatrix}^T \begin{bmatrix} H(x_{1p}, x_{2p}) \end{bmatrix} \begin{bmatrix} \Delta x_1 \\ \Delta x_2 \end{bmatrix} \tag{7.27}
$$

For *n-variables*, with X_p the current point and ΔX the displacement vector,

$$
f(X_p + \Delta X) = f(X_p) + \nabla f(X_p)^T \Delta X + \frac{1}{2} \Delta X^T H(X_p) \Delta X \tag{7.28}
$$

Equation (7.28) can be written in terms of the difference in function values as

$$
\Delta f = f(X_p + \Delta X) - f(X_p) = \nabla f(X_p)^T \Delta X + \frac{1}{2} \Delta X^T H(X_p) \Delta X \tag{7.29}
$$

$$
\Delta f = \delta f + \delta^2 f
$$

where $\delta f = \nabla f^T \Delta X$ is termed as the first or linear variation. $\delta^2 f$ is the second or quadratic variation and is given by the second term in the above equation.

Linearization: In linearizing the function f about the current value of design X_0, only the first variation is used. The neighboring value of the function can be expressed as:

$$\tilde{f}(X_0) = f(X_0) + \nabla f(X_0)^T \Delta X \qquad (7.30)$$

All the functions in the problem can be linearized similarly. It is essential to understand the difference between $f(X)$ and $\tilde{f}(X_0)$. This is illustrated in Figure 7.6 using the objective function of Example 7.1 expanded about the current design $x_1 = 3$, $x_2 = 2$. The curved lines $f(X)$ are the contours of the original function. The inclines lines are $\tilde{f}(X_0)$, the contours of the linearized function obtained through the following:

Original function: $f(x_1, x_2)$: $x_1^4 - 2x_1^2 x_2 + x_1^2 + x_1 x_2^2 - 2x_1 + 4$

Linearized function: $\tilde{f}(\Delta x_1, \Delta x_2) = 64 + 92 \Delta x_1 - 6 \Delta x_2$

where the coefficients represent the evaluation of the function and gradients at $x_1 = 3$, $x_2 = 2$ and substituted in (7.30). In Figure 7.6 several contours are shown

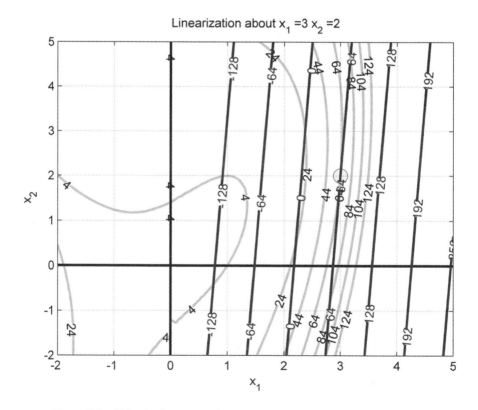

Figure 7.6 Objective function of Example 7.1 linearized about $x_1 = 3$; $x_2 = 2$.

though the ones of interest are in the neighborhood of the current design. Any nonlinear function can be expanded in the same manner. If another point was chosen then the slopes of the lines would be different.

MATLAB Code: Figure 7.6 is created using the code **LinearizationFigure.m.** You can observe that the nonlinear function and the linear representation about the point (3, 2). The error increases away from the point. You can rerun the code by selecting a different point to linearize the function.

Quadratic Expansion: The quadratic expansion of the function can be obtained by using (7.28). The expanded curves will be nonlinear (quadratic). Figure 7.7 illustrates the quadratic expansion for the same function about the same point. You can observe that function contours are elliptic and certainly approximate the original function significantly better around the point (3, 2). It is important to recognize the contours would be different if another point was chosen for expansion.

Original function: $f(x_1, x_2)$: $x_1^4 - 2x_1^2 x_2 + x_1^2 + x_1 x_2^2 - 2x_1 + 4$

Quadratic Expansion: $\tilde{f}(\Delta x_1, \Delta x_2) = 64 + [92\,\Delta x_1 - 6\,\Delta x_2]$

$$+ \left[0.5 \times 102 \times \Delta x_1^2 - 8 \times \Delta x_1 \times \Delta x_2 + 0.5 \times 6 \times \Delta x_2^2 \right]$$

MATLAB Code: Figure 7.7 is created using the code **QuadraticExpansion Figure.m.** You can observe that the nonlinear function and the quadratic expansion are coincident around the point (3, 2). The two functions diverge at a significant distance from the point. You can rerun the code by selecting a different point to expand the function.

In the following sections, four direct methods are discussed. All of them use one or both constructions illustrated. The first is the *Sequential Linear Programming (SLP)*, where the solution is obtained by successively solving the corresponding linearized optimization problem. The second, *Sequential Quadratic Programming (SQP)* uses the quadratic expansion for the objective function. Like the SUMT, the current solution provides the starting values for the next iteration. The third, *Generalized Reduced Gradient Method (GRG)* develops a sophisticated search direction and follows it with an elaborate one dimensional process. The fourth method, *Sequential Gradient Restoration Algorithm (SGRA)* uses a two-cycle approach working on feasibility and optimality alternately to find the optimum. There are several other methods, but they differ from those listed in small details.

Keep in mind that one method may not solve all problems, though it will have a particular strength in a class of problems. With the code and other procedures available from this section, it should not be difficult to program the additional techniques worth exploring. A note to the reader is not to discount simple methods

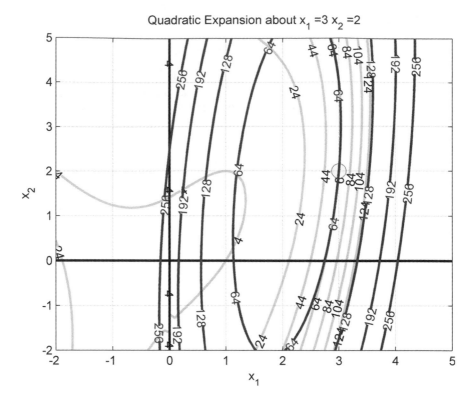

Figure 7.7 Quadratic Expansion of Objective function of Example 7.1 about $x_1 = 3; x_2 = 2$.

like the Scan and the Zoom in favor of the technically sophisticated methods in the following sections.

Except for special classes of problems in the following two chapters, this chapter is the principal reason for the book. It provides a resolution of the complete optimization problem as postulated. There are not many new or improved numerical techniques for regular constrained optimization. Many of the new advances in techniques are in the area of global optimization. There is a lot of applied optimization effort directed towards actual design applications and expanding computer aided design (CAE) to provide optimization along with analysis. The standard format of the Nonlinear Programming Problem (NLP) is reproduced here once again for convenience:

$$\text{Minimize} \quad f(x_1, x_2, \ldots x_n) \tag{7.1}$$

$$\text{Subject to:} \quad h_k(x_1, x_2, \ldots, x_n) = 0, \quad k = 1, 2, \ldots, l \tag{7.2}$$

$$g_j(x_1, x_2, \ldots, x_n) \leq 0, \quad j = 1, 2, \ldots, m \tag{7.3}$$

$$x_i^l \leq x_i \leq x_i^u \quad i = 1, 2, \ldots, n \tag{7.4}$$

In vector notation

$$\text{Minimize} \quad f(X), \quad [X]_n \tag{7.5}$$

$$\text{Subject to:} \quad [h(X)]_l = 0 \tag{7.6}$$

$$[g(X)]_m \leq 0 \tag{7.7}$$

$$X^{\text{low}} \leq X \leq X^{\text{up}} \tag{7.8}$$

7.3.3 Sequential Linear Programming (SLP)

In the SLP all the functions are expanded linearly[5]. If X_i is considered the current design vector, then the linearized optimal problem can be set up as

$$\text{Minimize:} \quad \tilde{f}(\Delta X) = f(X_i) + \nabla f(X_i)^T \Delta X \tag{7.31}$$

$$\text{Subject to:} \quad \tilde{h}_k(\Delta X): \quad h_k(X_i) + \nabla h_k^T(X_i) \Delta X = 0; \quad k = 1, 2, \ldots, l \tag{7.32}$$

$$\tilde{g}_j(\Delta X): \quad g_j(X_i) + \nabla g_j(X_i)^T \Delta X \leq 0; \quad j = 1, 2, \ldots, m \tag{7.33}$$

$$\Delta x_i^{low} \leq \Delta x_i \leq \Delta x_i^{up}; \quad i = 1, 2, \ldots, n \tag{7.34}$$

Relations (7.31) to (7.34) represents a *Linear Programming (LP)* problem (Chapter 3). All of the functions in (7.31) to (7.34), except for ΔX, have numerical values after substituting a numerical vector for X_i. Assuming an LP program code is available, it can be called repeatedly after the design is updated as

$$X_{i+1} = X_i + \Delta X$$

In order to include a search direction and stepsize calculation the ΔX in (7.31) to (7.34) is considered as the search direction at the current design S. The optimization subproblem becomes:

$$\text{Minimize:} \quad \tilde{f}(\Delta X) = f(X_i) + \nabla f(X_i)^T S \tag{7.35}$$

$$\text{Subject to:} \quad \tilde{h}_k(\Delta X): \quad h_k(X_i) + \nabla h_k^T(X_i) S = 0; \quad k = 1, 2, \ldots, l \tag{7.36}$$

$$\tilde{g}_j(\Delta X): \quad g_j(X_i) + \nabla g_j(X_i)^T S \leq 0; \quad j = 1, 2, \ldots, m \tag{7.37}$$

$$S_i^{low} \leq S_i \leq S_i^{up}; \quad i = 1, 2, \ldots, n \tag{7.38}$$

Stepsize computation strategy is no longer simple. If there are violated constraints alpha attempts to reduce this violation. If current solution is feasible the stepsize will attempt to reduce the function without causing the constraints to be excessively violated. Such strategies are largely implemented at the discretion of the algorithm or code developer.

Algorithm: Sequential Linear Programming (SLP): (A 7.4):

Step 1:   Choose X^1, $f(X^1)$, N_s (no. of iterations),
ε_i's (for convergence and stopping)
$q = 1$ (iteration counter)

Step 2:   Call LP to optimize (7.35)—(7.38)
Output: S
Use a constrained α^* calculation
$\Delta X = \alpha^* S$
$X^{q+1} = X^q + \Delta X$
$f^{q+1} = f(X^{q+1})$

Step 3:   Convergence for SLP
If $h_k = 0$, for $k = 1, 2, \ldots l$;
If $g_j \leq 0$, for $j = 1, 2, \ldots m$;
If all side constraints are satisfied
If $f^{q+1} > f^q$
Then Stop
(Other Stopping Criteria)
$\Delta X = X^{q+1} - X^q$
Else If $\Delta X^T \Delta X \leq \varepsilon_1$: Stop (design not changing)
Else If $q = N_s$: Stop (maximum iterations reached)
Continue
$q \leftarrow q + 1$
Go to Step 2

In algorithm (A 7.4), we have not really checked the KT conditions for the optimum since we are not calculating the multipliers. It is expected that the iterations are initially driven by the requirement of feasibility. For a current feasible design the iteration continues if the current objective is greater than the previous feasible one. We can certainly include the computation of the multipliers based on the current gradient values although it is not trivial. In this way we can formally establish the KT conditions for the minimum.

Example 7.1:   This example is used to illustrate the various algorithms

$$\text{Minimize} \quad f(x_1, x_2): \quad x_1^4 - 2x_1^2 x_2 + x_1^2 + x_1 x_2^2 - 2x_1 + 4 \qquad (7.9)$$

$$\text{Subject to:} \quad h(x_1, x_2): \quad x_1^2 + x_2^2 - 2 = 0 \qquad (7.10a)$$

$$g(x_1, x_2): \quad 0.25 x_1^2 + 0.75 x_2^2 - 1 \leq 0 \qquad (7.10b)$$

$$0 \leq x_1 \leq 4; \quad 0 \leq x_2 \leq 4 \qquad (7.10c)$$

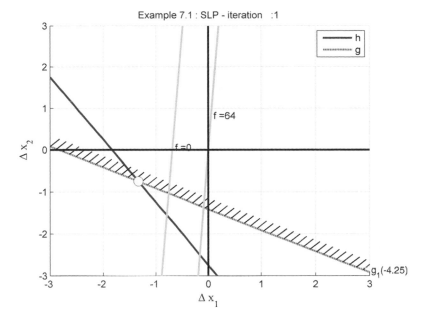

Figure 7.8 Example 7.1. SLP iteration 1.

The SLP algorithm is applied to Example 7.1 through a direct calculation of the linear subprogram expressed in (7.31) to (7.34). It is also graphically illustrated in Figures 7.8 to 7.10. No explicit code for solving the LP program is developed, although the MATLAB LP code from the optimization toolbox should work just fine.

Application of Sequential Linear Programming (SLP) to Example 7.1:

Step 1: $X^1 = [\, 3 \; 2\,]$

Step 2: Linearized sub problem

$$\text{Minimize:} \quad \tilde{f}(\Delta X) = 64 + \begin{bmatrix} 92 & -6 \end{bmatrix} \begin{bmatrix} \Delta x_1 \\ \Delta x_2 \end{bmatrix}$$

$$\text{Subject to:} \quad \tilde{h}(\Delta X): \quad 11 + \begin{bmatrix} 6 & 4 \end{bmatrix} \begin{bmatrix} \Delta x_1 \\ \Delta x_2 \end{bmatrix} = 0$$

$$g(\Delta X): \quad 4.25 + \begin{bmatrix} 1.5 & 3 \end{bmatrix} \begin{bmatrix} \Delta x_1 \\ \Delta x_2 \end{bmatrix} \leq 0$$

Figure 7.8 describes the problem graphically. The solution is identified on the figure. Note, two lines are drawn for linearized objective functions so that the direction of the minimum can be established easily. In all of the figures, the solution is at the intersection of the equality and the inequality constraints. From

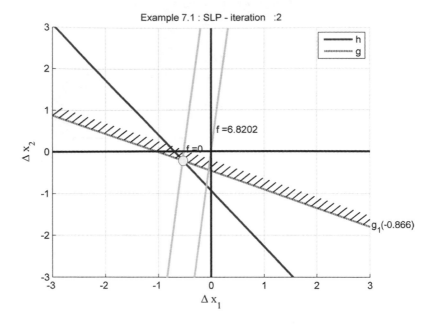

Figure 7.9 Example 7.1. SLP iteration 2.

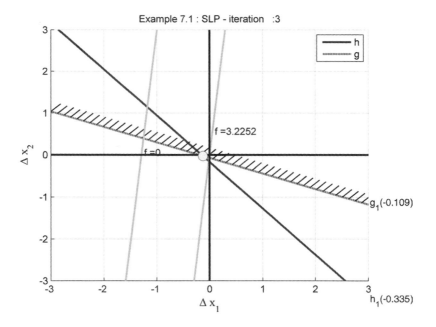

Figure 7.10 Example 7.1. SLP iteration 3.

Figure 7.8 the solution is $\Delta x_1 = 1.3333$, $\Delta x_2 = -0.75$. For the next iteration $X^2 = [1.6667\ 1.25]$. Step 2 is then repeated.

Step 2:   Linearized subproblem

$$\text{Minimize:}\quad \tilde{f}(\Delta X) = 6.8202 + \begin{bmatrix} 13.081 & -1.3889 \end{bmatrix} \begin{bmatrix} \Delta x_1 \\ \Delta x_2 \end{bmatrix}$$

$$\text{Subject to:}\quad \tilde{h}(\Delta X):\quad 2.3403 + \begin{bmatrix} 3.3333 & 2.5 \end{bmatrix} \begin{bmatrix} \Delta x_1 \\ \Delta x_2 \end{bmatrix} = 0$$

$$g(\Delta X):\quad 0.8663 + \begin{bmatrix} 0.8333 & 1.875 \end{bmatrix} \begin{bmatrix} \Delta x_1 \\ \Delta x_2 \end{bmatrix} \leq 0$$

From Figure 7.9 the solution is $\Delta x_1 = -0.5333$, $\Delta x_2 = -0.2250$. For the next iteration $X^3 = [1.1333\ 1.0250]$. Step 2 is repeated.

Step 2:   Linearized sub problem

$$\text{Minimize:}\quad \tilde{f}(\Delta X) = 3.2252 + \begin{bmatrix} 2.4934 & -0.2456 \end{bmatrix} \begin{bmatrix} \Delta x_1 \\ \Delta x_2 \end{bmatrix}$$

$$\text{Subject to:}\quad \tilde{h}(\Delta X):\quad 0.3350 + \begin{bmatrix} 2.2667 & 2.05 \end{bmatrix} \begin{bmatrix} \Delta x_1 \\ \Delta x_2 \end{bmatrix} = 0$$

$$g(\Delta X):\quad 0.10908 + \begin{bmatrix} 0.5667 & 1.5375 \end{bmatrix} \begin{bmatrix} \Delta x_1 \\ \Delta x_2 \end{bmatrix} \leq 0$$

From Figure 7.10 the solution is $\Delta x_1 = -0.1255$, $\Delta x_2 = -0.0247$. For the next iteration $X^4 = [1.0078\ 1.0003]$. We can run the code for additional iterations but the iterations have almost converged to the solution at $X = [1.0000\ 1.0000]$. It started at the infeasible point $X = [3\ 2]$. Note that the change in design variables are decreasing, as it should if it were approaching the solution.

MATLAB Code: The SLP algorithm applied to Example 7.1 is available in the m-file **SLP_Example7_1.m**. The students are encouraged to walk through the code as it is quite compact. It also creates the Figures 7.8 to 7.10. It uses symbolic calculation to obtain the solution to the linear subproblem. The information is printed to the Command window. The last iteration obtained by running the code is:

```
*****************************
Iteration:       3
*****************************
Linearized about [x1, x2]    :      1.1333     1.0250
Objective function value f(x1,x2) :      3.2252
```

```
Equality constraint value value h(x1,x2) :      0.3351
Inequality constraint value value g(x1,x2) :      0.1091


solution for  [delx1 delx2]  :      -0.1255   -0.0247


Linearized function
f(x1,x2): 3.2252 +2.4934*dx1+-0.24556 *dx2
Linearized equality constraint:
     h(x1,x2): 0.33507 +2.2667*dx1 +2.05 *dx2
Linearized inequality constraint:
     g(x1,x2): 0.10908 +0.56667*dx1 +1.5375 *dx2
New Design Values [x1, x2]   :      1.0078   1.0003
```

The SLP took three iterations to get close to the solution. This is very impressive. This may be true for those cases where the number of equality plus active inequality constraints is the same as the number of design variables–*a full complement*. The main disadvantage of the method appears if there is no full complement. In that case the side constraints (7.34) are critical to the determination of the solution as they develop into active constraints. In these situations the limits on the side constraints are called the ***move limits***. In case of active side constraints the move limits establish the values of the design changes. If it is the search direction (see algorithm) then both *value* and *direction* are affected. For these changes to be small as the solution is approached, the move limits *have to be adjusted (lowered) with each iteration*. The strategy for this adjustment will influence the solution.

To simulate the need for move limits consider Example 7.1 without the inequality constraint, but with the equality constraint. Figure 7.11 represents the linearization about the point.[3,2] The move limits in the figure (box) represent the side constraint limits

$$-2 \leq \Delta x_1 \leq 2; \quad -2 \leq \Delta x_2 \leq 2$$

MATLAB Code: The SLP iteration with ***move limits*** is illustrated in Figure 7.11. The figure and the values for the functions are calculated in the code **SLP_MoveLimits_Example7_1.m**. It is clear in Figure 7.11 that without the move limits the solution is unbounded, which would not be very helpful. Using the solution obtained in the figure, the next iteration can be executed. Once again the move limits will influence the solution. If the move limits are left at the same value, the solution will always be on the square with no possibility of convergence. These limits must be lowered with each iteration. Usually they are done geometrically through a scaling factor. This implies the convergence will depend on the strategy for changing the move limits, not an appealing situation. The SLP is not used extensively although it provides an excellent example of the ideas/properties of linearization. The next method avoids the move limits altogether.

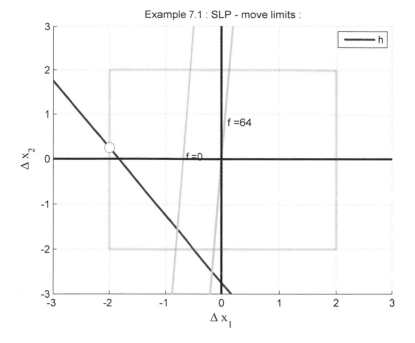

Figure 7.11 SLP with move limits.

7.3.4 Sequential Quadratic Programming (SQP)

The lack of robustness in the SLP can be countered by including an element of nonlinearity in the problem. The linearized subproblem is

$$\text{Minimize} \quad \tilde{f}(\Delta X) = f(X_i) + \nabla f(X_i)^T \Delta X \tag{7.31}$$

$$\text{Subject to:} \quad \tilde{h}_k(\Delta X): \quad h_k(X_i) + \nabla h_k^T(X_i)\,\Delta X = 0; \quad k = 1, 2, \ldots, l \tag{7.32}$$

$$\tilde{g}_j(\Delta X): \quad g_j(X_i) + \nabla g_j(X_i)^T\,\Delta X \le 0; \quad j = 1, 2, \ldots, m \tag{7.33}$$

$$\Delta x_i^{low} \le \Delta x_i \le \Delta x_i^{up}; \quad i = 1, 2, \ldots, n \tag{7.34}$$

One way is to include nonlinearity is to replace (7.34) by

$$\Delta X^T \Delta X \le c \text{ (constant)}$$

or require the search directions (remember in actual implementation ΔX is S) in (7.35)–(7.38) to be limited by the unit circle which can be described the constraint:

$$S^T S \le 1 \tag{7.39}$$

The stepsize determination will account for the actual change in the design vector. The problem expressed by (7.35) to (7.39) is a *quadratic programming (QP)* subproblem because of (7.39). An iterative solution of this subproblem can be one form of the *Sequential Quadratic Programming (SQP)* method. While the QP subprogram is a simple nonlinear problem with linear constraints it would still require methods of this chapter for solution. Fortunately, quadratic programming problems appear in many disciplines and there are efficient procedures and algorithms to address these problems.[6–10] One of these methods is an extension of the Simplex method of Linear Programming.

There is a more popular version of the Quadratic Programming that is developed below. The solution to the QP is well documented in the literature. Readers are encouraged to seek out the various references to gain appreciation for the implementation details. The QP method is also available in the *Optimization Toolbox* in MATLAB. The formal methods for QP problems are based on efficiency in various stages of the implementation. An inefficient, but successful solution to the problem can be obtained by using the ALM of the previous section. After all, computing resources are not a constraint today. In this book a more intuitive development of the SQP algorithm and the corresponding numerical technique is presented through Example 7.1. The author is confident that you will be capable of understanding the method and translating it into the necessary code. The translation of SQP into code is not trivial is usually the basis of commercial software. For example, SQP is the principal algorithm for NLP in the MATLAB Optimization Toolbox. A detailed discussion of the solution technique is also available in.[11]

The quadratic programming subproblem employed in this subsection, is based on expanding the objective function quadratically about the current design. The constraints are expanded linearly as in SLP.[8] This is called the *Sequential Quadratic Programming (SQP)*.

Minimize: $\tilde{f}(\Delta X) = f(X_i) + \nabla f(X_i)^T \Delta X + \dfrac{1}{2} \Delta X^T \nabla^2 f(X_i) \Delta X$

$$(7.40)$$

Subject to: $\tilde{h}_k(\Delta X)$: $h_k(X_i) + \nabla h_k^T(X_i) \Delta X = 0$; $k = 1, 2, \ldots, l$ (7.32)

$\tilde{g}_j(\Delta X)$: $g_j(X_i) + \nabla g_j(X_i)^T \Delta X \le 0$; $j = 1, 2, \ldots, m$

$$(7.33)$$

$\Delta x_i^{low} \le \Delta x_i \le \Delta x_i^{up}$; $i = 1, 2, \ldots, n$ (7.34)

In (7.40) $\nabla^2 f(X_i)$ is the *Hessian* matrix. In practical implementation, since Hessian computation is discouraged, a metric $[H]$ that is updated with each iteration is used. This is based on the success of the Variable Metric Methods (VMM) of the previous chapter. Several researchers have shown the BFGS update for the Hessian provides an efficient implementation of the SQP method.[6,9] This QP is a well posed and convex and should yield a solution. Solution to this subproblem is at the heart of SQP. A substantially large amount of research has been invested to develop efficient techniques to handle this subproblem. In a formal implementation of the SQP the ΔX in (7.40) and (7.32) to (7.34) must be replaced by the search direction S. The QP for S also modifies the constraint equations so that a feasible direction can be found with respect to the current active constraints. Figure 7.12 illustrates the concern for moving linearly to a point on the constraint. X^0, the current design is on the active constraint in the figure. If the search direction S^1 follows the linearized function at X^0 (the tangent to the function at that point), any stepsize, along the search direction, however small, will cause the constraint to be violated. In order to determine a neighboring point that will still satisfy the constraints, a search direction slightly less than the tangent is used. This introduces a lack of consistency, as the deviation from tangency becomes a matter of individual experience and practice. Experiments have suggested that 90 to 95 percent of the tangent is a useful figure although making it as close to 100% is recommended.[2] The search direction finding QP subproblem is

$$\text{Minimize} \quad \tilde{f}(S) = f(X_i) + \nabla f(X_i)^T S + \frac{1}{2} S^T [H] S \qquad (7.41)$$

$$\text{Subject to:} \quad \tilde{h}_k(S): \quad c\, h_k(X_i) + \nabla h_k^T(X_i)\, S = 0; \quad k = 1, 2, \ldots, l \quad (7.42)$$

$$\tilde{g}_j(S): \quad c\, g_j(X_i) + \nabla g_j(X_i)^T\, S \le 0; \quad j = 1, 2, \ldots, m \qquad (7.43)$$

$$S^T S \le 1 \qquad (7.39)$$

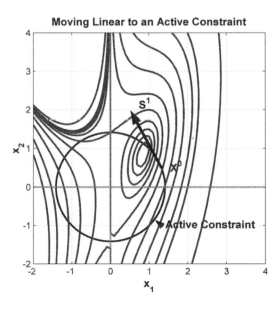

Figure 7.12   Moving linear to an active constraint.

The side constraints are not necessary as this is a well posed problem. The factor c is adjustment for moving tangent to the constraint. The suggested value for c is *0.9* to *0.95*. For the inequality constraint, the value of c is *1* if the constraint is satisfied but not active (should be able to move tangent to the constraint).

Stepsize Calculation: Once the search direction has been determined it is necessary to calculate the stepsize. Since the methods of this chapter have seen an enormous increase in the number of calculations, the stepsize calculations are based on simultaneously decreasing the objective as well as improving the constraint satisfaction. In order to accomplish this goal a suitable function must be developed and the stepsize determination is then based on minimizing this unconstrained function. Several authors refer to it as the *descent function*. The description of this function varies in the literature. Some define this as the Lagrangian function in which case the Lagrange multipliers must be determined. Many others include only the constraints with a penalty multiplier and adopt the EPF form for the function. Still others include only the maximum violated constraint. Two forms of the function are listed here:

$$X^{i+1} = X^{i-1} + \alpha S \tag{7.44}$$

Minimize: $$\phi(X^{i+1}) = f(X^{i+1}) + r \sum_{k=1}^{l} h_k(X^{i+1})^2$$

$$+ r \sum_{j=1}^{m} \max[\, g_j(X^{i+1}), 0\,]^2 \tag{7.45}$$

Minimize: $$\phi(X^{i+1}) = f(X^{i+1}) + \sum_{k=1}^{l} |\lambda_k h_k(X^{i+1})|$$

$$+ \sum_{j=1}^{m} \beta_j \, \max[\, g_j(X^{i+1}), 0\,] \tag{7.46}$$

Equations (7.45) or (7.46) can be used to calculate the stepsize α. The function in (7.46) requires an important observation. Should the Lagrange multipliers be calculated for each value of α. Imagine using the Golden Section method to solve (7.46) with the multipliers calculated from the satisfaction of the Kuhn-Tucker conditions for each new value of X. Several updating strategies for the multipliers in (7.46) can be found in the listed references. A simple one is to hold it constant at the previous iteration value.

Calculating the Multipliers: The popularity of the SQP method is that it attempts to update the Hessian based on the Lagrangian of the problem. Calculating the multipliers is necessary to verify the Kuhn-Tucker conditions at the solution. Calculating the multipliers, for the current value of the design vector is necessary for additional operations, including setting up the descent function in (7.46), as well as updating the metric replacement for the Hessian (follows next). The calculation uses the FOC based on the Lagrangian. The values of the gradients of the various functions are computed at the current design. The multipliers are then obtained as a solution to a set of linear equations. The solution to QP and SQP problems usually take advantage of an *active constraint set*. This includes all equality constraints and only those inequality constraints that are active. This implies that SQP iterations *start at a feasible point*. The multipliers corresponding to inactive constraints are set to zero. The remaining multipliers can be established, considering a set of appropriate number of linear independent equations from (7.12).

$$\frac{\partial F}{\partial x_i} = \frac{\partial f}{\partial x_i} + \lambda_1 \frac{\partial h_1}{\partial x_i} + \ldots + \lambda_l \frac{\partial h_l}{\partial x_i} + \beta_1 \frac{\partial g_1}{\partial x_i} + \ldots + \beta_m \frac{\partial g_m}{\partial x_i} = 0;$$

$$i = 1, 2, \ldots, n \tag{7.12}$$

Replacing Hessian with BFGS update: In (7.40) or (7.41), calculating the actual Hessian is strongly discouraged. Since the BFGS method converges to the Hessian, the BFGS metric is considered to be an excellent replacement of the Hessian.[9] The metric is based on the Lagrangian of the function.[12]

$$p = \nabla f(X^{q+1}) + \sum_{k-1}^{l} \lambda_k \nabla h_k(X^{q+1}) + \sum_{j-1 \forall j \in J}^{m} \beta_j \nabla g_j(X^{q+1})$$

$$- \nabla f(X^q) + \sum_{k=1}^{l} \lambda_k \nabla h_k(X^q) + \sum_{j=1 \forall j \in J}^{m} \beta_j \nabla g_j(X^q) \tag{7.47}$$

$$y = X^{q+1} - X^q$$

$$[H^{q+1}] = [H^q] + \frac{pp^T}{p^T y} - \frac{[H^q]^T [H^q]}{y^T [H^q] y}$$

$j \in J$ implies the set of active inequality constraints. It is recommended that the metric be kept positive definite as much as possible even if it is likely that it will not be positive definite at the solution. One way of ensuring this is to verify that

$$p^T y > 0$$

for each iteration.

Algorithm: Sequential Quadratic Programming (SQP)–(A 7.5):

Step 1: Choose X^1, N_s (no. of iterations),
         ε_i's (for convergence and stopping)
         $q = 1$ (iteration counter)

Step 2:   Call QP to optimize (7.41)—(7.43)
Output: **S**
Use a constrained α^* calculation (descent function)
$\Delta X = \alpha^* S$
$X^{q+1*} = X^q + \Delta X$

Step 3:   Convergence for SQP
If $h_k = 0$, for $k = 1,2, \ldots \ l$;
If $g_j \leq 0$, for $j = 1,2, \ldots \ m$;
If KT conditions are satisfied
Then Converged, Stop
Stopping Criteria:
$\Delta X = X^{q+1*} - X^{q*}$
Else If $\Delta X^T \Delta X < \varepsilon_1$: Stop (design not changing)
Else If $q = N_s$: Stop (maximum iterations reached)
Continue
Update Metric **[H]**
$q \leftarrow q + 1$
Go to Step 2

Example 7.1:   This example is used to illustrate the various algorithms

$$\text{Minimize} \quad f(x_1, x_2): \quad x_1^4 - 2x_1^2 x_2 + x_1^2 + x_1 x_2^2 - 2x_1 + 4 \tag{7.9}$$

$$\text{Subject to:} \quad h(x_1, x_2): \quad x_1^2 + x_2^2 - 2 = 0 \tag{7.10a}$$

$$g(x_1, x_2): \quad 0.25 x_1^2 + 0.75 x_2^2 - 1 \leq 0 \tag{7.10b}$$

$$0 \leq x_1 \leq 4; \quad 0 \leq x_2 \leq 4 \tag{7.10c}$$

Application of Sequential Quadratic Programming (SQP) to Example 7.1:
The SQP technique is applied to Example 7.1. There are several ways to accomplish this. The application can use graphical support, as in the demonstration of the SLP (an exercise). The application can use the QP program from MATLAB (an exercise). The application solves the QP problem by applying the KT conditions—which is the feature of this subsection. After all for this example since the functions are analytically defined we can harness the symbolic operations in MATLAB to calculate the derivatives and the actual Hessian. Similar to SLP, the actual change in design ΔX is computed (7.40, 7.32–7.34) rather then the search direction S for the QP subproblem.

MATLAB Code: `SQP_Example7_1.m` applies the basic idea behind SQP to Example 7.1 symbolically and analytically. It applies the KT conditions through two cases: Case (a) corresponds to $\beta = 0$ (*g must be less than zero*); Case (b) corresponds to $g = 0 (\beta > 0)$. The best output is then selected to establish the new design. The output from running the program is transcribed here for several iterations. Note that Algorithm (A 7.5) is not being applied.

Iteration 1:

$$X^1 = \begin{bmatrix} 3 & 2 \end{bmatrix}$$

$$f = 64; \quad \nabla f = \begin{bmatrix} 92 \\ -6 \end{bmatrix}; \quad [H] = \begin{bmatrix} 102 & -8 \\ -8 & 6 \end{bmatrix}$$

$$h = 11; \quad \nabla h = \begin{bmatrix} 6 \\ 4 \end{bmatrix}; \quad g = 4.25; \quad \nabla g = \begin{bmatrix} 1.5 \\ 3 \end{bmatrix};$$

The solution to the QP problem is **Case (a)**

$$\Delta x_1 = -1.0591; \quad \Delta x_2 = -1.1613$$

$$\tilde{h} = 0; \quad \tilde{g} = -0.8226;$$

$$\tilde{x}_1 = 1.9409; \quad \tilde{x}_2 = 0.8387;$$

Iteration 2:

$$X^2 = \begin{bmatrix} 1.9409 & 0.8387 \end{bmatrix}$$

$$f = 13.123; \quad \nabla f = \begin{bmatrix} 25.318 \\ -4.278 \end{bmatrix}; \quad [H] = \begin{bmatrix} 43.848 & -6.086 \\ -6.086 & 3.882 \end{bmatrix}$$

$$h = 2.470; \quad \nabla h = \begin{bmatrix} 3.881 \\ 1.677 \end{bmatrix}; \quad g = 0.4693; \quad \nabla g = \begin{bmatrix} 0.970 \\ 1.258 \end{bmatrix};$$

The solution to the QP problem is **Case (a)**.

$$\Delta x_1 = -0.6186; \quad \Delta x_2 = -0.0411$$

$$\dot{h} = 0; \quad \tilde{g} = -.18281;$$

$$\tilde{x}_1 = 1.3222; \quad \tilde{x}_2 = 0.7976;$$

Iteration 3:

$$X^3 = \begin{bmatrix} 1.3222 & 0.7976 \end{bmatrix}$$

$$f = 4.2127; \quad \nabla f = \begin{bmatrix} 6.3089 \\ -1.3875 \end{bmatrix}; \quad [H] = \begin{bmatrix} 19.789 & -3.6938 \\ -3.6938 & 2.6445 \end{bmatrix}$$

$$h = 0.3844; \quad \nabla h = \begin{bmatrix} 2.6445 \\ 1.5951 \end{bmatrix}; \quad g = -0.0858; \quad \nabla g = \begin{bmatrix} 0.6611 \\ 1.1963 \end{bmatrix};$$

The solution to the QP problem is **Case (a)**

$$\Delta x_1 = -0.2686; \quad \Delta x_2 = 0.2042$$

$$\tilde{h} = 0; \quad \tilde{g} = -0.0190;$$

$$\tilde{x}_1 = 1.0537; \quad \tilde{x}_2 = 1.0018;$$

Iteration 4:

$$X^4 = \begin{bmatrix} 1.0537 & 1.0018 \end{bmatrix}$$

$$f = 3.0685; \quad \nabla f = \begin{bmatrix} 1.5679 \\ -0.1092 \end{bmatrix}; \quad [H] = \begin{bmatrix} 11.315 & -2.2111 \\ -2.2111 & 2.1073 \end{bmatrix}$$

$$h = 0.1138; \quad \nabla h = \begin{bmatrix} 2.1073 \\ 2.0036 \end{bmatrix}; \quad g = 0.0302; \quad \nabla g = \begin{bmatrix} 0.5268 \\ 1.5027 \end{bmatrix};$$

The solution to the QP problem is **Case (b)**.

$$\Delta x_1 = -0.0523; \quad \Delta x_2 = -0.0018; \quad \beta = 0.9276$$

$$\tilde{h} = 0; \quad \tilde{g} = 0;$$

$$\tilde{x}_1 = 1.0014; \quad \tilde{x}_2 = 1.0000;$$

$$f = 3.0014; \quad h = 0.0027; \quad g = 0.0006$$

The solution is almost converged and there are two more iterations for the stopping criteria adopted in the code. The last iteration is copied from the Command window.

```
*****************************
Iteration:       6
*****************************
Linearized about [x1, x2]    :     1.0000    1.0000
Objective function value f(x1,x2) :     3.0000
Equality constraint value value h(x1,x2) :    1.8657e-006
Inequality constraint value value g(x1,x2) :    4.6642e-007


_____

QP-SUB PROBLEM
_____

Quadratic Objective function  f(x1,x2):
3 +1*dx1 +-1.8657e-006 *dx2 + 0.5*10 * dx1^2 +-2 * dx1*dx2
+ 0.5*2 * dx2^2
Linearized equality    h(x1,x2): 1.8657e-006 +2*dx1 +2 *dx2
```

```
Linearized inequality  g(x1,x2):  4.6642e-007 +0.5*dx1
+1.5 *dx2


Case a: beta = 0
Change in design vector:      -0.0625     0.0625
The linearized quality constraint:   1.3878e-017
The linearized inequality constraint:     0.0625
New design vector:       0.9375     1.0625
The objective function:     2.9671
The equality constraint:      0.0078
The inequality constraint:      0.0664


 Case b: g = 0
Change in design vector:    1.0e-006 *
   -0.9328    -0.0000
The linearized quality constraint:       0
The linearized inequality constraint:       0
New design vector:       1.0000     1.0000
The objective function:     3.0000
The equality constraint:    8.6997e-013
The inequality constraint:    2.1760e-013
Multiplier beta:       1.0000
Multiplier lamda:      -0.7500


&&&&&&&&&&&&&&&&&&&&&&&&&&&&&&
Stopped:  Design Not Changing
&&&&&&&&&&&&&&&&&&&&&&&&&&&&&&
```

You can see, for each iteration, there are values for the design variable and functions. The QP subproblem also formulated analytically. The two cases based on the KT conditions are applied and evaluated to determine the best solution. The solution has converged not only in the design variables but also in the multipliers, even though the multipliers are for the QP subproblem.

7.3.5 Generalized Reduced Gradient (GRG) Method

The Generalized Reduced Gradient (GRG) method is another popular technique for constrained minimization. The original method, the Reduced Gradient Method has seen many modifications through several researchers.[13,14] Like the SQP this is an active set method—*it deals with active inequalities*. The method is based on equality constraints only. The inequality constraints are transformed to equality constraints using a linear slack variable of the type used in LP problems. The general optimization problem is restructured as follows:

Minimize $f(x_1, x_2, \ldots x_n)$ $\qquad\qquad\qquad\qquad\qquad$ (7.48)

Subject to: $h_k(x_1, x_2, \ldots, x_n) = 0, \quad k = 1, 2, \ldots, l$ (7.49)

$$h_{l+j} = g_j(x_1, x_2, \ldots, x_n) + x_{n+j} = 0, \quad j = 1, 2, \ldots, m \quad (7.50)$$

$$x_i^l \le x_i \le x_i^u \quad i = 1, 2, \ldots, n \quad (7.51)$$

$$x_{n+j} \ge 0 \quad j = 1, 2, \ldots, m \quad (7.52)$$

The total number of variables is $n + m$. The active set strategy assumes that out of $(n + m)$ variables, it is possible to define or choose a subset of $(n - l)$ *independent* variables from the set of $(l + m)$ equality constraints. Once these variables are established, the *dependent variables* $(l + m)$ can be recovered through the constraints. The number of independent variables is based on the original design variables and the original equality constraints of the problem. The following development is adapted from.[2]

The set of the design variables $[X]$ is partitioned into the independent set $[Z]$ and dependent set $[Y]$.

$$X = \begin{bmatrix} Z \\ Y \end{bmatrix}$$

The constraints are accumulated as

$$H = \begin{bmatrix} h \\ g \end{bmatrix}$$

Relations (7.48) to (7.52) can be recast as

Minimize $f(Z, Y)$ (7.53)

Subject to: $H(Z, Y) = 0$ (7.54)

$$z_i^l \le z_i \le z_i^u; \quad i = 1, 2, \ldots, n - l \quad (7.55)$$

A similar set of side constraints for Y can be expressed. In actual implementation the inactive inequality constraints are not included in the problem. Also, the slack variables can belong to the independent set. In GRG, the linearized objective is minimized subject to the linearized constraints about the current design X^i.

Minimize $\tilde{f}(\Delta Z, \Delta Y) = f(X^i) + \nabla_Z f(X^i)^T \Delta Z + \nabla_Y f(X^i)^T \Delta Y$ (7.56)

Subject to:

$$\tilde{H}_k(\Delta Z, \Delta Y) = H_k(X^i) + \nabla_Z H_k(X^i)^T \Delta Z + \nabla_Y H_k(X^i)^T \Delta Y = 0 \quad (7.57)$$

If the GRG began at a feasible point, the $H(X^i) = 0$, and the linearized constraints are also zero, then the linearized equations in (7.57) determine a set of linear equations in ΔZ and ΔY. Considering the vector $[H]$ the linear equation can be written as follows:

$$[A]\,\Delta Z + [B]\,\Delta Y = 0 \tag{7.58}$$

where

$$[A] = \begin{bmatrix} \nabla_Z H_1^T \\ \nabla_Z H_2^T \\ \cdot \\ \nabla_Z H_{l+m}^T \end{bmatrix} \; ; \quad [B] = \begin{bmatrix} \nabla_Y H_1^T \\ \nabla_Y H_2^T \\ \cdot \\ \nabla_Y H_{l+m}^T \end{bmatrix}$$

From (7.58) the change in the dependent variable ΔY can be expressed in terms of the changes in the independent variables as

$$\Delta Y = -[B]^{-1}[A]\,\Delta Z \tag{7.59}$$

Substituting (7.59) into (7.56) reduces it to an unconstrained problem in ΔZ.

$$\begin{aligned}
\tilde{f}(\Delta Z) &= f(X^i) + \nabla_Z f(X^i)^T \Delta Z - \nabla_Y f(X^i)^T [B]^{-1}[A]\,\Delta Z \\
&= f(X^i) + \left\{ \nabla_Z f(X^i)^T - \nabla_Y f(X^i)^T [B]^{-1}[A] \right\} \Delta Z \\
&= f(X^i) + \left\{ \nabla_Z f(X^i) - \left([B]^{-1}[A]\right)^T \nabla_Y f(X^i) \right\}^T \Delta Z \\
&= f(X^i) + [G_R]^T \Delta Z \tag{7.60}
\end{aligned}$$

$[G_R]$ is the **reduced gradient** of the function $f(X)$. It provides the *search direction*. The stepsize α is found by requiring the changes in ΔY, corresponding to the changes in ΔZ, determine a feasible design. The changes in ΔZ is defined through

$$\Delta Z = \alpha S = \alpha(-G_R)$$

This is implemented through Newton's method as described next. Since the functions are linear there is no quadratic convergence associated with this choice. The iterations start with a feasible current design vector $[X^i]^T = [Z^i\,Y^{iT}]$. The stepsize computation is performed for two nonzero values of alpha and the corresponding values of the objective function are recorded. The default values of $\alpha = 0$ is available for the starting value of the f. A quadratic interpolation is then used to determine the best alpha and a final computation based on this optimum alpha is used to determine the end values for this iteration.

Stepsize Computation Algorithm: (GRG):

Step 1: Find G_R, $S = -G_R$; Select α
Step 2: $q = 1$
$\qquad \Delta Z = \alpha\,S$; $\quad Z = Z^i + \Delta Z$;
$\qquad \Delta Y^q = -[B]^{-1}[A]\,\Delta Z$

Step 3: $Y^{q+1} = Y^q + \Delta Y^q$

$$X^{i+1} = \begin{bmatrix} Z \\ Y^{q+1} \end{bmatrix}$$

If $[H(X^{i+1})] = 0$; Stop, Converged
Else $q \leftarrow q + 1$
$\Delta Y^q = [B]^{-1} [-H(X^{i+1})]$
Go to Step 3

In the above algorithm neither Z, $[A]$, $[B]$ nor α is changed. The singularity of $[B]$ is a concern and can be controlled by the selection of independent variables. The algorithm assigns the responsibility of optimization of the objective to ΔZ, while the feasibility is handled by ΔY. Since the generalized gradient is unscaled, the value of alpha used may be problem dependent and different from 1.

Algorithm: Generalized Reduced Gradient (GRG) (A 7.6):

Step 1:   Choose X^1 (must be feasible)
N_s (no. of iterations),
ε_i's (for convergence and stopping)
p = 1 (iteration counter)

Step 2:   Identify Z, Y
Calculate *[A]*, *[B]*
Calculate $[G_R]$
Calculate Optimum Stepsize α^* (see algorithm)
Calculate X^{p+1}

Step 3:   Convergence for GRG (note H should be zero)
If $h_k = 0$, for $k = 1,2, \ldots$ l;
If $g_j \leq 0$, for $j = 1,2, \ldots$ m;
If KT conditions are satisfied
Then Converged, Stop
Stopping Criteria:
$\Delta X = X^{p+1*} - X^{p*}$
Else If $\Delta X^T \Delta X \leq \varepsilon_1$: Stop (design not changing)
Else If $p = N_s$: Stop (maximum iterations reached)
Continue
$p \leftarrow p + 1$
Go to Step 2

Application of the Generalized Reduced Gradient (GRG) Method: The
GRG algorithm is applied to Example 7.1. Once again the illustration is through a set of calculations generated using MATLAB. For a start, Example 7.1 is rewritten using a slack variable for the inequality constraint.

$$\text{Minimize} \quad f(x_1, x_2, x_3): \quad x_1^4 - 2x_1^2 x_2 + x_1^2 + x_1 x_2^2 - 2x_1 + 4$$

Subject to: $H_1(x_1, x_2, x_3)$: $x_1^2 + x_2^2 - 2 = 0$

$H_2(x_1, x_2)$: $0.25\, x_1^2 + 0.75\, x_2^2 - 1 + x_3 = 0$

$0 \leq x_1 \leq 4$; $0 \leq x_2 \leq 4$; $x_3 \geq 0$

For this problem n = 2, l = 1, m = 1, n + m = 3, n − l = 1, l + m = 2

Example 7.1 is not a good vehicle to illustrate the working of the GRG method because the only feasible point where g is active is also the solution. This makes **Z**, and therefore **G$_R$**, a scalar. The example is retained for the uniformity of presentation. In the following implementation, $Z = x_1$ and $Y = [x_2\ x_3]^T$.

M$_{ATLAB}$ Code: The code for the GRG algorithm, applied to Example 7.1 is available in **GRG_Example7_1.m**. The calculations are based symbolic computation. x_1 is the independent variable. It sets up the **Z** and **Y** automatically and computes *[A]* and *[B]*. Experimenting with the code indicates that not all feasible points provide a nonsingular *[B]*, which is required to continue the algorithm. Two values of α are used to perform a quadratic interpolation, which includes iteration for **Δ*Y*** for feasibility. The final computation is based on the optimum stepsize and this completes one iteration of the method. The code is programmed to run several iterations based on user choice. The following is the listing of the first and the seventh (last) iteration. The program must start at a design where the equality constraints are met.

```
The initial design vector [ 1.3000      0.5568       0.3450]
***********************

GRG-Example 7.1
*******************


******************************
Iteration:      1
******************************
Current Design [x1, x2, x3]:     1.3000     0.5568     0.3450
Objective function value f(x1,x2)          :      4.4672
Equality constraint value value h1(x1,x2):      0
Equality constraint value value h2(x1,x2):      0
Inequality constraint value value g(x1,x2):     -0.3450


_____

Optimum alpha computation
_____

Generalized Gradient   :    11.3146
Step size              :     0.0250

No. of iterations for feasibility:      11
```

```
design vector:       1.0171      0.9826      0.0173
function and constraints (f h1 h2)
     3.0195     0.0001     0.0001
```

```
Generalized Gradient  :     11.3146
Step size             :      0.0500
```

```
No. of iterations for feasibility:     12
design vector:      0.7343     1.2087    -0.2305
function and constraints (f h1 h2)
     3.1307     0.0000    -0.0001
```

```
Generalized Gradient  :     11.3146
Step size             :      0.0357
```

```
No. of iterations for feasibility:     11
design vector:      0.8959     1.0943    -0.0988
function and constraints (f h1 h2)
     2.9712    -0.0000    -0.0001
```

Checking Stopping Criteria

```
Difference in design vector    -0.4041     0.5375    -0.4438
Difference in objective function    -1.4960
```

```
*******************************
Iteration:       7
*******************************
Current Design [x1, x2, x3]:      0.9278     1.0673    -0.0696
Objective function value f(x1,x2)            :       2.9656
Equality constraint value value h1(x1,x2):   -8.5728e-005
Equality constraint value value h2(x1,x2):   -2.1367e-005
Inequality constraint value value g(x1,x2):       0.0695
```

Optimum alpha computation

```
Generalized Gradient  :      0.0032
Step size             :      0.0250
```

```
No. of iterations for feasibility:      2
design vector:      0.9277     1.0674    -0.0696
function and constraints (f h1 h2)
     2.9656    -0.0001    -0.0000
```

```
Generalized Gradient   :    0.0032
Step size              :    0.0500


No. of iterations for feasibility:     3
design vector:      0.9276    1.0675   -0.0697
function and constraints (f h1 h2)
      2.9656   -0.0001   -0.0000


Generalized Gradient   :    0.0032
Step size              :    0.0073


No. of iterations for feasibility:     2
design vector:      0.9278    1.0673   -0.0696
function and constraints (f h1 h2)
      2.9656   -0.0001   -0.0000
```

```
Checking Stopping Criteria
```

```
Difference in design vector  1.0e-004 *
   -0.2380    0.3560   -0.3802
Difference in objective function  3.7873e 006


&&&&&&&&&&&&&&&&&&&&&&&&&&&&&
Stopped:  Design Not Changing
&&&&&&&&&&&&&&&&&&&&&&&&&&&&&&
```

This is far from satisfactory, as x_3, the slack variable, is *negative*. The stepsize and design change calculation must also be made sensitive to the side constraints. Running additional iterations did not improve the solution. Since $g(X)$ is in violation, the KT conditions were not computed. Once again, this is probably not a great example for illustrating the method because of the scalar independent variable.

7.3.6 Sequential Gradient Restoration Algorithm (SGRA)

This SGRA employs a two-phase strategy for finding a minimum to the constrained optimization problem. The two phases are the *Gradient* phase and the *Restoration* phase. Starting at a feasible design, the *Gradient* phase decreases the value of the objective function while satisfying the linearized active constraints. This will cause constraint violation for nonlinear active constraints. The *Restoration* phase brings back the design to feasibility which may establish a new and different set of active constraints. This cycle of two phases is repeated until the optimum is found.[15,16] The method incorporates a descent property with each cycle. Like other algorithms in this section it uses an active constraint strategy, basically ignoring the inequality constraints that are inactive and satisfied. In fact, the GRG of the previous section does something similar in a single iteration without seperate phases. The SGRA compares favorably with the GRG and the Gradient Projection method.[17]

The SGRA as originally introduced in the listed references assumes the inequality constraints of the form

$$g_j(X) \geq 0; \quad j = 1, 2 \ldots, m$$

To be consistent with the discussion in the book it is changed to the standard form of NLP problems defined in this text. In the following development of the algorithm only the salient features are indicated. They are also transcribed consistent with the other algorithms in this section. Please review the references for a more detailed description. The general problem is

$$\text{Minimize} \quad f(X), [X]_n \tag{7.5}$$

$$\text{Subject to:} \quad [h(X)]_l = 0 \tag{7.6}$$

$$[g(X)]_m \leq 0 \tag{7.7}$$

$$X^{low} \leq X \leq X^{up} \tag{7.8}$$

In the SGRA, only active inequality constraints are of interest. Equality constraints are always active. Active inequality constraints also include violated constraints. If v indicates the set of active constraints then (7.6) and (7.7) are combined into a vector of active constraints (ϕ):

$$\phi(X) = \begin{bmatrix} h(X) \\ g(X) \geq 0 \end{bmatrix}_v \tag{7.61}$$

The number of active inequality constraints is $(v - l)$. The Lagrangian for the problem can be expressed in terms of the active constraints alone (since the multipliers for the $g_j(X) < 0$ will be set to zero as part of KT conditions). The KT conditions are then expressed as

$$\nabla_X F(X, \lambda^v) = \nabla_X f(X) + [\nabla_X \phi] \lambda^v = 0 \tag{7.62a}$$

$$\phi(X) = 0 \tag{7.62b}$$

$$\lambda_{v-l} \geq 0 \tag{7.62c}$$

where $[\nabla_X \phi] = \begin{bmatrix} \nabla_X \phi_1 & \nabla_X \phi_2 & \cdots & \nabla_X \phi_v \end{bmatrix}$; $\quad \lambda^v = \begin{bmatrix} \lambda_1 & \lambda_2 & \cdots & \lambda_v \end{bmatrix}^T$

Gradient Phase: Given a feasible design X^i, find the neighboring point

$$\tilde{X}_g = X^i + \Delta X$$

such that $\delta f < 0$, and $\delta \phi = 0$. Imposing a quadratic constraint on the displacement ΔX, the problem can be set up as an optimization subproblem whose KT conditions determine the following:[16]

$$\Delta X = -\alpha \nabla F_X(X^i, \lambda^v) = \alpha S \qquad (7.63)$$

The interpretation of the search direction **S** is used to connect to other algorithms in this book. It is not part of the original development. The search direction is opposite to the gradient of the Lagrangian, which is a novel feature. To compute the Lagrangian the Lagrange multipliers have to be calculated. This is done by solving a system of v—*linear* equations

$$\left[\nabla_X \phi(X^i)\right]^T \nabla_X f(X^i) + \left[\nabla_X \phi(X^i)\right]^T \left[\nabla_X \phi(X^i)\right] \lambda^v = 0 \qquad (7.64)$$

Stepsize for Gradient Phase: The stepsize α calculation is based on driving the optimality conditions in (7.62a) to zero. Therefore, if

$$\tilde{X}_g = X^i - \alpha \nabla F_X(X^i, \lambda^v)$$

then optimum α^* is found by a quadratic or cubic interpolation of trying to satisfy

$$\nabla_X F(\tilde{X}_g, \lambda^v)^T \nabla_X F(X^i, \lambda^v) = 0 \qquad (7.65)$$

Care must be taken that this stepsize does not cause significant constraint violation. This is enforced by capping the squared error in the constraints by a suitable upper bound which is set up as

$$\phi(\tilde{X}_g)^T \phi(\tilde{X}_g) \leq P_{\max} \qquad (7.66)$$

Reference suggests that P_{max} is related to another performance index Q, which is the error in the optimality conditions[16].

$$Q = \nabla F_X(X^i, \lambda^v)^T \nabla F_X(X^i, \lambda^v) \qquad (7.67)$$

Restoration Phase: It is expected that at the end of the gradient phase the function will have decreased but there would be some constraint dissatisfaction (assuming there was at least one nonlinear active constraint at the beginning of the gradient phase). The restoration phase establishes a neighboring feasible solution. It does this by ensuring that the linearized constraints are feasible. Prior to this, the active constraint set has to be updated ($\bar{v}$) since previously feasible constraints could have become infeasible (and previously infeasible constraints could have become feasible) with the design changes caused by the gradient phase. The design vector and the design changes for the Restoration phase can be written as follows:

$$\tilde{X}_r = \tilde{X}_g + \Delta X_r \qquad (7.68)$$

The design changes for this phase ΔX_r is obtained as a least squared error in the design changes subject to the satisfaction of the linear constraints. Setting up a NLP subproblem this calculation for the design changes is

$$\Delta X_r = -\nabla_X \phi(\tilde{X}_g) \sigma^{\tilde{\nu}} \tag{7.69}$$

Here, σ is the $\tilde{\nu}$- vector Lagrange multiplier of the quadratic subproblem. The values for the multipliers are established through the linear equations

$$\mu \phi(\tilde{X}_g) - \nabla_X \phi(\tilde{X}_g)^T \nabla_X \phi(\tilde{X}_g) \sigma^{\tilde{\nu}} = 0 \tag{7.70}$$

The factor μ is a user-controlled parameter to discourage large design changes. The Restoration phase is iteratively applied until

$$\phi(\tilde{X}_r)^T \phi(\tilde{X}_r) \le \varepsilon_1 \tag{7.71}$$

where ε_1 is a small number.

At the conclusion of the Restoration phase the constraints are feasible and the next cycle of *Gradient—Restoration* phase can be applied.

Algorithm: Sequential Gradient Restoration Algorithm (SGRA)–(A 7.7):

Step 1: Choose X^1 (must be feasible)
N_s (no. of iterations),
ε_i's (for convergence and stopping)
p $= 1$ (iteration counter)

Step 2: Execute Gradient Phase
Calculate Stepsize using cubic/quadratic interpolation
Calculate X_g^{p+1}

Step 3: Execute Restoration Phase
X^{p+1}

Step 4: Convergence for SGRA
If $h_k = 0$, for $k = 1,2, \ldots\ l$;
If $g_j \le 0$, for $j = 1,2, \ldots\ m$;
If KT conditions are satisfied
Then Converged, Stop
Stopping Criteria:
$\Delta X = X^{p+1} - X^p$
Else If $\Delta X^T \Delta X \le \boldsymbol{\varepsilon}_1$: Stop (design not changing)
Else If p $= N_s$: Stop (maximum iterations reached)
Continue
$p \leftarrow p + 1$
Go to Step 2

Application of the Sequential Gradient Restoration Algorithm (SGRA):
The SGRA method is applied to Example 7.1. Symbolic and numeric computation is used to implement the algorithm. The code, at first glance, appears to be problem specific, but the sequence of steps is well commented so that it can be extended to other problems directly. It does consider a scalar equality and inequality constraint.

Example 7.1 This example is used to illustrate the various algorithms

$$\text{Minimize} \quad f(x_1, x_2): \quad x_1^4 - 2x_1^2 x_2 + x_1^2 + x_1 x_2^2 - 2x_1 + 4 \qquad (7.9)$$

$$\text{Subject to:} \quad h(x_1, x_2): \quad x_1^2 + x_2^2 - 2 = 0 \qquad (7.10a)$$

$$g(x_1, x_2): \quad 0.25 x_1^2 + 0.75 x_2^2 - 1 \le 0 \qquad (7.10b)$$

$$0 \le x_1 \le 4; \quad 0 \le x_2 \le 4 \qquad (7.10c)$$

MATLAB Code: The MATLAB code is available in **SGRA_Example7_1.m**. Note it is set up for only two variables with a single equality and an inequality constraint. It requires a feasible starting point. Three iterations or cycles are selected to be executed though the last does not produce any changes since the solution has been obtained after the second iteration. In the Gradient phase, the default value of α for scanning is set to 1. The value of μ in the Restoration phase is 0.5. The Performance indices are also printed. The output to the Command window of the first two cycles or iterations is as follows:

```
The initial design vector [     1.3000        0.5568]
***********************
SGRA-Example 7.1
********************
****************************
Iteration:     1
****************************
Current Design [x1, x2]          :    1.3000e+000  5.5678e-001
Objective function value f(x1,x2)         :    4.4672e+000
Equality constraint value value h1(x1,x2):        0
Inequality constraint value value g(x1,x2):   -3.4500e-001
_____

 Gradient Phase
_____

The Lagrange multipliers
 -1.9419e+000

alpha for quadratic interpolation
          0  6.2500e-002  1.2500e-001
```

```
function values for quadratic interpolation
  1.9843e+001  1.6944e+001  2.3516e+001

optimum alpha:    5.0381e-002

The design vector [    1.2116       0.7631]

function and constraints(f h1 h2):
   3.6651e+000  5.0367e-002 -1.9627e-001

The performance indices Q, P:
   1.6488e+001  2.5368e-003
```
_____

 Restoration Phase
_____

```
 Number of Restoration iterations:     10

The design vector [    1.1967       0.7537]

function and constraints(f h1 h2):
   3.6107e+000  9.8983e-005 -2.1597e-001

The performance index P:  9.7976e-009
```
_____

Checking Stopping Criteria
_____

```
Difference in design vector -1.0330e-001  1.9689e-001
Difference in objective function -8.5652e-001

******************************
Iteration:      2
******************************
Current Design [x1, x2]     :   1.1967e+000  7.5366e-001
Objective function value f(x1,x2)      :   3.6107e+000
Equality constraint value value h1(x1,x2):  9.8983e-005
Inequality constraint value value g(x1,x2):  -2.1597e-001
```
_____

 Gradient Phase
_____

```
The Lagrange multipliers
 -1.0594e+000

alpha for quadratic interpolation
          0  1.2500e-001  2.5000e-001
```

```
function values for quadratic interpolation
   9.8608e+000   6.3279e+000   1.0848e+001

optimum alpha:    1.1734e-001

The design vector [    1.0003        1.0655]

function and constraints(f h1 h2):
   3.0046e+000  1.3587e-001  1.0156e-001

The performance indices Q, P:   6.0690e+000   1.8460e-002
```
─────────────────────────
```
Restoration Phase
```
─────────────────────────
```
Number of Restoration iterations:     12

The design vector [    1.0000        1.0000]

function and constraints(f h1 h2):
   3.0000e+000  6.8435e-005  5.1161e-005

The performance index P).
   7.3008e-009
```

─────────────────────────────
```
Checking Stopping Criteria
```
─────────────────────────────
```
Difference in design vector -1.9670e-001   2.4637e-001
Difference in objective function -6.1068e-001
```

Note, the solution has been reached at the end of the restoration phase of the second cycle. This is an impressive accomplishment in comparison to the other techniques in this chapter. Also note that the objective function should decrease with each iteration. At the end of the Gradient phase, the constraints may be violated. At the end of the Restoration phase, they are satisfied. The SGRA is of the same caliber as the GRG (some comparisons show it to be better). It is difficult to compare different algorithms because of different coding structures and non standard implementation details. Nevertheless, the SGRA has considerable merit. It appears definitely better than the GRG as far as Example 7.1 is concerned (Section 7.3.5). This comparison may not be fair since the active constraint strategy was not implemented in the code for the GRG even though it is part of the algorithm.

7.4 ADDITIONAL EXAMPLES

Three additional examples are introduced here to illustrate and expand the use of these techniques. During the early development of these techniques, there was a significant emphasis on efficiency, robustness, and keeping computational resources as low as possible. Today, the search for global solutions (that are largely heuristic), coupled by simple calculations repeated endlessly, performed on desktop PCs with incredible computing power, through software that can provide a wide range of resources with a small learning curve, has shifted the focus to experimentation and creativity. The author feels such an effort must be coupled with knowledge and insight gained through a wide class of application and understanding the progress of numerical calculation. In spite of so many algorithms in the text, for many real design problems the solution is not automatic. Dealing with nonlinear equations is a challenge and often surprising. Trying to understand why the technique does not work is as much a learning experience as a routine solution to the problem. Some of the examples use the methods of the previous section, modified for the problem. These modifications are small but the student is encouraged to pay attention the reason for their need as well as the evolution of the algorithms to become more general. Students are urged to determine a consistent way to incorporate changes, as this will ensure economy of code and standardization.

7.4.1 Example 7.2 — Flagpole Problem

This is the same problem that was solved graphically in Chapter 2, analytically in Chapter 4, and now considered for numerical solution. In this subsection, the scaled version of the problem is solved. From Chapter 4, Example 4.3, it is reintroduced as:

Example 7.2

$$\text{Minimize} \quad f(x_1, x_2) = 6.0559\,E\,05\,(x_1^2 - x_2^2) \tag{7.72}$$

$$\text{Subject to} \quad g_1(x_1, x_2)\colon 2399x_1^2 + 128x_1 - 31174(x_1^4 - x_2^4) \le 0 \tag{7.73a}$$

$$g_2(x_1, x_2)\colon (250 + 7497\,x_1)(x_1^2 + x_1x_2 + x_2^2)$$
$$- 854140(x_1^4 - x_2^4) \le 0 \tag{7.73b}$$

$$g_3(x_1, x_2)\colon 2500x_1 + 800 - 65450(x_1^4 - x_2^4) \le 0 \tag{7.73c}$$

$$g_4(x_1, x_2)\colon x_2 - x_1 + 0.001 \le 0 \tag{7.73d}$$

$$0.02 \le x_1 \le 1.0; \quad 0.02 \le x_2 \le 1.0 \tag{7.74}$$

Analytical Solution: In Chapter 4, this example was discussed extensively, including looking at possible solutions flagged graphically. Finally a direct solution established the solution at:

$$x_1^* = 0.9963; \quad x_2^* = 0.9753 \tag{7.75a}$$

and the objective and constraint functions are:

$$f^* = 25073.7643; \quad g_1^* = -3118.563; \quad g_2^* = -39011079.31$$

$$g_3^* = -0.0188; \quad g_4^* = -0.02; \quad (7.75b)$$

Numerical Solutions: Numerical Solutions are obtained using the Constrained Scan and Zoom Method and the ALM method. The first is a zero-order method while the second is a gradient based indirect method.

MATLAB Code: Constrained Scan and Zoom Method: The MATLAB code is available in **Constrained_Scan_Zoom_Example7_2.m**. This is a wrapper file which calls the algorithm in **VConstrained_Scan_Zoom.m**.

The solution is tracked in Figure 7.13. The function m-files are **VOfun_Example7_2.m** for the objective function. **VGfun_Example7_2.m** for the constraint functions. **VHfun_Example7_2.m** just returns empty value. The algorithm remains the same but several changes are made to capitalize on array programming for function evaluation, and handling vector inequality constraints. Note the same function m-files will handle arrays and single pair of values. Once more the number of function evaluations per zoom cycle and the area scanned is increased

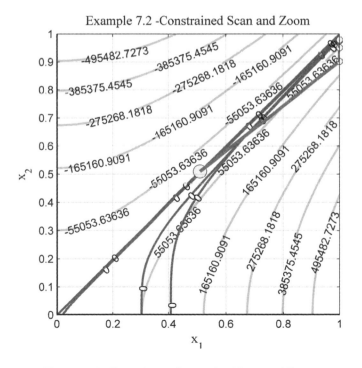

Figure 7.13 Example 7.2; Constrained Scan and Zoom.

if the design variables remain the same as the previous zoom cycle. While the iterations are run for the maximum number of cycles, the design does not change after the eleventh zoom cycle. The starting values and the values for the eleventh iteration are reproduced.

```
Example 7.2 (CONSTRAINED SCAN ZOOM)
*********************************

*****************
*STARTING VALUE:
*****************
Midpoint Design Vector (Xmid):     5.1000e-001  5.1000e-001
Largest Objective function  (flimit)  :    6.0535e+005
Largest Inequality Constraint value:   8.5526e+005
Lower Design Space Limit (X) :    2.0000e-002  2.0000e-002
Upper Design Space Limit (X) :        1           1

*******************
*ZOOM LEVEL:      11
*******************
Design Vector (X)    :     1.0000e+000  9.7907e-001
Objective function   :    2.5081e+004
f_limit   :    1.8917e+004
Inequality Constraints:  -1.7459e+000 -4.6527e+004
-2.0091e+003 -1.9927e-002
Max Inequality Constraints:    2.3640e+003
g_limit:    2.6727e+004

Lower Limit Design (X)    :     9.9234e-001  9.7142e-001
Upper Limit Design (X)    :     1.0077e+000  9.8673e-001
Number of scan steps:      61
```

In comparison to (7.75), the final values for the design vector and the various functions are:

$$x_1^* = 1.00000; \quad x_2^* = 0.97907$$

$$f^* = 25081;$$

$$g_1^* = -1.7459; \quad g_2^* = -46527; \quad g_3^* = -2009.1; \quad g_4^* = -0.019927$$

The design variables and the objective function are close in value. The first three constraint functions differ significantly in values, illustrating the high degree of nonlinearity. Note that the Constrained Scan and Zoom method can identify a solution on the boundary.

MATLAB Code: ALM Method: Example 7.2 is also solved using the ALM method, which uses the DFP algorithm to carry out the unconstrained minimization subproblem. The code is available in `ALM_Example7_2.m`, which is a wrapper file that calls `ALM.m`. Figure 7.14 tracks the solution through the ALM iterations. The function values are available in `ALM_Ofun_Example7_2.m`, `ALM_Gfun_Example7_2.m`, `ALM_Hfun_Example7_2.m`, and `FALM_Example7_2.m`. We could have used the function files from the Scan and Zoom method but that would require rewriting the ALM algorithm. The student is recommended to complete this exercise for consistency of code. The solution after 4 iterations is

$$x_1^* = 0.89353; \quad x_2^* = 0.86978$$

$$f^* = 25358;$$

$$g_1^* = -0.021; \quad g_2^* = -39407; \quad g_3^* = -122751; \quad g_4^* = -0.0237$$

The solution differs from both the analytical one and the numerical one from the Constrained Scan and Zoom method. It appears there are several local optimums to the augmented Lagrangian function. It was necessary to work with a

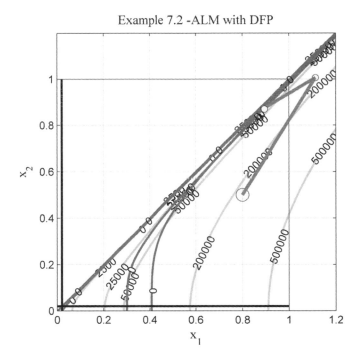

Figure 7.14 Example 7.2: ALM with DFP.

scaled version of the example in order to obtain decent solutions. This solution is actually rescaled from the solution of the scaled problem. The scaled problem was generated from (7.72) to (7.73) by dividing throughout by the largest coefficient in the various functions. This can be seen in the code. The reported solution makes the first constraint active, which is more evident in the solution of the scaled problem. The first constraint is almost active in the Constrained Scan and Zoom technique, too. The starting values and the values after the fourth iteration, for the scaled problem, are copied from the Command window.

```
* * * * * * * * * * * * * * * * * * * * * *
*ALM iteration number:      0
* * * * * * * * * * * * * * * * * * * * * *
Design Vector (X):     8.0000e-001  5.0000e-001
Objective function:   3.9000e-001
Inequality constraints:  -2.9456e-001 -3.3766e-001
-3.0432e-001 -2.9900e-001
Sum of Squared Error in inequality constraints(g):     0
Lagrange Multipliers (beta):      1     1     1     1
Penalty Multipliers (rg):      1


* * * * * * * * * * * * * * * * * * * * * * * * * * * * * *
*ALM iteration number:      4
* * * * * * * * * * * * * * * * * * * * * * * * * * * * * *
Design Vector (X):               8.9353e-001  8.6978e-001
Objective function:                         4.1873e-002
Inequality constraints:          6.7525e-007 -4.6136e-002
-1.8755e-002 -2.2747e-002
Sum of Squared Error in constraints (g):   4.5597e-013
Lagrange Multipliers (beta): 6.5975e-001            0
          0 -5.5511e-017
Penalty Multipliers (rg):                       1000
```

Like other implementations enforcing the side constraints are not part of the code. It should be as Figure 7.14 illustrates that the solution travels through the infeasible region. You can also run the code by changing the various parameters and experience how difficult it is to solve this nonlinear problem. It appears that change in every parameter has a significant influence on the solution.

7.4.2 Example 7.3–I-Beam Design

The example is the design of an I-beam for use in a particular structural problem. This is a traditional mechanical/civil engineering structural optimization problem. This is an example with four design variables, three inequality constraints, and no equality constraints. It appeared in Chapter 1 as Example 1.2. Many versions of this problem can easily be formulated. It can be reduced to a two-variable

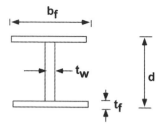

Figure 7.15 Example 7.3: Design variables.

problem if symmetry was imposed. Standard failure criteria with respect to combined stresses or principal stresses can also be included instead of the simplistic stress constraints considered here. If the cantilevered end is bolted then additional design functions regarding bolt failure needs to be examined. The original problem statement is given. Figure 7.15, same as Figure 1.3, is the description of the design variables for the problem. The original problem statement is from Chapter 1.

Example 7.2 Design a cantilevered beam, of minimum mass, carrying a point load F at the end of the beam of length L. The cross section of the beam will be in the shape of the letter I (referred to as an I-beam). The beam should be sufficiently strong in bending and shear. There is also a limit on its deflection.

Since the problem is formulated in Chapter 1, we directly proceed to the mathematical model associating x_1 with d, x_2 with t_w, x_3 with b_f, and x_4 with t_f, so that the design vector is $X = [x_1, x_2, x_3, x_4]$. The problem in standard format is:

$$\text{Minimize} \quad f(X): \quad \gamma L A_c \tag{7.76}$$

$$\text{Subject to:} \quad g_1(X): \text{FL}x_1/2\text{I}_c - \sigma_{\text{yield}} \le 0 \tag{7.77a}$$

$$g_2(X): \text{FQ}_c/\text{I}_c x_2 - \tau_{\text{yield}} \le 0 \tag{7.77b}$$

$$g_3(X): \text{FL}^3/3\text{EI}_c - \delta_{\text{max}} \le 0 \tag{7.77c}$$

$$0.01 \le x_1 \le 0.25; \quad 0.001 \le x_2 \le 0.05; \tag{7.77d}$$

$$0.01 \le x_3 \le 0.25; \quad 0.001 \le x_4 \le 0.05 \tag{7.77d}$$

The designer must ensure that the problem definition is also consistent with the unit system chosen for the parameters and variables. The parameters for this problem (value is given in the parenthesis) are F (10000 N); L (3 m); γ (steel: = 7860 kg/m^3); E (210 GPa); σ_{yield} (250 E+06 N/m$^{2)}$; τ_{yield} (145 E+06 N/m^2); and the maximum deflection δ_{max} (0.005 m).

We will include some geometric constraints among the design variables, as the program is not aware that x_1 must be greater than x_4, for example.

$$g_4(X): x_1 - 3x_3 \leq 0 \tag{7.78a}$$

$$g_5(X): 2x_3 - x_1 \leq 0 \tag{7.78b}$$

$$g_6(X): x_2 - 1.5x_4 \leq 0 \tag{7.78c}$$

$$g_7(X): 0.5x_4 - x_2 \leq 0 \tag{7.78d}$$

In the code for the various algorithms that we have developed, we always recognized the fact that that we did not explicitly handle the side constraints. One way to incorporate them directly in the problem formulation is to create linear inequality constraints from (7.77d). There will be eight such constraints:

$$g_8(X) = -x_1 + 0.01 \leq 0 \tag{7.79a}$$

$$g_9(X) = x_1 - 0.25 \leq 0 \tag{7.79b}$$

$$g_{10}(X) = -x_2 + 0.001 \leq 0 \tag{7.79c}$$

$$g_{11}(X) = x_2 - 0.05 \leq 0 \tag{7.79d}$$

$$g_{12}(X) = -x_3 + 0.01 \leq 0 \tag{7.79e}$$

$$g_{13}(X) = x_3 - 0.25 \leq 0 \tag{7.79f}$$

$$g_{14}(X) = -x_4 + 0.001 \leq 0 \tag{7.79g}$$

$$g_{15}(X) = x_4 - 0.05 \leq 0 \tag{7.79h}$$

For an analytical solution, you will have to consider 2^{15} cases. This will be extremely challenging and makes a numerical solution justified. To complete the model we will need the following relations based on the design variables:

$$A_c = 2x_3x_4 + x_1x_2 - 2x_2x_4$$

$$I_c = \frac{x_3x_1^3}{12} - \frac{(x_3 - x_2)(x_1 - 2x_4)^3}{12}$$

$$Q_c = 0.5x_3x_4(x_1 - x_4) + 0.5x_2(x_1 - x_4)^2$$

This section requires the use of the *Optimization Toolbox* from MATLAB. More specifically, it uses the *Quadratic Programming (Quadprog)* function to solve the quadratic programming subproblem in the SQP method. The problem is solved using the direct procedure illustrated through the example in Section 7.3.2. This section combines the *symbolic* capability of MATLAB with the resources available in the optimization toolbox to implement an original interpretation of the SQP.

MATLAB Code: The code for the Example is available in **SQP_Example7_2.m**. It is sufficiently commented so that the reader can follow the implementation. The reader is strongly recommended to review the code in the m-file for the following reasons:

- The power of symbolic manipulation is exploited further using vector definitions to make the implementation fairly easy. The symbolic calculations are used to generate the data for the *QP* subproblem.
- The symbolic calculations are effortlessly integrated with numerical computations (the solution to the *QP* is numerical).
- The implementation of the SQP is straightforward and direct (and simple).
- The solution is the same as the numerical solution by MATLAB's own SQP program (see Chapter 10).

The starting iteration is copied from the Command window as:

```
SQP Method-Example 7.3
************************
USES QuadProg FROM OPTIMIZATION TOOLBOX

Initial Design    [ 0.20000  0.01000  0.01000  0.02000 ]
Objective Function:        47.1600
Normal Stress Constraint:   2.6667e+003
Shear Stress Constraint:   -7.8667
Deflection Constraint:        249000
The normal stress and the deflection constraints are in
     violation. The final iteration, once again copied from
     the Command window is:
```

```
No change in design-exiting-iteration number        6
```

```
Final Design [m]:[ 0.25000  0.01211  0.12500  0.02422 ]
Objective Function: (kg)    200.3002

Constraints
```

```
Normal Stress Constraint-g1:   -3.5357e+004
Shear Stress Constraint-g2:    -143.9800
Deflection Constraint-  g3:  -5.0185e-012
Acceptability constraint-g4:     -0.1250
Acceptability constraint-g5:      0
Acceptability constraint-g6:     -0.0242
Acceptability constraint-g7:      0
Side constraint        -g8:     -0.2400
Side constraint        -g9:     -0.0111
```

```
Side constraint          -g10:       -0.1150
Side constraint          -g11:       -0.0232
Side constraint          -g12:        0
Side constraint          -g13:       -0.0379
Side constraint          -g14:       -0.1250
Side constraint          -g15:       -0.0258
```

From the results, constraints g_3, g_5, g_7, and g_{12} are active. The mass of the structure is around 201 kg. The reader is encouraged to explore the difference between this implementation and the SQP implementation in the *Optimization Toolbox*, and the influence of the Hessian on the solution.

7.4.3 Example 7.4–Box Design

This problem appeared as Example 2.5 in Chapter 2, where it was solved graphically. The simple problem is developed extensively there, and once again we borrow the mathematical model after the problem statement. We will apply the GRG method to numerically solve the problem with three design variables. We apply it using symbolic calculation in MATLAB.

Problem Definition: Today's concern over the waste, recycling, and the environment has manufacturers trying to adopt new packaging materials to deliver their products. One such case involves using bio-degradable cartons made from recycled materials. The cost of the material is based on the surface area of the rectangular container that will house the product. The cost per unit area is $1.5 per square meter. The different products can be accommodated by a single container. The container must hold a volume of 0.032 m^3. The perimeter of the base must be less than or equal to 1.5 m. Its sides are scaled geometrically to hold information labels. The width should not exceed three times the length. Its height must be less than two thirds the width. Its length and width are less than 0.5 m. Find the container of minimum cost.

The design variables are shown in Figure 2.12. The parameters for the example are ***Cost*** = $1.5/m^2[material cost], ***Vol*** = 0.032 m^3 [volume constraint], ***Per*** = 1.5 m [Perimeter Constraint]. The design variables are x_1 (length of container), x_2 (width of container), x_3 (height of container).

Mathematical Model: The functions in the model are

$$\text{Minimize} \quad f(x_1, x_2, x_3) = 2^*Cost^*(x_1x_2 + x_2x_3 + x_3x_1) \qquad (7.80)$$

$$\text{Subject to:} \quad h_1(x_1, x_2, x_3): x_1x_2x_3 - Vol = 0 \qquad (7.81a)$$

$$g_1(x_1, x_2, x_3): 2(x_1 + x_2) - Per \leq 0 \qquad (7.81b)$$

$$g_2(x_1, x_2, x_3): x_2 - 3x_1 \leq 0 \qquad (7.81c)$$

$$g_2(x_1, x_2, x_3): x_3 - (2/3)x_2 \leq 0 \tag{7.81d}$$

$$0 \leq x_1 \leq 0.5; \quad 0 \leq x_2 \leq 0.5 \tag{7.81e}$$

The graphical solution, difficult to see, was shown to be $x_1 = 0.3132\,\text{m}$, $x_2 = 0.3915\,\text{m}$, and $x_3 = 0.2610\,\text{m}$, with an objective function value of \$ 0.9196. For the GRG method, the inequality constraints are expressed as equality constraints using slack variables. This increases the total number of design variables to 6. There are two independent variables and four dependent variables.

MATLAB Code: The code for the problem is available in **GRG_Example7_4.m**. There are 7 iterations before the program stops because the design variables are not changing. Once again the first and the last iterations are copied from the Command Window.

```
*********************
GRG-Example 7.4
*********************
The initial design vector    0.2000    0.2500    0.6400
0.6000      0.3500    -0.4733
*****************************
Iteration:      1
*****************************
Current Design [x1, x2, x3, x4, x5, x6]  :      0.2000
0.2500    0.6400    0.6000    0.3500    -0.4733
Objective function value f(X)            :      1.0140
Equality constraint values   h(X):      0    0    0    0
Inequality constraint values g(X):    -0.6000 -0.3500  0.4733


*****************************
Iteration:      7
*****************************
Current Design [x1, x2, x3, x4, x5, x6]  :      0.3169 0.3191
     0.3164    0.2280    0.6317    -0.1037
Objective function value f(X)            :      0.9071
Equality constraint values   h(X):    1.0e-004 *
    -0.0419    -0.8282    0.0191    0.2039
Inequality constraint values g(X):    -0.2281 -0.6317 0.1037


Difference in design vector   1.0e-004 *
    0.0349    0.1905    -0.1119    -0.2255    -0.0429    0.1754
Difference in objective function   2.1508e-005
&&&&&&&&&&&&&&&&&&&&&&&&&&&&&&
```

```
Stopped:  Design Not Changing
&&&&&&&&&&&&&&&&&&&&&&&&&&&&&&&&
```

On first glance, it appears that we were successful in applying the method. However, the last inequality constraint is violated, or the corresponding slack variable is negative. This means the solution is not feasible. Starting from several different starting designs, we seem to converge to the same solution. This suggests that the last inequality constraint may not be justified. It was just introduced to increase the number of constraints for illustration. Students are asked to obtain the solution by extensive evaluation in the exercise at the end of the chapter. This example was included to examine the GRG method because the previous example was limited by the number of constraints.

REFERENCES

1. Fiacco, A.V., and G.P. McCormick. *Nonlinear Programming, Sequential Unconstrained Minimization Techniques*, Wiley, New York: 1968.
2. Vanderplaats, G.N., *Numerical Optimization Techniques for Engineering Design*, McGraw-Hill, New York: 1984.
3. ANSYS.
4. Bertekas, D.P. *Constrained Optimization and Lagrange Methods*, Academic Press, New York: 1982.
5. Kelley, J.E. "The Cutting Plane Method", *Journal of SIAM*, Vol. 8, pp. 702–712, 1960.
6. Gill, P.E., W. Murray, and M.H. Wright. *Practical Optimization*. Academic Press, New York: 1981.
7. Boot, J.C.G., *Quadratic Programming, Studies in Mathematical and Managerial Economics*. H. Theil (ed.), Vol. 2., North-Holland: 1964.
8. Biggs, M.C. *Constrained Minimization Using Recursive Quadratic Programming: Some Alternate Subproblem Formulations, Towards Global Optimization*. L.C.W. Dixon and G. P. Szego (eds.), pp. 341–349, North–Holland, 1975.
9. Powell, M.J.D. *A Fast Algorithm for Nonlinear Constrained Optimization Calculations*, No. DAMPTP77/NA 2, University of Cambridge, England: 1977.
10. Han, S.P. "A Globally Convergent Method for Nonlinear Programming", *Journal of Optimization Theory and Applications*, Vol. 22, p. 297, 1977.
11. Arora, J.S. *Introduction to Optimum Design*, McGraw-Hill Inc., New York: 1989.
12. Branch, M, A., and A. Grace A. *Optimization Toolbox, User's Guide*, The MathWorks Inc., 1996.
13. Wolfe, P. "Methods of Nonlinear Programming," *Recent Advances in Mathematical Programming*, R.L. Graves, and P. Wolfe (eds.), McGraw-Hill, New York: 1963.
14. Gabriele, G.A., and K.M. Ragsdell. "The Generalized Gradient Method: A Reliable Tool for Optimal Design," *ASME Journal of Engineering and Industry, Series B*, Vol. 99, (May) 1977.

15. Miele, A., H.Y. Huang, and J.C. Heideman. "Sequential Gradient Restoration Algorithm for the Minimization of Constrained Functions–Ordinary and Conjugate Gradient Versions," *Journal of Optimization Theory and Applications*. 1969, vol. 4, No. 4, (1969).

16. Levy, A.V., and S. Gomez. "Sequential Gradient-Restoration Algorithm for the Optimization of a Nonlinear Constrained Function," *Journal of the Astronautical Sciences, Special Issue on Numerical Methods in Optimization*, Dedicated to Angelo Miele, K.H.Well (sp. ed.), vol. XXX, No. 2, (1982).

17. Rosen, J.B. "The Gradient Projection Method for Nonlinear Programming, Part II: Nonlinear Constraints," *SIAM Journal of Applied Mathematics*. vol. 9, no. 4, (1961).

PROBLEMS

You can confirm your solution by using the Optimization Toolbox to solve the problems—see Chapter 10

7.1 Modify `GraphicalPlot_EPF_Example7_1.m` to review contour values and plot automatically selected contours within range for decent plot.

7.2 (uses the Optimization Toolbox) Develop the SLP by using the LP solver from the Optimization Toolbox and apply to Example 7.1.

7.3 Solve Example 7.1 by solving the QP problem graphically.

7.4 Develop the code to find a feasible initial design vector for constrained problems.

7.5 Develop a program to calculate the multipliers for the NLP problem at a given design.

7.6 Develop a program for a constrained stepsize calculation procedure.

7.7 (uses Optimization Toolbox) Use the QP program from the Toolbox and develop your own SQP implementation.

7.8 (uses Optimization Toolbox) Use the SQP program from MATLAB to solve Example 7.1.

7.9 Create a new version of SQP where the Hessian matrix is maintained as the identity matrix for all iterations.

7.10 Solve Example 7.1 with $Z = [x_2]$ and $Y = [x_1 x_3]^T$.

7.11 Modify the GRG code to include consideration of active constraints and compare the performance.

7.12 Build in KT condition calculation into the SGRA.

Solve the following problems by at least two numerical techniques, one direct and one indirect. Use your own sensible choice of parameters if they are not included in problem definition. Check your solution using the Optimization Toolbox (optional).

7.13 Solve Problem 1.11.

7.14 Solve Problem 1.12.

7.15 Solve Problem 1.13.

7.16 Solve Problem 1.14.

7.17 Solve Problem 1.16.

7.18 Solve Problem 1.17.

7.19 Solve Problem 1.18.

7.20 Set up the analytical conditions for Problem 1.19.

7.21 Solve Problem 1.20.

7.22 Solve Problem 1.21.

7.23 Solve Problem 1.23.

7.24 Solve Problem 1.25.

7.25 Solve Problem 1.26.

7.26 Setup the analytical conditions for Problem 1.27.

7.27 Solve Problem 1.35.

7.28 Solve Problem 1.36.

7.29 Solve Problem 1.39.

7.30 Solve Problem 1.40.

7.31 Find the solution to Example 7.4 using direct evaluation and verify if constraint 7.81d is possible.

8

DISCRETE OPTIMIZATION

This chapter introduces some basic ideas, concepts, methods and algorithms associated with discrete optimization. The subject of *discrete optimization* cannot be constrained by a single chapter in a course on optimization, as the topic is capable of spawning several courses. Discrete optimization is very different, difficult, diverse, and continues to develop even today. Today, almost all types of communication, signals, and controls have moved to the digital format where information is coded, usually in binary form. This is essentially discrete information or values. In the coming year, the TV signal will be digital. Dealing with discrete values will be essential for the engineer. Discrete optimization problems are largely combinatorial and are computationally more time intensive and expensive than continuous problems. The algorithms are also simpler. The problems addressed by the discrete optimization research community are mainly in the area of operations research usually characterized by linear models.

Since the last edition of the book there have been only a few books in the area of discrete, integer, or combinatorial programming. There are still fewer in the area of nonlinear discrete problems. Engineering optimization, by contrast, mostly incorporates nonlinear relations. From a real perspective, discrete design variables should be fundamental to engineering optimization. For example, in the beam design problem in Chapters 1 and 7, a practical solution should involve identifying an "off the shelf," beam as the rolling mill will probably charge prohibitively for a limited production of the unique beam that was identified as the solution to the continuous optimization problem. In the example of identifying the different placement machines, the solution for the number of different placement machines of different kinds was expected to be integers (*discrete value*). In other problems, which may require the identification of diameters, lengths, washers, valves, men, components, stock sizes and so on, engineers avoid high

403

procurement costs associated with nonstandard components unless cost recovery is possible because of large volumes. These choices involve identifying a component from a *set of available* designs—*a discrete selection*. Continuous optimization can be justified in *one of a kind* design or if the item is to be completely manufactured in-house (no off-the-shelf component is necessary).

Practical engineering design requires that some design variables will belong to an ordered set of values—*discrete variables*. A typical catalog will order the design based on some property of the component or device. Washers can be organized in terms of increasing sizes, materials, or thickness. Designing or selecting washers is a discrete optimization problem. Still today there are very few engineering problems that are currently solved as a discrete optimization problem. First, discrete optimization algorithms are difficult to apply. Second, they are time consuming. Third, many of the algorithms and code address linear mathematical models. Discrete optimization in engineering will necessarily involve adaptations from the currently available techniques used by the decision making community. Examples of such adaptations are uncommon. It is rare for books on design optimization, both past and contemporary, to include the subject of nonlinear discrete optimization; notwithstanding the fact that it is enormous in extant. A modest effort is being made in this book with the aim of providing the opportunity to acquaint the reader with the subject area.

The typical approach for incorporating discrete design variables in engineering is to solve the corresponding continuous optimization problem and adjust the optimal design to the nearest discrete values (*this is similar to the rounding process to arrive at an* integer *number*). According to Fletcher, there is no guarantee that this process is correct or that a good solution can be obtained this way[1]. Very often, this *rounding* may result in infeasible designs. Given that the alternate is to solve the discrete optimization problem itself, a systematic approach to this rounding process has become acceptable. Papalambros and Wilde povide rounding process that is based on maintaining a feasible design.[2]

This chapter and the remaining ones will differ significantly from the previous chapters, both in their content and organization. Their primary focus is on presenting new ideas and content rather than developing a technique or assisting in code development. Discrete problems require significantly different concepts than the traditional continuous mathematical models of the earlier chapters. Simple examples are used to bring out the difference. For example, *derivatives, gradients, Hessian*, and so on, do not apply in discrete problems. By extension, *search directions*, and *one-dimensional step size* computation lose their relevance. While there are several excellent references on discrete optimization, almost all of them deal with linear mathematical models. There may be few, if any, about nonlinear discrete optimization applied to engineering problems.

Discrete optimization problems *(DP)* are implied when the design variables are not continuous. In engineering, it will be very rare if there were an unlimited number of discrete values for the design variable. These values will be explicit and function as natural side constraints. It makes sense that this set of values for the design variable be ordered–typically in increasing value. To be generous, we can have each design variable with its own distinct set, with no constraints on the spacing of values within the set. There are also several categories of discrete variables. If the set is made up only integers, then the problem is characterized as an *integer optimization* problem. If the mathematical model is linear such problems are then identified as *Integer Programming (IP)* problem as compared to the *Linear Programming (LP)* problem in Chapter 3. A large class of *IP* problems require the design variables (decision variables) to have values of either "0" or "1". These are also known as binary values and we can therefore recognize a *binary IP* problem. These are also referred as *Zero-One IP* problem (ZIP – this book only). All IP problems can be reduced to Zero-One (0–1) IP problems by representing each integer by its binary equivalent, and associating a 0-1 design variable to each binary value location. For example,

integer value of 8 ⇒ binary value of 1000

Therefore if the design variable x is restricted to integers between $0 \leq x \leq 16$, then x can be replaced by five 0-1 (binary) design variables $[y_1, y_2, y_3, y_4, y_5]$ from which x can be assembled as

$$x = y_1 \left(2^4\right) + y_2 \left(2^3\right) + y_3 \left(2^2\right) + y_4 \left(2^1\right) + y_5 \left(2^0\right)$$

Such a transformation is not recommended if there are a large number of integer variables. Engineering design problems can be expected to contain both continuous and discrete variables. These are termed as *Mixed Programming (MP)* problems if the mathematical model is linear. In this book the classification is extended to nonlinear problems, too.

The focus in this chapter is to present only basic ideas on discrete variables and their special handling. Three methods are presented in this chapter on discrete optimization. This keeps the chapter manageable. These methods are only representative and in no way address the topic of discrete optimization sufficiently, let alone completely. The third one has seen limited use, but with computational resources not being a hurdle today, they might start to look attractive. The order of presentation does not represent any priority, though the methods are among those that have been applied often in engineering. The methods of the next chapter have evolved through their initial application to discrete optimization. They could also belong to this chapter but have been kept distinct because they are the driving force in the search for global optimum today. The methods of this chapter are (i) Exhaustive Enumeration, (ii) Branch and Bound (Partial or Selective Enumeration), and (iii) Dynamic Programming.

8.1 CONCEPTS IN DISCRETE PROGRAMMING

We introduce a contrived simple unconstrained minimization example to introduce some key ideas and expectations in discrete programming. This way the feasibility issues do not intrude in presenting the important concepts in discrete optimization. We also explore the feature of finding a continuous solution followed by adjusting the discrete design variables to neighboring discrete values.

Example 8.1:   Minimize the objective function f, where x_1 is a continuous variable, and x_2, x_3 are discrete variables. x_2 must have a value from the set [0.5 1.5 2.5 3.5] and x_3 must have a value from the set [0.22 0.75 1.73 2.24 2.78].

$$\text{Minimize } f(x_1, x_2, x_3) = (x_1 - 2)^2 + (x_1 - x_2)^2 + (x_1 - x_3)^2 + (x_2 - x_3)^2$$
$$(8.1)$$

The side constraints on the design variables can be set up as

$$x_1 \in \mathbf{R} \tag{8.2a}$$

$$x_2 \in [0.5 \quad 1.5 \quad 2.5 \quad 3.5] \in \mathbf{X}_{2d} \tag{8.2b}$$

$$x_3 \in [0.22 \quad 0.75 \quad 1.73 \quad 2.24 \quad 2.78] \in \mathbf{X}_{3d} \tag{8.2c}$$

The symbol $\in$ identifies that the variable on the left can have one of the values on the right side. In (8.1a) the $\mathbf{R}$ stands for a real value. This represents a standard use of the symbol $\in$ and represents an effective way to express the idea that the discrete values can only have selected values.

8.1.1   Problem Relaxation

If this were a problem in continuous variables the solution can be obtained by setting the $\nabla f = 0$ and solving the three equations for the values of $x_1, x_2,$ and x_3. Alternately, for a continuous problem, inspection of the objective function suggests that the best value of the function is 0. This can be obtained by the following solution:

$$x_1^* = 2; \quad x_2^* = 2; \quad x_3^* = 2; \quad f^* = 0 \tag{8.3}$$

This solution violates the side constraints in (8.2). For the original problem $\partial f / \partial x_1$ is defined, but not $\partial f / \partial x_2, \partial f / \partial x_3$ since x_2 and x_3 are discrete values. Derivatives are defined by taking the limit of the ratio of change of objective function f to the change in the value of the design variable, as the change in the variable approaches zero. The value of f is only defined at selected combination of x_2, x_3 in (8.2) and is not defined elsewhere. Small/infinitesimal changes in the objective function or in the discrete design variables are not defined in Example 8.1. Derivatives with respect to the discrete variables do not exist. This conclusion is of major significance as it makes the previous body of work in the book of limited usefulness in the pursuit of the solution to the DP Problem. To recollect, the necessary and sufficient conditions, which drove the algorithms, were based on the gradients and their derivatives.

Two areas in the previous chapters escape with a limited impact due to the absence of derivatives. They are Linear Programming (*LP*) and Zero-order methods for numerical solution to unconstrained nonlinear optimization. It is not surprising that both of them play significant role discrete optimization problems. It is therefore possible to conclude that solution to Example 8.1 as established in (8.3) is incorrect (unless there is a coincidence).

In DP the solution (8.3) represents the solution to a ***relaxation*** problem. Problem *relaxation* can take several forms. Mostly it is applied to the relaxation or the *weakening* of the constraints or the objective functions. There are no explicit constraints in Example 8.1. In this instance, the *relaxation* refers to the removal of the restriction of *discreteness* of the variables. This is identified as ***continuous relaxation***. The relaxed problem or *relaxation* has several advantages.[3]

- If a constraint relaxation is infeasible, so is the problem/model it relaxes.
- Constraint relaxation expands the set of feasible solutions. The relaxed optimal value must improve or equal the best feasible solution to the original problem/model.
- The optimal value of any relaxed model provides a lower bound on the optimal solution if it is a minimization problem. Similarly, it establishes an upper bound for maximization problems.
- Many relaxation models produce optimal solutions that are easily rounded to good feasible solutions of the original problem/model. This appears to drive discrete optimization for engineering design problems.

Applied to Example 8.1, the values established in (8.3) is a solution to the relaxation of Example 8.1. It was easy to obtain this solution while a technique for the solution to the discrete unconstrained optimization is still to be developed—the Kuhn-Tucker conditions are no longer useful. The solution in (8.3) is not acceptable as x_2^* and x_3^* are not elements of the permissible set (8.2). The value of $f^* = 0$ will be better then the best discrete solution— lower bound on the solution to the original problem, as this is a minimization example.

8.1.2 Discrete Optimal Solution

A standard approach to solving discrete optimization problem, particularly in engineering, is to use the *continuous relaxation* of the mathematical model and first generate a continuous solution. For Example 8.1, this is given in (8.3). More than likely, the discrete variables will not belong to the predefined discrete sets. These variables are then changed (or rounded) to the nearest discrete values. For problems with constraints, feasibility is checked. The best feasible solution is then chosen. Unlike continuous problems, there are no necessary and sufficient conditions to satisfy.

In the case of Example 8.1, (8.2) and (8.3) provide the relevant information to establish the solution. For convenience (8.2) and (8.3) are reproduced here:

$$x_1 \in R \tag{8.2a}$$

$$x_2 \in [0.5 \quad 1.5 \quad 2.5 \quad 3.5] \in X_{2d} \tag{8.2b}$$

$$x_3 \in [0.22 \quad 0.75 \quad 1.73 \quad 2.24 \quad 2.78] \in X_{3d} \tag{8.2c}$$

$$x_1^* = 2; \quad x_2^* = 2; \quad x_3^* = 2; \quad f^* = 0 \tag{8.3}$$

The continuous variable x_1 is chosen to have a value of 2.0. Figure 8.1 indicates four sets of discrete values for x_2, and x_3 around their continuous solution. By one strategy the least value of the objective function at these points will be considered the solution. Evaluating the values of the objective function at those points:

$$x_1^* = 2; \quad x_2^* = 1.5; \quad x_3^* = 1.73; \quad f^* = 0.3758 \tag{8.4a}$$

$$x_1^* = 2; \quad x_2^* = 1.5; \quad x_3^* = 2.24; \quad f^* = 0.8552 \tag{8.4b}$$

$$x_1^* = 2; \quad x_2^* = 2.5; \quad x_3^* = 1.73; \quad f^* = 0.9158 \tag{8.4c}$$

$$x_1^* = 2; \quad x_2^* = 2.5; \quad x_3^* = 2.24; \quad f^* = 0.3752 \tag{8.4d}$$

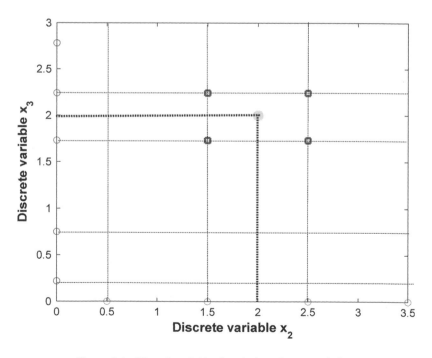

Figure 8.1   Discrete neighborhood of continuous solution.

From this exercise, the least value of the objective function is available in (8.4d) and would be regarded as the adjusted optimum solution to the original *MP* problem.

In fact, the best solution is obtained at

$$x_1^* = 1.7433; \quad x_2^* = 1.5; \quad x_3^* = 1.73; \quad f^* = 0.1782 \qquad (8.5)$$

The difference between (8.5) and (8.4a), only in the variable x_1, suggests that additional continuous optimization needs to be performed once the discrete values are selected, confirming the earlier observations by Fletcher[1]. This optimization should be easier, as the order of the mathematical model will be reduced by the number of discrete variables since they have been assigned numerical values. This simple example recommends at the very least a three-step procedure for the solution to the MP:

Step 1: A *continuous relaxation* identifies several combination of discrete variables for further exploration.

Step 2: For each such set of discrete variable combination, a continuous optimization is performed to establish a new optimum value of the continuous variables and the corresponding objective function. If all the variables are discrete, then only the function and constraints need to be evaluated at each of the set of variables.

Step 3: A simple comparison of the solutions/values in Step 2 identifies the optimum solution to the discrete problem.

This unconstrained optimization example in three variables has demonstrated that discrete optimization is a lot of work compared to continuous optimization. We are looking at four single-variable continuous optimization problems for the discrete variables in (8.4) after continuos relaxation. This work expands significantly if the number of variables increases or if the mathematical model is enhanced by the inclusion of constraints. Another essential feature in the exploration is that no new mathematical conditions were necessary for establishment of the discrete solution beyond a simple comparison of the objective function. The nature of discrete variables and discrete functions precludes any sophisticated mathematical conditions established by derivatives of the functions involved in the model. Trapping and branching based on comparison of values is the mainstay of discrete algorithms. The techniques are classified as **heuristic methods**. This encourages unique and personal implementations of the search for discrete optimization that can be tailored for your class of problems.

8.2 DISCRETE OPTIMIZATION TECHNIQUES

There are three discrete optimization techniques in this section. The first one is *Exhaustive Enumeration*. This involves identifying solution to the mathematical model for all possible combinations of the discrete variables. This is suggestive of the Zero-Order numerical techniques of Chapter 6. Those methods lacked sophistication as they involved only evaluation of the functions at a phenomenal number of points. They were able to take advantage of plentiful computer resources today. Imagine operating in the peer-to-peer computing environment afforded by "MP3" or "Gnutella" and solving the problem using all the PCs in the world. The second method is the *Branch and Bound* method. This is based on partial enumeration where only part of the combinations are explored. The remaining are pruned from consideration because they will not contain the solution. This is currently the most popular method for discrete optimization for engineering designs. The last method is *Dynamic Programming*, an elegant approach to optimization problems, which did not gain favor because it involved significantly large amounts of computation than competitive methods even for problems of reasonable size. It is restricted to problems that require a sequential selection of the design variables. Today, such resource limitations are disappearing in the world of powerful personal PCs and hence, the method deserves to be considered.

Standard Format: Discrete Optimization: There is no standard format for the discrete optimization problem. It subsumes the format of the corresponding continuous relaxation problem. In this book, the following format for the mixed optimization problem is used.

$$\text{Minimize} \quad f(X, Y), \quad [X]_{n_c} ; [Y]_{n_d} \tag{8.6}$$

$$\text{Subject to:} \quad h(X, Y) = [0]; \quad [h]_l \tag{8.7}$$

$$g(X, Y) \leq [0]; \quad [g]_m \tag{8.8}$$

$$x_i^l \leq x_i \leq x_i^u; \quad i = 1, 2, \ldots n_c \tag{8.9}$$

$$y_i \in Y_{d_i}; \quad \left[Y_{d_i} \right]_{p_i}; \quad i = 1, 2, \ldots n_d \tag{8.10}$$

X represents the set of n_c continuous variables. Y represents the set of n_d discrete variables. f is the objective function. h is the set of l equality constraints. g is the set of m inequality constraints. Expression (8.9) represents the side constraints on each continuous variable. Expression (8.10) expresses the side constraints on the discrete variables. Each discrete variable y_i must belong to a pre-established set of p_i discrete values Y_{di}. If $n_d = 0$ then it a continuous optimization problem. If $n_c = 0$ then it is a discrete optimization problem. If both are nonzero then it is a *Mixed Problem*.

Continuous Relaxation: The continuous relaxation of the mixed optimization problem (8.6–8.10) is identical to the problem expressed in (8.6–8.10) with the discrete constraint (8.10) being replaced by a continuous side constraint for the discrete variables of the form of (8.9). Since the set of allowable values for each discrete variable is expected to be ordered, the continuous limits can be identified as the first and the last value of the set for each variable. This is not explicitly developed since (8.5–8.9) and a modified (8.10) is fairly representative of the continuous mathematical models you have come across in the previous chapters.

Reduced Model: The reduced model is important in subsequent discussions. It is the continuous relaxation model solved after a set of discrete variables are set at some allowable discrete values. This removes those variables from the problem as their values have been set. The mathematical model is then defined by the remaining discrete variables, as well as the original continuos variables. If Z is the remaining discrete variable that needs to be solved then n_z

$$\text{Minimize} \quad \tilde{f}(X, Z), \quad [X]_{n_c}; \; [Z]_{n_z} \tag{8.11}$$

$$\text{Subject to:} \quad \tilde{h}(X, Z) = [0]; \quad [h]_l \tag{8.12}$$

$$\tilde{g}(X, Z) \leq [0]; \quad [g]_m \tag{8.13}$$

$$x_i^l \leq x_i \leq x_i^u; \quad i = 1, 2, \ldots, n_c \tag{8.14}$$

$$z_i \in Z_{d_i}; \quad \left[Z_{d_i}\right]_{\zeta_i} \tag{8.15}$$

Note that Z_{di} is not a new set. It reflects the remaining discrete variable and the appropriate Y_{di}. In *Exhaustive Enumeration* n_z is zero, as all the discrete variables are set to some allowable value.

8.2.1   Exhaustive Enumeration

This is the simplest of the discrete optimization techniques. It evaluates an optimum solution for all combinations of the discrete variables. The best solution is obtained by scanning/comparing the list of feasible solutions in the previous investigation. The total number of evaluations are

$$n_e = \prod_{i=1}^{n_d} p_i \tag{8.16}$$

If either n_d or $p_i's$ (or both) are large then this is represents a lot of work. It also represents an exponential growth in the calculations with the number of discrete variables. In a mixed optimization problem this would involve n_e continuous optimum solution. If the mathematical model and its computer calculations are easy to implement this is not a serious burden. If the mathematical model requires extensive calculations—for example finite element or finite difference calculations—then some concern may arise. In such cases, limiting the elements of the discrete variables to smaller clusters may represent a useful approach.

Algorithm: Exhaustive Enumeration — A8.1:

Step 1: $f^* = \inf, X = [0 \ 0 \dots 0]$
For every allowable combination of $(y_1, y_2, \dots y_{n_d}) \Rightarrow (Y_b)$
Solve Optimization Problem (8.11)–(8.15) (Solution X^*)
If $h(X^*, Y_b) = [0]$ and
If $g(X^*, Y_b) \leq [0]$ and
If $f(X^*, Y_b) < f^*$
Then $f* \leftarrow f(X^*, Y_b)$
$X \leftarrow X^*$
$Y \leftarrow Y_b$

Example 8.1:   Minimize the objective function f

Minimize $f(x_1, x_2, x_3) = (x_1 - 2)^2 + (x_1 - x_2)^2 + (x_1 - x_3)^2 + (x_2 - x_3)^2$

$$(8.1)$$

The side constraints on the design variables can be set up as

$$x_1 \in R \tag{8.2a}$$

$$x_2 \in [0.5 \quad 1.5 \quad 2.5 \quad 3.5] \in X_{2d} \tag{8.2b}$$

$$x_3 \in [0.22 \quad 0.75 \quad 1.73 \quad 2.24 \quad 2.78] \in X_{3d} \tag{8.2c}$$

MATLAB Code: The exhaustive enumeration of Example 8.1 is available in **Exhaustive_Enumeration_Example8_1.m**.[*] Table 8.1a is the optimum value of the continuous variable x_1 for every allowable combination of the discrete variables x_2, and x_3. Table 8.1b is the corresponding value of the objective function for each of the cases. Since this is an unconstrained problem no discussion of feasibility is involved. There are 20 values and the best solution for Example 8.1 is (8.5).

$$x_1^* = 1.7433; \quad x_2^* = 1.5; \quad x_3^* = 1.73; \quad f^* = 0.1782$$

Table 8.1a   Optimal value of continuous variable x_1 for discrete combination of x_2 and x_3

$x2/x3$	0.2200	0.7500	1.7300	2.2400	2.7800
0.5	0.9067	1.0833	1.4100	1.5800	1.7600
1.5	1.2400	1.4167	1.7433	1.9133	2.0933
2.5	1.5733	1.7500	2.0767	2.2467	2.4267
3.5	1.9067	2.0833	2.4100	2.5800	2.7600

*Files to be downloaded from the web site are indicated by boldface courier type.

Table 8.1b **Optimal value of objective function for discrete combination of x_2 and x_3**

$x2/x3$	0.2200	0.7500	1.7300	2.2400	2.7800
0.5	1.9107	1.3542	2.7915	4.8060	7.8840
1.5	3.3240	1.3542	0.1782	0.8327	2.4707
2.5	8.0707	4.6875	0.8982	0.1927	0.3907
3.5	16.1507	11.3542	4.9515	2.8860	1.6440

There is some good news, though. First each of the 20 cases is essentially a *single variable* optimization problem. The number of design variables in the original model (3) is reduced by the number of discrete variables (2). Model reduction is an important feature in enumeration techniques.

Example 8.2 illustrates the use of algorithm A8.1 with feasibility requirements.

Example 8.2: The Aerodesign Club has initiated fundraising activities by selling pizza during lunch in the student activity center. Over the years it has collected data to identify a mathematical model that will reduce its cost which translates to higher profits. Only two types of pizza will sell—pepperoni (x_1) and cheese (x_2). The local pizza chain will deliver a maximum of 26 pizzas. Factoring into account wastage and the time available to make the sale the club members arrive at the following mathematical model:

$$\text{Minimize} \quad f(x_1, x_2) = (x_1 - 15)^2 + (x_2 - 15)^2 \tag{8.17}$$

$$\text{Subject to:} \quad g_1(x_1, x_2): \quad x_1 + x_2 \leq 26 \tag{8.18a}$$

$$g_2(x_1, x_2): \quad x_1 + 2.5x_2 \leq 37 \tag{8.18b}$$

$$0 \leq x_1 \leq 30; \quad 0 \leq x_2 \leq 30; \quad both\ integer \tag{8.18c}$$

Matlab Code: The file: Exhaustive_Enumeration_Example8_2.m will solve Example 8.2 both through exhaustive enumeration and continuous relaxation. The latter is obtained by applying the KT conditions.

The discrete solution is

$$x_1^* = 12; \quad x_2^* = 10; \quad f^* = 34; \quad g_2\ is\ active \tag{8.19a}$$

And the continuous relaxation solution is

$$x_1^* = 12.862; \quad x_2^* = 9.6552; \quad f^* = 33.1379; \quad g_2\ is\ active \tag{8.19b}$$

In this example, the rounding off to a neighboring discrete value works well. Programming exhaustive enumeration is straight forward and particularly easy with MATLAB. Given the processing speed, large memory availability on the desktop, easy programming through software like MATLAB, *exhaustive enumeration* is probably a very good idea today. What recommends it even further is that the solution is a ***global optimum***. Furthermore, unlike the earlier chapters where the concepts of derivatives, matrices, linear independence etc. were essential to understand and implement the techniques of optimization, lack of such knowledge is no disadvantage to solving discrete optimization problem using exhaustive enumeration. There is only essential skill—translating the mathematical model into program code. With the experience provided in this book, this is definitely not an issue.

Exhaustive Enumeration for Continuous Optimization Problems: The Scan and Zoom method that was introduced in earlier chapters can be regarded as a multistage exhaustive enumeration for continuous variables. This consideration is based on the availability of serious desktop computing power, the promise of global optimization, and the ease of implementation. Exhaustive enumeration, after all is a *search strategy*, similar to the ones used before.

A higher stage in multistage enumeration involves enumerating the design on decreased intervals centered around the solution from the previous stage. The values of the design variables used for enumeration will incorporate progressively improved tolerance. The strategy must take into account the possibility that the initial interval might overlook global optimal solutions that do not appear attractive in a particular stage. Hence, more than one minimum can be tracked in each stage. Each of these minimum can be considered an *enumeration cluster*. Each such cluster is multistaged until a prescribed tolerance is achieved. A comparison of the solution will determine the best one. In engineering, manufacturing tolerances of the order of 0.0001 are achievable today. From a programming perspective, it just adds an extra loop to the implementation. The final solution obtained represents a continuous solution within the final tolerance implemented by the technique.

Such an exercise, of tracking multiple minimums, is left to the student. As with all implementation, the merit of this approach lies in the tractability of the mathematical model and the need for a global optimum solution. Since the method is largely heuristic, there is no guarantee that the solution is actually a global optimum (of course assuming the functions involved are not convex).

8.2.2 Branch and Bound

Exhaustive or complete enumeration is possible only for a limited set of variables because of the exponential growth of effort required to examine all possible solutions. Using *partial, or incomplete or selective* enumeration to arrive at the solution to the discrete optimization problem would provide the necessary advantage in handling large problems. Partial enumeration works by making decisions about groups of solutions rather then every single one. The ***Branch and Bound-nobreak (BB)*** algorithm is one of the successful algorithms to use selective/partial enumeration. They do so by using *relaxation models* instead of the original complete discrete mathematical models. The Branch and Bound (BB) algorithms use a *tree* structure to represent *completions* of *partial solutions*. The tree structure uses *nodes* and *edges/links* to represent the trail of partial solutions. This is called *fathoming* the partial solution. The terminology is borrowed from operation research literature and is explained with respect to Example 8.1 which is reproduced here for convenience[3].

$$\text{Minimize} \quad f(x_1, x_2, x_3) = (x_1 - 2)^2 + (x_1 - x_2)^2 + (x_1 - x_3)^2 + (x_2 - x_3)^2 \tag{8.1}$$

$$x_1 \in \mathbf{R} \tag{8.2a}$$

$$x_2 \in [0.5 \quad 1.5 \quad 2.5 \quad 3.5] \in X_{2d} \tag{8.2b}$$

$$x_3 \in [0.22 \quad 0.75 \quad 1.73 \quad 2.24 \quad 2.78] \in X_{1d} \tag{8.2c}$$

Partial Solution: A solution in which some of the discrete design variables are fixed and others are left *free* or undetermined. Variables that are not fixed are represented by the symbol f (please avoid confusion with objective function). For example,

$$X = [f, \ 0.5, \ f] \tag{8.20}$$

represents the optimization problem with the discrete variable x_2 set at a *discrete value* of 0.5 and x_1 and x_3 yet to be determined. The partial solutions are also termed as the nodes of the tree. The free variables are usually solved using a *continuous relation* of the model.

Completions (of Partial Solution):   Each partial solution or node can give rise to additional partial solution/nodes directly under it. This is called **branching** or *expansion* of the node. In this new set of nodes, another one of the *free* discrete variables is fixed and the remaining continue to be free. For example, under the node represented in (8.20), five additional nodes are possible:

$$X = [f, 0.5, 0.22]; \quad X = [f, 0.5, 0.75]; \quad X = [f, 0.5, 1.73];$$

$$X = [f, 0.5, 2.24]; \quad X = [f, 0.5, 2.78]$$

These completions branch out and hence form a treelike structure. There is a hierarchical structuring of the nodes. Each branch leads to a tier in which some variable is held at its prescribed discrete value.

Fathoming the Partial Solution: Expanding the node all the way till the end is termed *fathoming* a node. During the fathoming operation, it is possible to identify those partial solutions at a node that do not need to be investigated further. Only the best solution at the node is **branched**. Those partial solutions/nodes that will not be investigated further are **pruned** or **terminated** in the tree.

Active Partial Solution: A node that has not been terminated or pruned and has not been completed is referred to as an *active partial solution*.

Edges/Links: These are node to node connections in a in a tree. They identify the value of the new variable that is fixed.

Root: The BB algorithm searches along this tree. It starts at the **root** where all the design variables are free. Branch and Bound search stops when all partial solutions in the tree have been branched or terminated. The search is depth first—completion at a node is preferred to jumping across nodes in the same hierarchy.

Incumbent Solution: The incumbent solution is the best feasible solution known so far with all discrete variables assuming discrete values. When the Branch and Bound search finally stops, the final incumbent solution, if one exists, is a global optimum solution. In the following, the incumbent solution is denoted by f^{**}. The incumbent design variables are X^{**}.

Algorithm: Branch and Bound — A8.2:   For even a small set of discrete variables, actually drawing the BB tree structure is most likely a difficult enterprise. The tree structure merely identifies the search process. Example 8.1 does not have constraints, so feasibility is not examined. With constraints present, choosing the partial solution to complete based on minimum value of the objective function is not a justifiable strategy. The best feasible solution is the candidate for completion. Also, an active feasible partial solution that has a better value than the current incumbent solution is also a candidate for completion. The following algorithm captures the essence of the Branch and Bound algorithm.

Step 0.   Establish the root node with all discrete variables free.
          Initialize solution index s to 0.
          Set current incumbent solution to $f^{**} = \infty$, $X^{**} = [\infty]$, (minimization problem).
Step 1.   If any active feasible partial solutions remain (that is better then the current incumbent solution), select one as $X^{(s)}$ and go to Step 2.
          If none exists and there exists an incumbent solution, it is optimal.
          If none exists and there is no incumbent solution, the model s infeasible.
Step 2.   Obtain the completion of the partial solution $X^{(s)}$ using a continuous relaxation of the original model.
Step 3.   If there are no feasible completions of the partial solution $X^{(s)}$, terminate $X^{(s)}$, increment $s \leftarrow s + 1$ and return to Step 1.

Step 4. If the best feasible completion of partial solution $X^{(s)}$ cannot improve the incumbent solution then terminate $X^{(s)}$, increment $s \leftarrow s + 1$ and return to Step 1.

Step 5. If the best feasible completion is better then the incumbent, update the incumbent solution

$$f^{**} \leftarrow f^{(s)}, \quad X^{**} \leftarrow X^{(s)}$$

terminate $X^{(s)}$, increment $s \leftarrow s + 1$ and return to Step 1.

Example 8.1 — Using Branch and Bound (BB): Here the BB search is applied to Example 8.1. The BB tree is in Figures 8.2 and 8.3.

Start: Node 0. All the variables are free. The solution to the continuous relaxation of the objective function (solved as a continuous problem) is identified at the node.

Nodes 1–4: This is the first **tier** of the BB tree. The discrete variable x_2 is assigned each of the permissible discrete values at the nodes in this tier. This is indicated on the edges between the nodes in Figure 8.2. The remaining variables

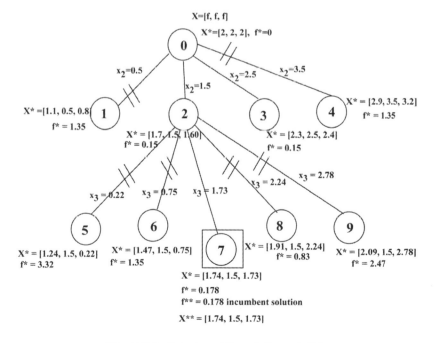

Figure 8.2 Branch and Bound: Example 8.1.

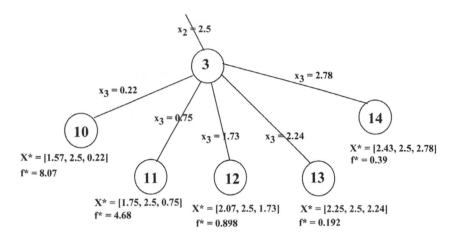

Figure 8.3   Branch and Bound: Example 8.1 — Fathoming Node 3.

are free/continuous. The optimum solution at the various nodes is displayed on the tree. The minimum solution is the one that will be fathomed at the next step. Here Nodes 2 and 3 are candidates for expanding the branches. Nodes 1 and 4 will not be fathomed—(unless their optimum value is better than the incumbent solution)—they will be terminated or pruned as the objective function has a higher value is. The continuous relaxation assures us a lower bound on the optimum value of the objective at the node. Node 2 is taken up for fathoming based on node ordering.

Completions of Node 2:   This expands the Node 2 by setting the discrete variable x_3 to its permissible set of discrete values. It creates branches Nodes 5–9. It is also the second tier of the tree. The solution is indicated at the various nodes. In all of them, the two discrete variables are set to indicated values while x_1 is free, which is a continues variable by definition. Since all the discrete variables have discrete values, this Node is also ***fathomed***. Node 7 indicates the best solution. Since ***all*** the discrete variables are assigned discrete values the solution at this node is the ***current incumbent solution***. The remaining nodes will be pruned.

The current incumbent solution ($f^{**} = 0.1782$) is higher then the solution at the node 3 ($f^* = 0.15$). This requires Node 3 to be fathomed.

Completions of Node 3:   Figure 8.3 illustrates the branching at Node 3. The process is similar to completions of Node 2 above. The best solution is at Node 13 with a value of $f^* = 0.1927$. This is larger than the current incumbent solution. All the nodes can be terminated.

No further nodes are available for expansion. The current incumbent solution is also the *global optimal solution*. This is the solution to the optimization problem. In this example, 15 partial solutions were necessary. Each partial solution can be considered an enumeration. In exhaustive enumeration, 20 enumerations were required for Example 8.1. There is a net gain in 5 enumerations. The BB search reduces the number of total enumerations needed to solve the problem.

***MATLAB* Code:** The Branch and Bound algorithm is applied symbolically in `Branch_Bound_Example8_1.m`. The Tier 1 and Tier 2 calculations are written to the Command window. These results are used in Figures 8.2 and 8.3. Please note that the code is problem specific.

Example 8.2: The Branch and Bound algorithm is now applied to Example 8.2. The search is applied through the BB tree. The mathematical model for Example 8.2, where x_1, and x_2 are integers.

$$\text{Minimize} \quad f(x_1, x_2) = (x_1 - 15)^2 + (x_2 - 15)^2 \qquad (8.17)$$

$$\text{Subject to:} \quad g_1(x_1.x_2): \quad x_1 + x_2 \leq 26 \qquad (8.18\text{a})$$

$$g_2(x_1.x_2): \quad x_1 + 2.5x_2 \leq 37 \qquad (8.18\text{b})$$

$$0 \leq x_1 \leq 30; \quad 0 \leq x_2 \leq 30; \quad \text{both integer} \qquad (8.18\text{c})$$

The solution from exhaustive enumeration is

$$x_1^* = 12; \quad x_2^* = 10; \quad f^* = 34; \quad g_2 \text{ is active} \qquad (8.19\text{a})$$

Using Branch and Bound: For Tier 1 we will allow x_1 to range through its discrete values and x_2 will be continuous and determined by applying the KT conditions to the model in (8.17)–(8.18). For each value of x_1, there are only three cases, since setting g_1 and g_2 equal to zero simultaneously will determined an inconsistent values of x_2. The three cases can be resolved as

$$
\begin{aligned}
\beta_1 &= 0; \ \beta_2 = 0: \ x_2 = 15; \\
g_1 &= 0; \qquad\qquad x_2 = 26 - x_1 \\
g_2 &= 0; \qquad\qquad x_2 = \frac{37 - x_1}{2.5}
\end{aligned}
\qquad (8.21)
$$

From Tier 1 calculations, three best results are selected corresponding to the value of discrete x_1. The Tier 2 calculations are then performed for the three values of x_1 where x_2 is allowed to range through its discrete values. The best values are recorded and the final solution is determined. Note, the solution is obtained by evaluating and comparing the functions.

***MATLAB* Code:** The Branch and Bound algorithm is applied symbolically in `Branch_Bound_Example8_2.m`. The Tier 1 and Tier 2 calculations are written to the Command window. These results are used in Figure 8.4. Please note that

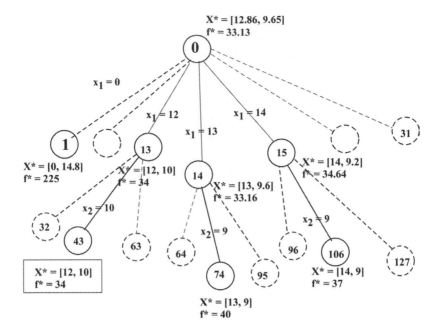

Figure 8.4 Branch and Bound tree for Example 8.2.

the code is problem specific. The three solutions, after Tier 2 calculations, are copied from the Command window below.

Tier -2: x1 and x2 are both discrete

```
Design Variables:      13      9
Objective function:     40
Constraints:    -4.0000    -1.5000
Design Variables:      12      10
Objective function:     34
Constraints:     -4      0
Design Variables:      14      9
Objective function:     37
Constraints:    -3.0000    -0.5000
```

The best solution matches the exhaustive enumeration in (8.19a).

The BB tree is shown in Figure 8.4. The data for the tree is calculated through. The tree is branched as follows:

Node 0: This is the root node. Both the variables are free. The continuous relaxation of the problem is solved. The solution is indicated on the tree. Currently this is the only available active partial solution and hence needs to be completed.

Nodes 1 to 31: This is the Tier 1 of the tree. The partial solution at Node 0 is completed allowing variable x_1 to assume its set of discrete values. Since it is likely that more than one node in this tier may be scrutinized, we choose the best three solutions at this level. The remaining nodes are pruned—shown by dotted lines. The three branches and the optimal values are indicated on the tree. The values are indicated at the nodes. For this problem they are also feasible. There are now three active partial solutions.

Node 13: The solution at Node 13 is also a feasible solution of the original discrete optimization problem since the design variables are discrete and belong to the permissible discrete set. Node 13 is therefore fathomed. This provides the first *incumbent solution*. This means that Nodes 32 to 63 are not necessary. This is not implemented in the code, as it would require identifying the discreteness of the value, and hence Nodes 32 and 63 are checked.

There are three available active feasible partial solutions (Nodes 13, 14, and 15) for completion. Since the solutions at the nodes are all feasible, only the best active feasible partial solution is picked up for completion.

Completions of Node 13, 14, 15: The completion of these nodes identifies three fathomed solutions that are feasible and the variables are discrete. The best solution is then selected from among these solutions. This is boxed on the tree.

This completes the BB search applied to Example 8.2. There were 127 enumerations as compared to 961 for the exhaustive enumeration. Note that for nodes 1 to 31, the evaluation of KT conditions were necessary, which is not necessary with exhaustive enumeration. Possibly, this is not a good comparison for performance.

In this example, the active partial solutions at the root and in the first tier were all feasible. What would be the strategy if the root node or other nodes have a infeasible active partial solution? In that case the BB search process must complete all active partial solutions and not only feasible active partial solutions. That is a modification that can be easily made to the Branch and Bound algorithm above.

Other Search Techniques: The two search techniques in this section, namely *Exhaustive Enumeration*, and *Branch and Bound* provide the additional feature of discovering *global optimum* solutions. They also belong to the class of methods used to address global optimum. The easy availability of large and faster computing resources have brought renewed emphasis on globally optimal designs. *Simulated annealing* and *genetic algorithms* are among the leading candidates for global optimization. They are covered in the next chapter and can be also characterized as enumeration techniques, like the ones in this section. Both of them can and are used for discrete optimization. They are not discussed in this chapter. The application of most of these techniques is still heuristic and often depends on experience based on numerical experiments among classes of problems. Some applications are problem specific and require user intervention and learning. Standard implementations in these methods, especially for nonlinear problems, are yet to coalesce.

8.2.3 Dynamic Programming

Discrete dynamic programming *(DDP)* is an exciting technique for handling some special class of problems. Examples 8.1 and 8.2 are not direct members of this class. Richard Bellman was responsible for first introducing the concept and the algorithm.[4] It is an optimizing procedure based on Bellman's *Principle of Optimality*. The principle is based on a sequence of decisions based on partial solutions, so that when certain variables have been determined, the remaining variables establish an optimum completion of the problem.[5] Another expression of this principle is obtained from Syslo[6]. An optimal sequence of decisions has the property that at the current time, whatever the initial state and the previous decisions, the remaining decisions must be an optimal set with regard to the state resulting from the first decision. Dynamic programming, is about a *sequence of decisions* (often in time) and is sometimes termed as a *sequential decision problem*. Problems like Examples 8.1 and 8.2 do not directly fit this classification but they can be transformed to fit this requirement[5]. Very often, DDP problems can be described as an optimal path determination problem. Such *path* problems are solved using directed graphs *(digraphs)*. Once again, these problems may actually have no connection to physical paths or distances. Like the BB tree they enable better appreciation/application of the algorithmic procedure and are not strictly required. DP requires definition of *states* and *stages*, and only the former is discussed here. The equations used to establish the principle of optimality are called *functional equations*. Example 8.3 will be used to define these terms.

Example 8.3: An established university is interested in developing a new College for Information Technology during the current year. Table 8.2 illustrates the necessary data to calculate the cost of recruiting faculty for this enterprise for the first year of operation. The college will operate during four quarters (fall, winter, spring, and summer) in a year indicated by the columns. Decisions are associated with the beginning of the quarters. The first row indicates the new faculty required to implement the courses that will be offered. The second row is the recruitment cost for new faculty each quarter, which is significant because of the competitive demand for qualified people in this area. The third row is the

Table 8.2 Faculty hiring policy data for Example 8.3

Item	Fall	Winter	Spring	Summer
Required Faculty	5	10	8	2
Recruitment Cost	10	10	10	10
Unit Faculty Cost	1	2	2	1
Unit Retaining Cost	2	2	2	2

unit cost of new faculty in terms of the standard institute faculty cost. Course specialization and availability make this cost different in different quarters. The last row is the unit cost for retaining extra faculty hired in an earlier quarter (to minimize cost over the year). What should be the hiring policy for the university to minimize cost?

This example is clearly illustrative of *sequential decision making*. Decisions are required to be made every quarter. For example, it is possible to recruit all needed faculty during the first quarter and pay the retaining cost, or it may cost less to hire during each quarter, or recruit for three quarters and hire again in the fourth. The problem will be solved using an *optimal/shortest path* method, even though there are no routes involved in the problem. One of the best way to understand the shortest path algorithm is to draw a directed graph between the *nodes/states* in the problem.

States: States represent points/conditions in DP where decisions have to be considered. This is sometimes referred to as *policy* decisions or just policies. This has a sequential connection because current decisions cannot be considered until the previous decisions have been made. Clearly in Example 8.3 they represent the beginning of each quarter. The number of states is usually one more than the number of decision states to incorporate the effect of the last policy/decision. *Nodes* in the digraph represent the *states* in the DDP problem. The **digraph** represents a connection between each pair of states that is permissible. This can be made clearer by stating that if a node is going to be affected by a decision at any previous node then there is a direct connection (line/arc) between the nodes and the cost of the connection is indicated on the connection itself. An arrow can be used to indicate the direction of traverse along this line/arc segment. These lines/arcs therefore represent *decisions*. Once the digraph for the DDP problem is available, the *optimum solution corresponds to the shortest path from the beginning to the ending state of the digraph*[3].

Figure 8.5 is the digraph for the hiring policy. There are five states with the first four representing the quarters where the decisions must be made. Node 1 represents the fall quarter, Node 2 represents the winter quarter and so on. Along the arcs on the digraph is the cost or the decision or policy that will originate at the current node and meet all the requirements at the latter node. In order to use the information in the digraph effectively, and to develop the mathematical model, the quarters are assigned indices from 1–4. The following abbreviations are made: Required Faculty *(RF)*; Recruitment Cost *(RC)*; Unit Faculty Cost *(FC)*; Unit Retaining Cost—Holding Cost)—*(HC)*. The policy decisions or cost between the start node i and the end node j is $C(i, j)$. For example $C(1, 4)$ represents the policy to hire in the fall quarter, which will meet the staffing requirements for the winter and spring quarters, thus paying the appropriate retaining penalties:

$$C(1, 4) = RC(1) + [RF(1) + RF(2) + RF(3)]^* FC(1)$$
$$+ \{[RF(2) + RF(3)]^* HC(1) + [RF(3)]^* HC(2)\}$$

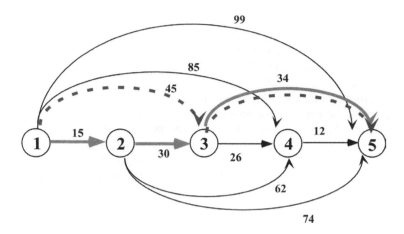

Figure 8.5 Digraph for Example 8.3.

Functional Equation for Dynamic Programming: The functional equation is the essential part of the DDP algorithm or method. It employs a recursive procedure to identify the best solution at each state. Let *Value (i)* be the best value of the decision at node i, taking into account all possible arcs/paths that lead to node i. Let *Node(i)* be the node that led to the best policy at the current state. The recursion relations for n states can be set up as follows:

Value (1) = 0

Value (k) = min {*Value* (l) + C(l, k); $\quad 1 \le l < k$}, $\quad k = 2, \ldots n$ $\quad$ (8.22)

For example, in Example 8.3

$$Value\ (4) = \min \{Value\ (1) + C(1, 4); \ Value\ (2) + C(2, 4); \ Value\ (3)$$

$$+ C(3, 4)\}$$

Value (4) = *min* {[0 + 85]; [15 + 62]; [45 + 26]} = 71

Node (4) = 3 *(which established the minimum value)*

These calculations are embedded in the code.

MATLAB Code: `DynamicProgramming_Example8_3.m` is the m-file that generates the data for the digraph for Example 8.3 and identifies the solution by solving the *functional equations*. It also backtracks to identify the optimal sequence. Note, for different problems only the cost calculation will differ. The functional equations and backtracking are elements of the DDP method and should be the same. The output in the Command window is:

Example 8.3: Dynamic Programming

Cost between pairs of nodes/states C(i,j)

0	15	45	85	99
0	0	30	62	74
0	0	0	26	34
0	0	0	0	12
0	0	0	0	0

Optimal path values at various nodes-Values(i)

0	15	45	71	79

Best previous node to reach current node -Node(i)

0	1	2	3	3

Additional nodes for same optimal value if any

0	0	0	0	0
0	0	0	0	0
0	1	0	0	0
0	0	0	0	0
0	0	0	0	0

The *Value* (4) is 71 and *Node* (4) is 3. The optimum value for the problem is *Value* (5), which is equal to 79.

Shortest Path: The *shortest path* is obtained by backtracking from the end state using the *Node (i)* information—which is the row of data above.

At Node 5, the best node from which to reach Node 5 is Node 3:

$$0 \quad 1 \quad 2 \quad 3 \quad \mathbf{3}$$

At Node 3 the best node from which to reach Node 3 is Node 2:

$$0 \quad 1 \quad \mathbf{2} \quad 3 \quad 3$$

At Node 2 the best node from which to reach Node 2 is Node 1 (this is the starting node):

$$0 \quad \mathbf{1} \quad 2 \quad 3 \quad 3$$

This solution is indicated by the thick solid line in the digraph on Figure 8.5.

The actual hiring policy according to this solution is to hire just the required faculty for the fall quarter in fall quarter. Do the same for the winter quarter. For the spring quarter, however, hire both for the spring and summer quarters.

Alternate Paths: There is an alternate solution for the *same optimal value*. The output from **DynamicProgramming_Example8_3.m** includes this information though it was not included above. It suggests that at the third node, *Node (3)* can also have the value of 1—that is Node 1 also provides the same value to reach Node 3. This will cause the *Node (i)* information to change as

$$0 \quad 1 \quad 1 \quad 3 \quad 3$$

The *shortest path* can then be interpreted as follows:
At Node 5 the best node from which to reach Node 5 is Node 3:

$$0 \quad 1 \quad 1 \quad 3 \quad \mathbf{3}$$

At Node 3 the best node from which to reach Node 3 is Node 1:

$$0 \quad 1 \quad \mathbf{1} \quad 3 \quad 3$$

In Figure 8.5 this is indicated by the thick dotted arcs (or in blue color). The hiring policy now would be to hire for fall and winter in fall. Then hire for spring and summer in spring.

Algorithm: Dynamic Programming (A 8.3): The following algorithm is identified in for paths that do not cycle[3]. The original Bellman algorithm includes an iterative refinement that allows for cyclic paths. In this example there are no cyclic paths. This is implemented in **DynamicProgramming_Example8_3.m**.

Step 0: Number nodes in sequential order. Arcs are defined for $j\rangle i$. Set the start node s optimal value as

$$\text{Value (s)} = 0$$

Step 1: Terminate if all *Value (k)* have been calculated. If p represents the last unprocessed node the
Value $(p) = \min\{\text{Value}(i) + C(i, p) : \text{if arc}(i, p) \text{ exists } 1 \le i < p\}$
Node $(p) = $ the number of Node$i$ that achieved the minimum above
Return to Step 1

This was a simple example of dynamic programming (DP). It is important to note that the solution in final *Value (i)* includes only the best solution at all the states/nodes. Like other techniques there are several different categories of problems addressed by Dynamic Programming, including special discrete problems like the Knapsack problems. It is also possible to bring Examples 8.1 and 8.2 under the framework of Dynamic Programming, though to impart a sequential nature in the process of selecting design variables in this case may involve a total transformation of the problem. Even then, it may not be efficient to solve it using DDP. In practice, DDP has always been regarded as computationally complex and prohibitive. It is seldom used for problems with large number of states, or if the estimation of cost is intensive. Another criticism of DDP is that it carries out a lot of redundant computation for continuous problems. Today, these are less of issue than in the past. For stochastic and discrete problems DDP may still be an effective tool. Bellman, has also indicated that it can be used to solve a set of unstable systems of equations. DDP also computes design sensitivity as part of its solution. This section was primarily meant to introduce Dynamic Programming as an useful tool for some categories problems. Readers are encouraged to consult the references for more information on the subject.

8.3 ADDITIONAL EXAMPLES

In this section, two additional examples are presented. The first one is an engineering example from mechanical design and illustrates how the discrete version of a four-variable problem very effectively *reduces to a single-variable problem* that can be easily addressed by enumeration. This is a stunning reduction in complexity yielding a global optimum solution through elementary calculations. It definitely invites scrutiny about the effort in solving continuous nonlinear optimization problems. The second one is the application of the Branch and Bound search to a 0-1 integer programming problem. Much of the application of discrete optimization is in the area of decision making, and decision—for or against, yes or no, are arrived at using a 0-1 linear integer programming model.

8.3.1 Example 8.4 — I-Beam Design — Single Variable?

This example is a structural design problem. This is the same as Example 1.2, which was visited as Example 7.3. The formal statement of the problem is reproduced from Example 1.2. The mathematical model is also reproduced without comment.

Example 1.2: Design a cantilevered beam, of minimum mass, carrying a point load F at the end of the beam of length L. The cross-section of the beam will be in the shape of the letter I (referred to as an I-beam). The beam should be sufficiently strong in bending and shear. There is also a limit on its deflection.

Since the problem is formulated in Chapter 1, and revisited as Example 7.3, we directly proceed to the mathematical model associating x_1 with (beam depth) d, x_2 with (web thickness) t_w, x_3 with (flange width) b_f, and x_4 with t_f (flange thickness), so that the design vector is $X = [x_1, x_2, x_3, x_4]$. Figure 8.6 describes the layout of the design variables. The problem in standard format is:

$$\text{Minimize} \quad f(X) : \gamma L A_c \tag{8.23}$$

$$\text{Subject to:} \quad g_1(X) : F L x_1 / 2 I_c - \sigma_{\text{yield}} \leq 0 \tag{8.24a}$$

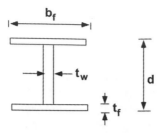

Figure 8.6 Design variables: Example 8.4.

$$g_2(X) : FQ_c/I_c x_2 - \tau_{\text{yield}} \leq 0 \tag{8.24b}$$

$$g_3(X) : FL^3/3EI_c - \delta_{\max} \leq 0 \tag{8.24c}$$

$$0.01 \leq x_1 \leq 0.25; \quad 0.001 \leq x_2 \leq 0.05; \tag{8.24d}$$

$$0.01 \leq x_3 \leq 0.25; \quad 0.001 \leq x_4 \leq 0.05 \tag{8.24e}$$

The designer must ensure that the problem definition is also consistent with the unit system chosen for the parameters and variables. The parameters for this problem (value is given in the parenthesis) are F (10,000 N); L (3 m); γ (steel: = 7860 kg/m^3); E (210 GPa); σ_{yield} (250 E+ 06 N/m$^{2)}$; τ_{yield} (145 E+ 06 N/m^2); and the maximum deflection δ_{max} (0.005 m).

We will include some geometric constraints among the design variables, as the program is not aware that x_1 must be greater than x_4 etc.

$$g_4(X) : x_1 - 3x_3 \leq 0 \tag{8.25a}$$

$$g_5(X) : 2x_3 - x_1 \leq 0 \tag{8.25b}$$

$$g_6(X) : x_2 - 1.5x_4 \leq 0 \tag{8.25c}$$

$$g_7(X) : 0.5x_4 - x_2 \leq 0 \tag{8.25d}$$

In the code for the various algorithms that we have developed, we always recognized the fact that that we did not explicitly handle the side constraints. One way to incorporate them directly in the problem formulation is to create linear inequality constraints from (7.77d). There will be eight such constraints developed:

$$g_8(X) = -x_1 + 0.01 \leq 0 \tag{8.26a}$$

$$g_9(X) = x_1 - 0.25 \leq 0 \tag{8.26b}$$

$$g_{10}(X) = -x_2 + 0.001 \leq 0 \tag{8.26c}$$

$$g_{11}(X) = x_2 - 0.05 \leq 0 \tag{8.26d}$$

$$g_{12}(X) = -x_3 + 0.01 \le 0 \tag{8.26e}$$

$$g_{13}(X) = x_3 - 0.25 \le 0 \tag{8.26f}$$

$$g_{14}(X) = -x_4 + 0.001 \le 0 \tag{8.26g}$$

$$g_{15}(X) = x_4 - 0.05 \le 0 \tag{8.26h}$$

To complete the model, we will need the following additional relations among the design variables.

$$A_c = 2x_3x_4 + x_1x_2 - 2x_2x_4$$

$$I_c = \frac{x_3x_1^3}{12} - \frac{(x_3 - x_2)(x_1 - 2x_4)^3}{12}$$

$$Q_c = 0.5x_3x_4(x_1 - x_4) + 0.5x_2(x_1 - x_4)^2$$

The solution to the continuous relaxation model is

$$x_1^* = 0.25\,[m]; \quad x_2^* = 0.0121\,[m]; \quad x_3^* = 0.125[m];$$

$$x_4^* = 0.02422\,[m]; \quad f^* = 200.3\,[kg]$$

The Discrete Problem: The discrete problem shares the same mathematical model except the values for the design variables will belong to a set of discrete values. It is possible to use either the complete enumeration or BB search to solve the problem. Before proceeding further in this direction, it is useful to recognize that the reason for discrete optimization is to choose an off-the-shelf I-beam which will keep the cost and production time down. Several steel mills provide information on standard rolling stock they manufacture, and these are well designated in the industry[7]. A partial data set of wide flange and standard beams is incorporated into the code. Any beam from this list has an explicit value for the set design variables. This becomes a *single variable problem*, the variable being the particular beam from the data set. In this case, therefore, no *geometrical constraints* are necessary and neither are the *side constraints*. However, just for completeness, the geometric and acceptability constraints are part of the problem formulation. A complete enumeration can be performed on the candidate beams from the stock list, and the best beam can easily be identified. We should compare this enumeration to solving the nonlinear mathematical model as Example 7.3. When such an opportunity arises, exploiting the discrete optimization problem is highly recommended. On the other hand, consider a different problem where a plastic I-beam will be molded in the manufacturing facility. Here, the continuous problem needs to be solved for the minimum cost.

MATLAB Code: The file: **DiscreteEnumeration_Example8_4.m** will solve this apparent one-dimensional optimization problem through direct enumeration. It uses **BeamPropertiesSI.m** to identify the properties of structural steel, and to create a structure called **RolledSteelBeamSI** that contains 32 beams.[7] The programming is straightforward. The output in the Command window will recognize feasible beams and identify the best beam for the problem—which minimizes the mass of the beam while satisfying all of the constraints.

There was a problem in solving the example as the constraints 12 (8.26e) and 14 (8.26g) were very conservative. It was decided to rerun the example ignoring these constraints but recording their values, nevertheless. The output copied from the Command window:

```
*********************************
Example 8.4 (Optimum Rolled Beam)
*********************************

Rolled Beam Designation: S310 X 74
Depth[m] Width[m]  Web Thickness[m] Flange Thickness [m]
0.30500  0.01740   0.13900                 0.017


Objective Function: (kg)     223.5384


Constraints
_____

Normal Stress Constraint-g1:         -53850
Shear Stress Constraint-g2:    -309.6380
Deflection Constraint-  g3:        -126900
Acceptability constraint-g4:      -0.1120
Acceptability constraint-g5:      -0.0270
Acceptability constraint-g6:      -0.0077
Acceptability constraint-g7:      -0.0090
Side constraint        -g8:      -0.2950
Side constraint        -g9:      -0.0164
Side constraint       -g10:      -0.1290
Side constraint       -g11:      -0.0157
Side constraint       -g12:       0.0550
Side constraint       -g13:      -0.0326
Side constraint       -g14:      -0.1110
Side constraint       -g15:      -0.0333
```

Note that the solution is significantly higher then the one obtained in the continuous relaxation, once again illustrating that the continuous relaxation provides a lower bound on the solution. Nevertheless, the solution here is the most economical one based on practical considerations.

8.3.2 Zero — One Integer Programming

Many LP programs/models have special features that are usually exploited through special algorithms. An important class of LP problems requires the variables be binary—that is, they can have one of two values. These problems can be set up so that values for the variables are zero or one—defining a Zero—One Integer Programming Problem (ZIP). The general form of a ZIP is

$$\text{Minimize} \quad f(X) : \sum_{j=1}^{n} c_j x_j$$

$$\text{Subject to} \quad \sum_{j=1}^{n} a_{ij} x_j \leq b_i, \quad i = 1, 2, \ldots m$$

$$x_j = 0 \ or \ 1, \quad j = 1, 2, \ldots, n$$

Except for the binary side constraints on the design variables, this represents the mathematical model for an LP problem (the standard model requires equality constraints only). Most algorithms assume $c'_j s$ are positive. If any c_j is negative, then $1 - x_j$ is substituted for x_j in the problem.

Several varieties of optimization/decision problems involve binary variables. They include the knapsack—a pure ILP with a single main constraint; capital budgeting—a multidimensional knapsack problem, assembly line balancing, matching, set covering and facility location and so on.

Example Knapsack Problem: Minimize the number of pennies (x_1), nickels (x_2), dimes (x_3) and quarters (x_4) to provide correct change for b cents

$$\text{Minimize:} \quad x_1 + x_2 + x_3 + x_4$$

$$\text{Subject to:} \quad x_1 + 5x_2 + 10x_3 + 25x_4 = b$$

$$x_1, x_2, x_3, x_4 \quad \text{are integers}$$

In this problem, the unknown variables are not binary variables.

Example Capital Budgeting: The College of Engineering is considering developing a unique Ph.D. program in Microsystems in the next three years. This expansion calls for significant investment in resources and personnel. Six independent focus areas have been identified for consideration, all of which cannot be developed. Table 8.3 presents the yearly cost associated with creating the various groups over the three years. The last line in the table represents the available budget (in millions of dollars) per year allotted to this development. The last column is the value (in \$100,000) the group is expected to generate in the next 5 years.

The capital budgeting problem uses six binary variables ($x_1, \ldots x_6$), one for each item in the table. The ZIP problem can be formulated as

$$\text{Maximize} \quad 20x_1 + 3x_2 + 30x_3 + 5x_4 + 10x_5 + 5x_6$$

$$\text{Subject to:} \quad 3x_1 + 2x_2 + x_3 + 2x_4 + 5x_5 \leq 10$$

Table 8.3 Capital Budgeting Example

Item	Group	Year1	Year2	Year3	Value
1	Systems	3	1	0	20
2	Microsystems	2	5	0	3
3	Software	1	3	1	30
4	Modeling	2	3	0	5
5	Materials	5	0	1	10
6	Photonics	0	2	2	5
Budget	10	8	3		

$$x_1 + 5x_2 + 3x_3 + 3x_4 + x_6 \le 8$$

$$x_3 + x_5 + 2x_6 \le 3$$

$$x_1, \ldots x_6 \in [0 \, or \, 1]$$

Note that this is not a minimization problem as required by the standard format.

Several algorithms have been developed for solving ZIP models. They all use implicit/selective enumeration through Branch and Bound search, together with continuous relaxation. However, the binary nature of the variables is exploited when completing the partial solutions or traversing the tree. Balas Zero—One Additive Algorithm/Method is one of the recommended techniques for this class of problems.[6] The algorithm from the same reference is included here.

Algorithm: Balas Zero–One Additive Algorithm (A 8.4): The enumeration is tracked by two vectors U and X. Both are of length n. If a partial solution consists of $x_{j1}, x_{j2}, \ldots, x_{js}$ assigned in order, the values of X corresponding to the assigned variables are 0 or 1. The remaining variables are free and have the value -1. The components of U are

$$u_k = \begin{cases} j_k & \text{if } x_{jk} = 1 \text{ and its complement has not been considered} \\ -j_k & \text{if } x_{jk} = 1 \text{ or } 0 \text{ and its complement has been considered} \\ 0 & \text{if } k > s \end{cases}$$

Step 1: If $b \ge 0$, optimal solution is $X = [0]$. Otherwise $U = [0]$. $X = [-1]$. $f^* = \infty$. $X^* = X$ (incumbent solution)

$$y_i = b_i - \sum_{j \in J} a_{ij} x_{ij}, \quad J \text{ is the set of assigned variables}$$

Step 2: Calculate $\bar{y} = \min y_i, \quad i = 1, 2, \ldots, m$

$$f = \sum_{j \in J} c_j x_j$$

If *If* $\bar{y} \ge 0$ *and* $f < f^*$ *then* $f^* = f,$ *and* $X^* = X$

go to Step 6

Otherwise continue

Step 3: Create a subset T of free variables x_{j}. defined as

$$T = \{j : f + c_j < f^*, a_{ij} < 0 \quad \text{for } i \text{ such that } y_i < 0\}$$

If $T = \emptyset$ (empty set) then go to Step 6, else continue

Step 4: Infeasibility test:

If i such that $y_i < 0$ and

$$y_i - \sum_{j \in T} \min(0, a_{ij}) < 0$$

go to Step 6, else continue.

Step 5: *Balas branching test:* For each free variable x_j create the set M_j:

$$M_j = \{i : y_i - a_{ij} < 0\}$$

If all sets M_j are empty go to Step 6. Otherwise calculate for each free variable x_j:

$$v_j = \sum_{i \in M_j} (y_i - a_{ij})$$

where $v_j = 0$ if the set M_j is empty. Add to the current partial solution the variable x_j that maximizes v_j. go to Step 2.

Step 6: Modify U by setting the sign of the rightmost positive component. All elements to the right are 0. Return to Step 2.

If there are no positive element in U. the enumeration is complete. The optimal solution is in the incumbent vector X^*. The objective is in f^*. If f^* is ∞ there is no feasible solution.

Balas.m is the m-file that implements the Balas Algorithm. It was translated from a Pascal program[6]. It prompts the user for the problem information $-n, m, A, b, c$. The test problem is included below[6].

Example 8.5:

$$\text{Minimize} \quad f(X) : 10x_1 + 14x_2 + 21x_3 + 42x_4$$

$$\text{Subject to:} \quad -8x_1 - 11x_2 - 9x_3 - 18x_4 \leq -12$$

$$-2x_1 - 2x_2 - 7x_3 - 14x_4 \leq -14$$

$$-9x_1 - 6x_2 - 3x_3 - 6x_4 \leq -10$$

$$x_1, x_2, x_3, x_4 \in [0 \ 1]$$

MATLAB Code: The code for Example 8.5 is available in the wrapper file **Zip_Balas_Example8_5.m**. The Balas algorithm is available in **zipBalas.m**. The output is copied from the Command window as

```
*******************************
Example 8.5 (ZIP-Balas)
*******************************
```

```
Solution exists
---------

The values for Zero-One design variables are:
     1     0     0     1

The objective function: 52.00

Constraints (RHS):-Actual Value-at solution
   -26    -16    -15

Constraints (RHS):-Permissible Values
   -12    -14    -10

Total cpu time (s)=   0.0625
```

This corresponds to the solution identified by Syslo et al.[6]

Complete Enumeration: The Balas algorithm performs partial enumeration. The *complete* or *exhaustive* enumeration (n_e) for n binary variables is given by this relation

$$n_e = 2^n$$

Typically, enterprise decision making involve several hundred variables and linear models. However, engineering problems characterized by a nonlinear model may have about 20 binary variables requiring $n_e = 1,048,576$ enumerations. This is no big deal with desktop resources today. Keep in mind that complete enumeration avoids any relaxation solution (purely discrete problems only).

CallSimplex.m: This is a brief description of an additional resource included among the code files to be downloaded. This is an function m-file that can be used to solve LP programs. It can be used to obtain solutions to LP relaxation models during the BB search. It is based on the algorithm available in Syslo. It is translated from a Pascal program. The mathematical model is

$$\text{Minimize}\quad c^T x$$

$$\text{Subject to}\quad [A][X] = [b]$$

$$x \geq 0$$

Greater than equal to constraints can be set up with a negative slack variables, as the program will automatically execute a dual-phase simplex method. Another alternative is to the use the LP programs provided by MATLAB. The function statement is

```
function [xsol, f]  = CallSimplex(A,B,C,M,N)
```

To use the function, A. B, C, M, N have to be defined. M is the number of constraints and N is the number of design variables.

REFERENCES

1. Fletcher, R. *Practical Methods of Optimization*, New York: John Wiley & Sons, Inc.,1987.
2. Papalambros, P.Y., and J.D. Wilde. *Principles of Optimal Design–Modeling and Computation*, New York: Cambridge University Press, 1988.
3. Ronald L. Rardin. *Optimization in Operations Research*, Englewood Cliffs, NJ: Prentice Hall, 1998.
4. Bellman, R. *Dynamic Programming*, Princeton University Press, 1957.
5. White, D.J. *Dynamic Programming*, Holden-Day Inc., 1969.
6. Syslo, M.M.,N. Deo, and J.S. Kowalik. *Discrete Optimization Algorithms*, Englewood Cliffs, NJ: Prentice Hall, 1983.
7. Beer, F. P., E.R. Johnston, and J.T. DeWolf. *Mechanics of Materials, 4th ed*, New York: McGraw-Hill, 2006.

PROBLEMS

8.1 Translate Example 1.1 into a discrete optimization problem and solve by exhaustive enumeration.

8.2 Translate Example 1.2 into a discrete optimization problem and solve by exhaustive enumeration.

8.3 Translate Example 1.3 into a discrete optimization problem and solve by exhaustive enumeration.

8.4 Implement a multistage exhaustive enumeration for the continuous optimization of Example 1.1.

8.5 Translate Example 1.1 into a discrete optimization problem and solve by Branch and Bound method.

8.6 Translate Example 1.2 into a discrete optimization problem and solve by Branch and Bound method.

8.7 Translate Example 3.1 into a discrete optimization problem and solve by Branch and Bound method.

8.8 Identify appropriate stock availability for the flagpole problem in Chapter 4 and solve the corresponding discrete optimization problem.

8.9 Solve Example 1.3 using the stock information available in **BeamPropertiesSI.m.**

8.10 You are the owner of an automobile dealership for *Brand X* cars. Set up a DP for monthly inventory control based on cars expected to be sold, high ordering costs, unit costs for the car, and inventory costs for excess cars. Solve it using the code in **DynamicProgramming_Example8_3.m.**

8.11 Solve the following using Balas Algorithm—use complete enumeration.

$$\text{Minimize} \quad f(X) = x_1 + x_2 + x_3$$
$$\text{Subject to:} \quad 3x_1 + 4x_2 + 5x_3 \geq 8$$
$$2x_1 + 5x_2 + 3x_3 \geq 3$$
$$x, x_2, x_3 \in [0 \ 1]$$

8.12 Program an exhaustive enumeration technique for Zero–One IP using MATLAB. Solve Problem 8.10.

8.13 Is it possible to solve the Capital Budgeting example? If so what is the solution using complete enumeration.

8.14 Solve Capital Budgeting example using Balas program.

8.15 Create an example that involves making binary choices regarding your participation in course and extra curricular activities to maximize your grades. Solve it.

8.16 Solve Problem 1.36 with the number of tubes as a discrete variable.

8.17 Solve Problem 1.18 with the number of coils as a discrete variable.

9

GLOBAL OPTIMIZATION

Contemporary literature in the area of optimization is dominated by methods and applications of **global optimization**. The methods of the previous chapters, to a large extent, are mostly capable of discovering locally optimum solutions. For many problems, executing these same methods, from many different initial guesses, from across the design space, can determine globally optimal solutions. We will not be able to claim so with certainty. The popular methods for global optimization are *Branch and Bound*, *Clustering Methods, Evolutionary Algorithms, Hybrid Methods, Simulated Annealing, Statistical Methods*, and *Tabu Search*.[1] The reader is strongly recommended to visit the survey site, as it is quite brief and provides each method with an overview, recognizes the application domain, identifies any available software, provides additional references and other links. Among these methods, we have seen the Branch and Bound method in the previous chapter. Exhaustive enumeration is also a global optimization technique for discrete variables. Endowing the Scan and Zoom method with heuristic search can adapt the method to seek a global solution. Matlab resources now include a *Genetic Algorithm and Direct Search Toolbox* for exploring problems in global optimization.[2] It works in conjunction with the Optimization Toolbox.

This chapter introduces two of the most popular techniques for finding the best value for the objective function, also referred to as the *global optimum* solution. They are *Simulated Annealing (SA)* and the *Genetic Algorithm (GA)*. The latter belongs to the category of evolutionary algorithms. The names are not accidental since the ideas behind them are derived from *natural processes*. SA is based on cooling of metals to obtain defined crystalline structures based on minimum potential energy. GA, is based on the combination and recombination of the genes in the biological system leading to improved DNA sequences or species selection. In actual application, they are more like continuation of

437

Chapter 8, because these methods can be applied to both the continuous and discrete categories of the optimization problems. They share an important characteristic with other optimization techniques in that they are primarily *search* techniques—they identify the optimum by searching the design space for the solution. Unlike the gradient-based methods, these techniques are largely heuristic and generally involve large amounts of computation. There is also a significant influence of statistical reasoning. Global optimization is being pursued seriously today, attracting a significant amount of new research to the area of applied optimization after a long hiatus. A significant drawback to many of these techniques is that they require *empirical tuning* based on the class of problems being investigated. There is also no easy way to determine these technique/problem sensitive parameters to implement an automatic optimization technique.

Global optimum is not really different from *local optimum*. The latter suggests that only a part of the design space (as opposed to the complete space) is viewed in the neighborhood of a particular design point. The standard Kuhn–Tucker conditions, established and used in earlier chapters (for continuous problems), will identify local optimum only. Most numerical investigations of continuous problems discover optimum values close to where they are started (initial guess). Although it makes sense to look for the global optimum, there is no method to identify if such an optimum exists for all classes of problems. At the present time, only continuous **convex problems** are guaranteed to have a global solution. It is not difficult to argue that such a class represents only a miniscule set of useful problems as most real optimization problems are not necessarily convex. In the literature, both the SA and GA methods are expected to produce solutions close to the global optimum solution, rather the global solution themselves. This is not to suggest that investigation of global solution is a waste of time if the possibility of one cannot be established with certainty. Enough justification for a global solution can be advanced through competitive and financial gains to warrant an investigation, even without the guarantee of useful results. Global optimization techniques are simple to program while requiring long run times for some calculations. This can potentially soak up all the unused time on the personal desktop computing resource, lending it immediate attractiveness.

The techniques and their discussion are very basic following the theme espoused in Chapter 8. The reader is encouraged to visit journal publications to monitor the maturing of these techniques. In this chapter, first some issues regarding global optimization are presented, followed by the presentation of Simulated Annealing and Genetic Algorithm.

9.1 PROBLEM DEFINITION

The following development on global optimization, in relation to the two methods of this chapter, is based on *unconstrained* continuous optimization. Constrained problems are usually solved by transforming the problem to an equivalent unconstrained form. Direct solution of constrained problems is still evolving and is not discussed here. The global optimum problem can be defined as

$$\text{Minimize} \quad f(X); \quad [X]_n \tag{9.1}$$

$$\text{Subject to} \quad x_i^l \leq x_i \leq x_i^u \tag{9.2}$$

If this problem were *unimodal* (only one minimum), the optimization could easily be established by any of the appropriate techniques used in the book, since the local minimum is also the global minimum. To investigate the global minimum, it is therefore expected that there would be many *local* minimums, several of them having different values for the objective. The least among them is then the *global* solution. Local minimums, in the case of continuous problems, are those that satisfy the KT conditions, and will be numerically determined if the starting guess was in the neighborhood of the particular minimum. This definition may be applied to discrete optimization problems that use continuous relaxation for obtaining the solution. For discrete problems using complete enumeration, the global optimization will be discovered by the process. Therefore, the discussion here applies to discrete problems that use only incomplete enumeration.

9.1.1 Global Minimum

The global minimum for the function $f(X)$ exists, if the function is continuous on a nonempty feasible set S, that is closed and bounded.[3]

The set S defines a region of the design space that is feasible. Any point in S is a candidate design point X. In the *n-dimensional* Cartesian space, the point X corresponds to vector of length $||X||$. The boundaries of the region are part of the set S (*closed*). The length of the vector X should be a finite number (*bounded*). The definition for the global minimum X^*, can be stated as

$$f(X^*) \leq f(X); \quad X \in S \tag{9.3}$$

The conditions for the existence of the global minimum as given above are attributed to Weierstrass. Although the global minimum is guaranteed if the conditions are satisfied, they do not negate the existence of the global solution if conditions are not satisfied. Considering the broad range and type of optimization problems, the conditions are fairly limited in their usefulness. The conditions for *local minimum* are the same as (9.3), except the set S is limited to a small region around X^*.

Convexity: If an optimization problem can be shown to be *convex*, then the local minimum will also be a global minimum. A convex optimization problem requires the objective function to be convex. If constraints are present, then all the *inequality constraints* must be convex, while the *equality constraints* must be linear.

Convex Set: This usually relates to the design vector (or design points) of the problem. Consider two design points X_1 and X_2. As before, the set S defines a design region and, by implication, both X_1 and X_2 belong to the set S. The line joining the points X_1 and X_2 contains a whole series of new design points. If all of these points also belong to S, then the set S is a *convex* set. A closed set S requires that the points X_1 and X_2 can also be on the boundary. If however, the line joining the points X_1 and X_2 contains some points that do not belong to S, then S is not a convex set. For a two-dimensional design problem, convex and nonconvex sets are illustrated in Figure 9.1.

Convex Functions: A *convex function* is defined only for a *convex set*. Let S be a *convex set* from which two design points X_1 and X_2 are selected. Let f be the function whose convexity is to be established. Consider any point X_a on a line joining X_1 and X_2. The function f is convex if $f(X_a)$ is less then the value of the corresponding point on the line joining $f(X_1)$ and $f(X_2)$. In *n-dimensions*, this is not easy to see but a one-dimensional representation is illustrated in Figure 9.2. Condition for convexity in the figure is

$$f(x_a) \leq f^+ \tag{9.4}$$

A mathematical condition of convexity is that the Hessian matrix (matrix of second derivatives) of the function f must be positive semi-definite at all points in the set S. Note: This corresponds to the KT conditions for *unconstrained minimization* problems. From a practical viewpoint, the *eigenvalues* of the function must be greater than or equal to zero at all points in S.

It is important to realize that a global optimum may still exist even if the convexity conditions are not met. In real optimization problems, it may be difficult to establish convexity. In practice, convexity checks are rarely performed. An assumption is made that a global optimum exists for the problem. This is a subjective element reinforced by experience, insight, and intuition. This is also

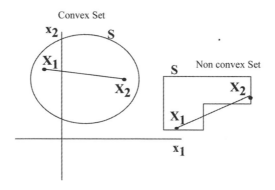

Figure 9.1 Convex and nonconvex sets.

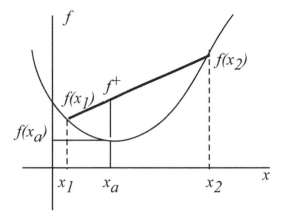

Figure 9.2 Convex function.

often justified by the possibility that a better solution than the current one can be found, and it is worth looking for one.

9.1.2 Nature of the Solution

We will assume that the global optimum solution will be obtained numerically. A good understanding of the solution to this problem is possible with reference to continuous design problems. Extension to discrete problems is possible by considering that any continuous solution corresponds to a specific combination of the set of discrete values for the design variables. Figure 9.3 is an illustration

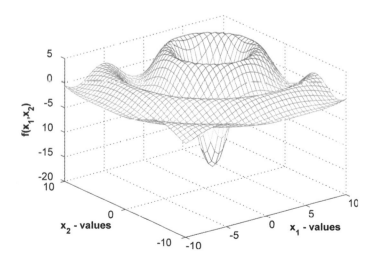

Figure 9.3 Surface plot of Example 9.1.

of a figure with several minima and maxima with a strong global minimum at $x_1 = 4$, $x_2 = 4$. The function is described by

$$f(x_1, x_2) = -20 \frac{\sin\left(0.1 + \sqrt{(x_1 - 4)^2 + (x_2 - 4)^2}\right)}{0.1 + \sqrt{(x_1 - 4)^2 + (x_2 - 4)^2}} \tag{9.5}$$

Figure 9.4 is the corresponding contour plot. The contour plot can be better appreciated in color and using the colorbar on the side for the values. The MATLAB code **DrawFigure9_3.m**[*] will create the Figures 9.3 and 9.4. From the relation (9.5) it is clear the ripples will subsidize as x_1 and x_2 move much further away from where the global minimum occurs, because of the influence of the denominator. In the limit, the ripples should start to become a plane.

We will use Figure 9.3 to outline some requirements of a global optimizer. In Figure 9.3 it is easy to spot the global solution by observation. Surrounding the global minimum are rings of local extrema, suggesting an infinite number of each of them. Numerically, it is a challenge to negotiate all the bumps and dips. A simple characteristic of the standard algorithms for minimization is that they attempt to find a ditch (or valley) to fall into. The KT conditions will lead the solution to be trapped in any one of the local minimum rings, particularly

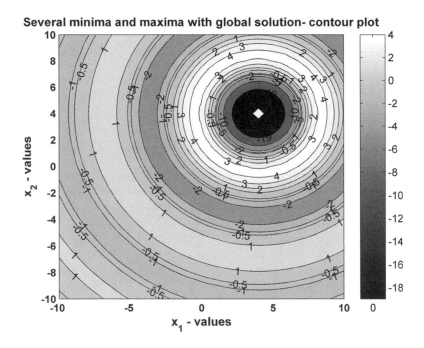

Figure 9.4 Contour plot of Example 9.1.

[*]Files to be downloaded from the web site are indicated by boldface courier type.

close to the initial guess for the iteration. If the iterations are started close to at $x_1 = 4$, $x_2 = 4$ then it is more than likely the global optimum will be discovered. Staring at $x_1 = -5$, $x_2 = -5$, using standard numerical techniques will probably not lead to the global minimum.

This is a simple illustration, but note there are infinite local minimums and maximums and a single global minimum. In real design problems it is not readily apparent how many local minimums exist, and if the global minimums are really that distinct, even if they are postulated to exist. What about the *complete enumeration* technique? There is no way the global minimum can escape the sifting due to this technique. There is also no technique that is simpler than complete enumeration. The problem, however, is to finish searching the infinite points of the design space to discover the solution.

The solution to global optimum must therefore lie between standard numerical techniques and complete enumeration. The former will not find the solution unless it is fortunate. The latter cannot find the solution in the time to be useful. Global optimization techniques tend to favor the latter approach because of the simplicity of implementation. To be useful, the solution must be discovered in a reasonable time.

9.1.3 Elements of a Numerical Technique

A simple approach to develop an algorithm for global minimum is to evolve it from traditional optimization methods since a lot of effort has preceded their development. The algorithm must incorporate some additional characteristics discussed here. In most standard numerical techniques, the local minimum serves as a strong magnet or a beacon for the solution. In most of these methods, design changes that do not improve the solution or the objective are not accepted. If standard numerical techniques have to be adapted to look for global minimum, then they must encourage solutions that, in the short run, may increase the objective value so that they find another valley at whose bottom is the global minimum. They must achieve this without sacrificing the significant potency of the local minimum, because a local minimum solution is better than no solution at all. This idea is illustrated using Figure 9.5.

In Figure 9.5, the iterations start at the point A. In a standard numerical technique, a search direction is found that leads to point B (lower objective). At B, a conventional numerical technique will discover the search direction leading to point C. A search direction leading to point D will be strongly discouraged. The iterations will continue to the local minimum x_1^*. But, if point D was somehow reachable, then it could move on to E and then F and continue to the global minimum at x_2^*. The main idea in finding the *global minimum* is to encourage solutions that will escape the attraction of the local minimum. This means that at point E, it is also quite possible that the search direction may lead to point G, and finally to x_3^*. Allowing a large number of iterations with the possibility of escape will cause the valley holding the global minimum to be revisited. From

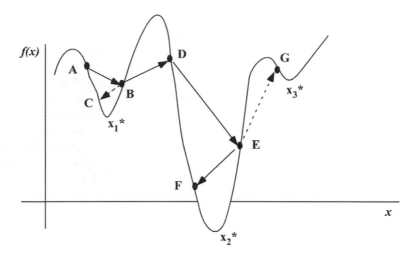

Figure 9.5 Strategy for global optimization search.

this brief illustration, the characteristics for a technique used to find a global minimum can be summarized:

- Short-term increases in the objective function must be encouraged.
- Traditional gradient-based techniques will not be particularly useful because they are based on search directions that decrease the objective function at the current iteration. They will not permit increase of the objective function value.
- If descent-based search directions are *not* part of the technique, then any direction that decreases the objective should be accepted to promote attraction by the local minimum.
- As the location of the global minimum is not known, search must be permitted over a large region with no bias, and for a long time.
- Since acceptable directions can both decrease or increase the objective function value, a heuristic determination of the search direction (or design change) must be part of the technique.
- These features resemble the *zero-order* technique for unconstrained optimization and therefore will require a large number of iterations to be effective.
- The large number of iterations will force the technique to be simple.
- The solution can only be concluded based on the number of iterations, as necessary conditions (available for continuous problems only) cannot distinguish between local and global optimum.

Without doubt, the best technique is complete or exhaustive enumeration. It is therefore necessary to assume that complete enumeration *is not an option* in further discussion.

The problem description, illustration, and discussion have dealt with *unconstrained minimization*. Constraints can be included by addressing these characteristics to apply to feasible solutions only. It is not easy to heuristically generate a large number of feasible design changes for consideration. Should short-term constraint violation be permitted to obtain the global optimum? This question is yet to be answered by the research community.

9.2 NUMERICAL TECHNIQUES AND ADDITIONAL EXAMPLES

This section introduces two of the popular methods for global optimization in the literature. Both of them can be are derived from natural physical process that help identify optimum states. The first one, **Simulated Annealing (SA)**, refers to a physical process for materials that establish different crystalline structures, each of which can be associated with different minimum potential energy. This is accomplished by reheating and cooling of the materials/metals while controlling the rate of cooling. The second, **Genetic Algorithm (GA)**, works through mutation of candidate solutions until the best solution can be established. This process can be related to natural selection or biological evolution, and therefore the algorithm is termed as an *evolutionary algorithm*. These techniques are illustrated with respect to unconstrained problems in this section.

9.2.1 Simulated Annealing (SA)

The first application using SA for optimization was presented by Kirkpatric et al.[4] The examples covered problems from electronic systems, in particular, component placement driven by the need for minimum wiring connections. The algorithm itself was based on the use of statistical mechanics, demonstrated by Metropolis et al. to establish thermal equilibrium in a collection of atoms[5] The name itself is more representative of the process of cooling materials, particularly metals, after raising their temperature to achieve a definite crystalline state after cooling. As mentioned earlier, the extension to optimization is mainly heuristic. The simple idea behind SA can be illustrated through the modified basic algorithm of SA due to Bohachevsky et al.[6] A third method, the Scan and Zoom method, a special development of this book introduced in Chapters 6 and 7, is also applied to the examples for comparison. The Scan and Zoom method should ideally seek out the global optimum. The method is not modified in any form to handle the global optimum from its initial appearance.

Algorithm: Basic Simulated Annealing (A 9.1):

Step 1. Choose starting design X_0. calculate $f_0 = f(X_0)$
(Need stopping criteria)

Step 2. Choose a random point on the surface of a unit *n-dimensional hypersphere* (it is just a sphere in n-dimensions) to establish a search direction S.

Step 3. Using a step size α, calculate
$$f_1 = f(x_0 + \alpha S); \quad \Delta f = f_1 - f_0$$

Step 4. **If** $\Delta f \leq 0$; **then** $p = 1$
Else $p = e^{-\beta \Delta f}$

Step 5. A random number $\mathbf{r}$, $0 \leq r < 1$, is generated.
If: $r \leq p$, then the step is accepted and the design vector is updated.
Else: no change is made to the design.
Go to Step 2.

If implemented, the algorithm has to be executed for a reasonably large number of iterations for the solution to drift to a global minimum. The simplicity of the algorithm is very appealing. A careful observation will indicate that α and β are important parameters for the algorithm and they are likely to be problem dependent. Generally, the parameter β can be related to the Boltzmann probability distribution. In statistical mechanics, it is expressed as $(-k/T)$. T is considered the annealing temperature; k is the Boltzmann's constant. It dictates the conditional probability that a worse solution can be accepted. For the algorithm to be effective, the value for p is recommended to be in the range $0.5 \leq p \leq 0.9$. In formal SA terminology, β represents the annealing temperature. The following ideas in the algorithm are apparent:

- The algorithm is heuristic.
- The algorithm is suggestive of a biased random walk.
- Descent of the function is directly accommodated.
- Directions that increase the value of the function are sometimes permitted. This is important to escape local optimum. This is the *potential* to discover the global optimum through the algorithm.

This basic algorithm can definitely be improved. During the initial iterations, it is necessary to encourage solutions to escape the local optimum. After a reasonably large number of iterations, the opportunity to escape local optimum (window of opportunity) must be continuously reduced so that the solution does not wander all over the place and cancel the improvements to the solution that have already taken place. This can be implemented by scheduling changes in the parameter β as the iterations increase. In SA terminology, this is referred to as the *annealing schedule*. These modifications do not in any way decrease the problem/user dependency of the algorithm. This problem dependence is usually unattractive for automatic or standard implementation of the SA method. Researchers are engaged in construction of algorithms that do not depend on user specified algorithm control parameters.[5]

The SA algorithm for continuous variables implemented in the book includes many of the basic ideas from the original algorithm. First, the negative sign in the probability term is folded into β itself. A simple procedure is included for obtaining the value of β. It is based on the initial value of X_0 and f_0, β

is calculated to yield a value of $p = 0.7$ for a Δf equal to 50 percent of the initial f_0. The algorithm favors a negative value of β by design. Since the process for establishing an initial β can determine a positive value, the algorithm corrects this occurrence. It is later illustrated through numerical experiments that a small magnitude of the parameter β allows the solution to wander all over the place, while a large value provides a strong local attraction. Instead of the SA step being incorporated when the objective increases, it is triggered when the objective is likely to increase based on the slope. In the following algorithm, the control parameter β is kept constant at its initial value. The calculation of α is borrowed from the traditional one-dimensional step size calculation using the Golden Section method (Chapter 6) if the objective is likely to decrease. If the slope of the objective is positive or zero, a fixed α is chosen. This value is a maximum of one or the previous value computed through the Golden Section calculations. Note, this does not ensure that the function is actually going to increase as required in the basic algorithm. Nevertheless, it appears to work as discussed. Stopping criteria are not built in and is left to the user. The algorithm is effective for Example 9.1. It has yet to be tested on a wide range of problems. It is also shown to work for Example 9.2.

Algorithm: Simulated Annealing (A 9.2):

Step 1: Choose X_0. Calculate $f(X_0)$, $p = 0.7$, $\beta = -\ln(p)/0.5^* f(X_0)$
If $|\beta| < 1$, $\beta = 1$, If $\beta > 0$, $\beta = -1^*\beta$
$i = 1$

Step 2: Choose a random point on the surface of a unit n-dimensional hypersphere (it is just a sphere in n-dimensions) to establish a search direction S.

Step 3: Calculate $\nabla f(x_0)$;
If $\nabla f(x_0) < 0$, then compute 1D α^*
$\Delta X = \alpha^* S$; $\tilde{X} = X_0 + \Delta X$; $X_0 \leftarrow \tilde{X}$; $i \leftarrow i + 1$
If $\nabla f(x_0) \geq 0$,
$\Delta f = f(x_0 + \tilde{\alpha} S) - f(x_0)$; $\tilde{\alpha} = \max(\alpha^*, 1)$
$p = e^{\beta \Delta f}$
A random number r; $0 \leq r < 1$ is generated.
If r $\leq$ p, then the step is accepted and the design vector is updated
$\Delta X = \tilde{\alpha} S$; $\tilde{X} = X_0 + \Delta X$; $X_0 \leftarrow \tilde{X}$; $i \leftarrow i + 1$
Else continue $i \leftarrow i + 1$
Go to Step 2

Example 9.1 This is the function used in Section 9.1.2.

$$\text{Minimize} \quad f(x_1, x_2) = -20 \frac{\sin\left(0.1 + \sqrt{(x_1 - 4)^2 + (x_2 - 4)^2}\right)}{0.1 + \sqrt{(x_1 - 4)^2 + (x_2 - 4)^2}} \quad (9.5)$$

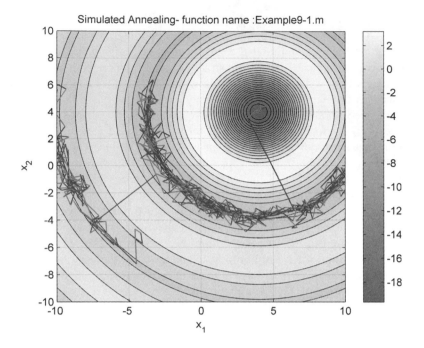

Figure 9.6 Simulated annealing: Example 1 — converged solution.

The global minimum is at $X^* = [4, 4]$ with an objective value $f^* = -19.6683$. In the figures, the contours suggest a strong global minimum encircled by many local maximums and minimums. For most of the runs, the starting value was far from the global minimum.

MATLAB Code: The code for Example 9.1 is executed through the wrapper file **SimulatedAnnealing_Example9_1.m**. The actual algorithm is programmed in **SimulatedAnnealing.m**. As the multi-variable Golden Section method is being used for the one-dimensional step size search, both **UpperBound_nVar.m** and **GoldSection_nVar.m** are required to exist in the same directory. The example function in (9.5) is coded in **Example9_1.m.**

Figure 9.6 captures the first of the many runs for Example 9.1. It escapes the starting relative minimum fairly easily to get trapped in a different local minimum. It is able to escape to the global minimum where it is starting to wander in the vicinity of the global minimum.

The following information is copied from the Command window:

```
*************************************
Example 9.1: SIMULATED ANNEALING
*************************************
```

```
Data for the Program
```

```
Number of iterations:              2000
Initial Guess for Design:    -5    -5


_____


Results from SA

_____

beta0 used for the program:    -2.7567
Number of useful iterations:      1111
Final Design Variables:         3.9988    3.9998
Final Objective function:      -19.6683


Total cpu time (s)=  1.8750
```

The useful number of iteration was when the objective function was updated. The beta used for the example is also shown. It is likely that the user will not be able to reproduce this result since the *rand* function is being used. It would be a singular coincidence if any of the results presented here are duplicated. Running numerical experiments is simply a mater of calling the SA algorithm and taking a look at the results (especially the plot). The design changes are illustrated using colors: red, representing decreasing slopes, and blue, for positive slopes that lead to design changes. The entrapment in the local minimum and the escape are clearly illustrated on the contour plots. The contour plots are filled color to identify areas on the figure where the function has a local minimum or maximum. Darker colors represent minimums and lighter colors the maximum. The examples were run usually for 2,000 iterations. Large number of iterations are useful to indicate if the solution escapes the global minimum once having obtained it. Although it was not evident in this example, it is likely problem dependent, since in this example the global minimum was significantly larger than the local minimums. Several features about the SA algorithm are illustrated using some numerical experiments.

Using Optimization Toolbox for Example 9.1:   The Optimization Toolbox provides the function **fminunc** to find a minimum of a scalar function of several variables, starting at an initial estimate. It can be executed in the Command window (without any fuss) as

```
>> fminunc(@Example9_1,xinit)
```

The result in the Command window is:

```
Optimization terminated: relative infinity-norm of gradient
less than options.TolFun.
ans =
   -5.9438    -5.9438
```

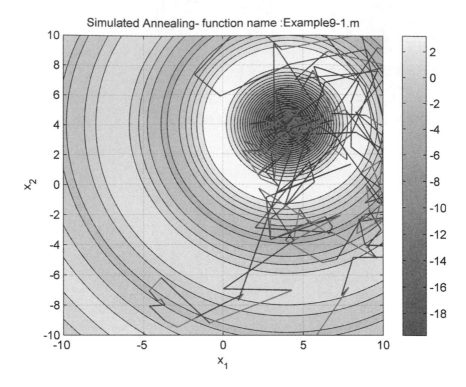

Figure 9.7 Simulated annealing — low β.

The solution $[-5.9438 \ -5.9438]$ is a local minimum as the global solution is at $[4 \ 4]$. This reinforces the previous statement that the standard algorithms will only determine a local minimum.

Numerical Experiments with Example 9.1: The second run requires a portion of the code in **SimulatedAnnealing.m** to be commented. Starting at the *solution*, using a low value of β, it can be seen that the solution wanders all over the design space including reaching much beyond the display region in Figure 9.7. The final statements are copied from the Command window as follows. There is no pattern in the limited points seen on the figure, except the suggestion that the search is encouraged to escape. This can be tied to the low magnitude of β.

```
*************************************
Example 9.1: SIMULATED ANNEALING
*************************************
_____

Data for the Program
_____
```

```
Number of iterations:                  2000
Initial Guess for Design:        4     4
_____

Results from SA
_____

beta0 used for the program:      -0.0363
Number of useful iterations:           1983
Final Design Variables:       -142.0848  186.8549
Final Objective function:         -0.0855
Total cpu time (s)=  3.4375
```

The third run is to restore the commented statements in **SimulatedAnnealing.m** and start at the solution. The value of β is -1. Figure 9.8 illustrates that the solution does not escape the global minimum region. The statements from the Command window is

```
************************************
Example 9.1: SIMULATED ANNEALING
************************************
_____

Data for the Program
_____

Number of iterations:                  2000
Initial Guess for Design:        4     4
_____

Results from SA
_____

beta0 used for the program:      -1
Number of useful iterations:           1088
Final Design Variables:          4.0306     4.0308
Final Objective function:        -19.6621
Total cpu time (s)=  1.9375
```

The choice of the control parameter was arbitrary, and several examples have to be studied to see if this value is problem independent. The value of 1 is typically a good choice for optimization algorithms, particularly for a well scaled problem. It appears that that a negative value, of $\beta = -1$ implements a local attractor.

The fourth run is a repeat of the first one with the maximum number of iterations as 500. Figure 9.9 conveys the effort at identifying the global minimum. It appears that 500 iterations are insufficient to determine the global minimum. The solution has not yet escaped the local minimum. Since the valleys (local minimums) in this example are circular, note that the solution moves around in the valley but has not managed to escape.

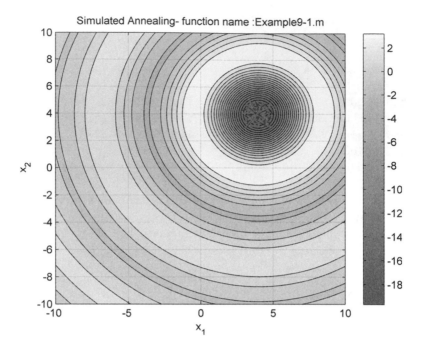

Figure 9.8   Simulated annealing: Increasing β — decreased opportunity to escape.

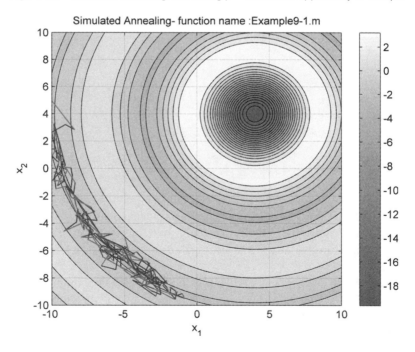

Figure 9.9   Simulated annealing: Non convergence — insufficient iterations.

Here are the statements copied from the Command window:

```
**************************************
Example 9.1: SIMULATED ANNEALING
**************************************
_____

Data for the Program
_____

Number of iterations:        500
Initial Guess for Design:    -5    -5
_____

Results from SA
_____

beta0 used for the program:     -2.7567
Number of useful iterations:     330
Final Design Variables:         -5.7981    21.8577
Final Objective function:        -0.9806


Total cpu time (s)=  1.2969
```

Discussion on the Numerical Experiments: The following observations are based on the experience of applying SA (A 9.2) to Example 9.1. They correspond to the reported experience with SA in the general literature:

- Annealing schedule (changing β) is an important consideration in SA. The following is not implemented but can be inferred from the experiments. Starting with a low value the control parameter β must be increased with the number of iterations. Not too fast, or it will get trapped in a local minimum. (This is not implemented.)
- Annealing schedule is developed through trial and error, insight that is not obvious, and based on probable behavior of the search.
- Length of time (or number of iterations) that the problem needs to be searched is difficult to determine. Although this can be based on monitoring, given the stochastic nature of the search, it cannot be asserted that the global minimum has been established. The SA algorithm itself can be expected to converge while this property is too slow for practical use.[8] Larger number of iterations is more likely to yield the global solution.
- SA search does not subscribe to iterative improvement. It is an example of a stochastic evolutionary process. The implementation used here represents a memoryless search.
- Restarting from different design must still be implemented to overcome strong local influences.
- Restarting from the same initial design is also important for establishing convergence properties as this is a stochastic search process.

Scan and Zoom Method for Example 9.1:   The Scan and Zoom method is applied to Example 9.1 through `Scan_Zoom_Example9_1.m`. It's performance is truly impressive and finds the global minimum in a single level. The statements in the Command window are

```
Example 9.1
_____

 Zoom Level ( 0):
x1mid =   0.0000000
x2mid =   0.0000000

 Zoom Level ( 1):
x1* =   4.0000000
x2* =   4.0000000
 f* =  -19.6683294

 Zoom Level ( 2):
x1* =   4.0000000
x2* =   4.0000000
 f* =  -19.6683294

 Change in x1 <= 1.000e-008
 Change in x2 <= 1.000e-008
Total time (s)=   0.0000
```

Example 9.1 was an unconstrained example. For other types of optimization problems the following suggestions will be useful.

Constrained Problems: This chapter does not deal with constraints or discrete problems. There are two ways of handling constraints:

1. If the new design point is not feasible, it is not accepted and no design change takes place. This is quite restrictive for random search directions.
2. The function that drives the algorithm during each iteration is the one with the largest violation. If the current point is feasible, then the objective function will drive the search. This makes sense as the process is largely stochastic and is expected to wander around the design space.

Constrained problems can be expected to take a significantly larger number of searches. There is currently no procedure to identify convergence property of the SA with respect to constrained optimization.

Discrete Problems: In discrete problems, the SA algorithm will use a discrete design vector determined stochastically from the available set. Each element of the design vector is randomly chosen from the permissible set of values. The remaining elements of the SA algorithm are directly applicable. If the objective decreases then the design change is accepted. If it increases, then a conditional probability of acceptance is determined allowing the solution to escape the minimum. This is left to the reader. The significant effort is the identification of the random design vector.

Example 9.2 This is a function of two variables that MATLAB uses for illustrating the graphics capability of the software. Typing `peaks` at the Command prompt will draw the surface plot. Typing `type peaks.m` at the Command prompt will provide the equations that were used to obtain the curve. This function used as the objective function for this example.

$$\textit{Minimize} \quad f(x_1, x_2) = 3(1 - x_1)^2 e^{-x_1^2} - (x_2 + 1)^2 - 10 \left(\frac{x_1}{5} - x_1^3 - x_2^5 \right)$$

$$\times e^{(-x_1^2 - x_2^2)} - \frac{1}{3} e^{(-(x_1+1)^2 - x_2^2)} \tag{9.6}$$

MATLAB Code: The wrapper file for Example 9.2 is **SimulatedAnnealing_ Example9_2.m**. The starting design and number of iterations are available in the output in the Command window, which is included here. Figure 9.10 traces

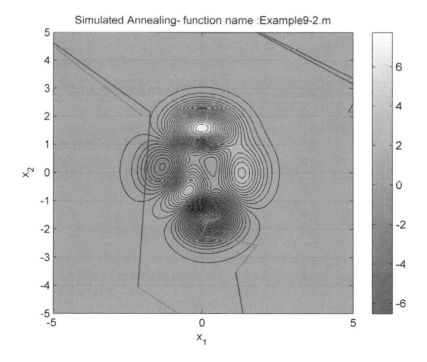

Figure 9.10 Simulated annealing: Example 9.2.

the iterations for this example. It is surprising to see the solution wander all over without being trapped by any of the global minimums in the figure. The final design is also the global optimal solution. It will be difficult to duplicate Figure 9.10 because of the use of the *rand* function.

```
*************************************
Example 9.2: SIMULATED ANNEALING
*************************************
```

Data for the Program

```
Number of iterations:              2000
Initial Guess for Design:     -2    2
```

Results from SA

```
beta0 used for the program:    -2.7567
Number of useful iterations:   438
Final Design Variables:        0.2283   -1.6255
Final Objective function:      -6.5511
```

```
Total cpu time (s)=  1.1875
```

Using Optimization Toolbox for Example 9.2:   The Optimization Toolbox function **fminunc** was used to find the solution minimum to Example 9.2.

```
>> fminunc(@Example9_2,[-2 2])
Optimization terminated: relative infinity-norm of gradient
less than options.TolFun.
ans =
   -1.3474     0.2045
```

This is a local minimum close to the initial guess in the design space.

Scan and Zoom Method for Example 9.2:   The Scan and Zoom method is applied to Example 9.2 through **Scan_Zoom_Example9_2.m**. Once more its performance is very impressive and arrives near the global minimum in 4 zoom levels. The statements copied from the Command window are

Example 9.2

```
 Zoom Level ( 0):
```

```
x1mid =   0.0000000
x2mid =   0.0000000

 Zoom Level ( 1):
x1* =   0.0000000
x2* = -2.0000000
 f* = -4.7596121

 Change in x1 <= 1.000e-008
 Zoom Level ( 2):
x1* =   0.0000000
x2* = -1.7500000
 f* = -5.9728893

 Change in x1 <= 1.000e-008
 Zoom Level ( 3):
x1* =   0.2500000
x2* = -1.6250000
 f^ = -6.5466445

 Zoom Level ( 4):
x1^ =   0.2500000
x2* = -1.6250000
 t* = -6.5466445

 Change in x1 <= 1.000e-008
 Change in x2 <= 1.000e-008
Total time (s)=   0.0313
```

9.2.2 Genetic Algorithm

This section presents a basic introduction to genetic algorithm (GA). It is largely unchanged from the previous edition. The actual methods and applications have multiplied, but the core ideas have remained the same. GA is an important part method in *evolutionary computation*. These can be largely considered as direct search processes and are naturally useful for discovering optimum solutions. Holland, was the first to use the technique but their use as an optimization tool began in earnest in the late 1980s, developed momentum in the mid-1990s, and continues to attract serious interest today.[9] The term *evolutionary* is suggestive of the natural process associated with biological evolution—primarily the Darwinian rule of the selection of the fittest. The natural process takes place through the constant *mutation* and *recombination* of the chromosomes in the population to yield a better gene structure. All of these terms appear in the discussions related to evolutionary computation. This basic process is also folded into the various numerical techniques. The discussions can become clear if the term chromosomes can be associated with **design variables**. Over the years, *evolutionary computation* has included **Genetic Programming**: that develop algorithms that mimic the evolutionary process; *Genetic* Algorithms (GA): optimization of combinatorial/discrete problems; **evolutionary programming**: optimizing continuous functions without recombination; **evolutionary strategies**: optimizing continuous functions with recombination.[10] This classification is quite loose. For example GA have been used, and continue to be used effectively to solve continuous optimization problems. Evolutionary Algorithms have been successfully applied to optimization problem in several areas: engineering design, parameter fitting, knapsack problems, transportation problem, image processing, traveling salesman, scheduling, and so on. As can be inferred from the list, a significant number are from the area of discrete or combinatorial programming.

This section includes only GA, though all of the evolutionary algorithms share many common features. One of the important characteristic of GA is the use stochastic information in their implementation and therefore can be considered global optimization techniques. They are exceptionally useful for handling ill-behaved, discontinuous, and non-differentiable problems. The SA can be proven to converge to the global optimum, but the same can be only weakly established for the GA. The GA's generate and use a *population* of solutions or *design vectors* (*X*) unlike the SA where a search direction (*S*) was used. This distinction is essential to understand the difficulty of handling continuous problems in GA. For example, using random numbers to generate a value for x_1, the values of 3.14, 3.141, or 3.1415 will all be distinct and have no relation to each other. The fact that they identify the same solution with different degrees of precision is ignored in GA. In SA, with the algorithm demonstrated in the previous section, getting close to the minimum is possible because of the step size computation. In GA typically, there may be no minor adjustments for neighboring solution. Such a predicament does not arise in the consideration of discrete problems. Hence, GA is more widely used for discrete problems. A basic GA can be assembled as follows:[11]

Generic Genetic Algorithm (GGA) (A 9.3):

Step 1: Set up an initial **population** P(0)—an initial set of solution or chromosomes.
 Evaluate the initial solution for fitness–differentiate, collate and rate solutions.
 Generation index $t = 0$

Step 2: Use **genetic operators** to generate the set of **children** (crossover, mutation).
 Add a new set of randomly generated population (**immigrants**).
 Reevaluate the population—fitness.
 Perform **competitive** selection—which members will be part of next generation.
 Select **population** $P(t + 1)$—same number of members.

If not converged $t \leftarrow t + 1$
Go to Step 2.
We will attempt to illustrate and explain many of the terms that appear in the algorithm.

Chromosomes: Although it is appropriate to regard each chromosome as a design vector, there is an issue of representation of the design vector for handling by genetic operators. It is possible to work with the design vector directly (as in the included example) or use some kind of mapping, real (*real encoding*) or binary (*binary encoding*). Earlier work in GA used binary encoding. In this book, the design vector is used directly for all transactions. Real encoding is recommended. A piece of the chromosome is called an *allele*.

Fitness and Selection: The fitness function can be related to the objective function. The population can be ranked/sorted according to their objective function value. A selection from this population is necessary for (i) identifying parents, and (ii) identifying if they will be promoted/evolve to the next generation. There are various selection schemes. A fraction of the best individuals can be chosen for reproduction and promotion, the remaining population being made up by new immigrants. This is called *tournament* selection. A modification to this procedure is to assign a probability of selection for the ordered set of the population for both reproduction and promotion. This probability can be based on *roulette wheel, elitist selection, linear*, or geometric ranking.[12] The implementation in this book uses ranking without assigning a probability of selection. It uses the objective function value to rank solutions.

Initial Population: The parameter here is the number of initial design vectors (*chromosomes*) for the starting generation, N_p. This number is usually kept the same in successive generations. Usually a normal random generation of the design vectors is recommended. In some instances, **doping**, adding some good solutions to the population is performed. It is difficult to establish how many members must be considered to represent the population in every generation.

Evaluation: The various design vectors have to be tested for fitness so that they can be used for creating children that can become part of the next generation. Although there are several fitness attributes, for the case of unconstrained optimum, the objective function $f(X)$ is a good measure. Since the global minimum is unknown, a progressive measure may be used.

Genetic Operators: These provide ways in defining new populations from existing ones. This is also where the evolution takes place. Two types of operation are recognized. The first is *crossover* or *recombination*, and the second is *mutation*.

Crossover Operators: There are several types of crossover strategies. A **simple crossover** is illustrated in Figure 9.11. Here, a piece of the chromosome—*allele*—is exchanged between two parents P_1, P_2 to produce children C_1, C_2. The length of the piece is usually determined randomly. Mathematically, if the parents are $X = [x_1, x_2, \ldots x_n]$ and Y (defined similarly), r is randomly selected truncation index, then the children $U (= C_1)$ and $V (= C_2)$, by simple crossover, are defined as follows:[13]

$$u_i = \begin{cases} x_i, & \text{if } i < r \\ y_i, & \text{otherwise} \end{cases}$$

$$v_i = \begin{cases} y_i, & \text{if } i < r \\ x_i, & \text{otherwise} \end{cases} \tag{9.7}$$

Arithmetic crossover is obtained by a linear combination of two parent chromosomes to yield two children.[11] If λ_1, and λ_2 are randomly generated numbers, then the children can be defined as follows:

$$C_1 = \lambda_1 X + \lambda_2 Y;$$

$$C_2 = \lambda_2 X + \lambda_1 Y; \tag{9.8}$$

A convex combination would arise if

$$\lambda_1 + \lambda_2 = 1; \quad \lambda_1, \lambda_2 > 0 \quad - \quad convex \quad combination$$

If there are no restrictions on $\lambda's$, it is an **affine combination**. For real $\lambda's$ it is *a* **linear combination**. An important crossover operator is a **direction-based crossover** operator, if $f(Y)$ is less than $f(X)$ and r is unit random number

$$C = r^*(Y - X) + Y$$

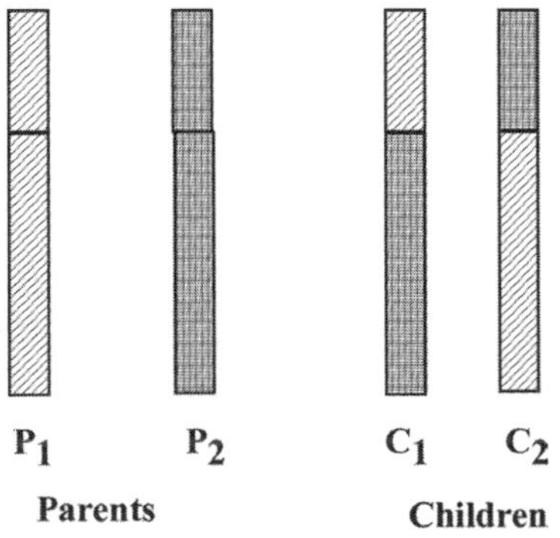

$$P_1 \qquad P_2 \qquad C_1 \qquad C_2$$

Parents   **Children**

Figure 9.11 Genetic Algorithm: Simple Crossover.

Mutation: **Mutation** refers to the replacement of a single element of a design vector by a randomly generated value. The element is also randomly chosen. For the design vector X, for a randomly selected element k

$$C = \begin{cases} x_i, & i \neq k \\ x'_k, & i = k \end{cases}$$

There are other types of mutation, including *non uniform mutation, directional mutation*, and so on.[11]

Immigrants: This is the addition of a set of randomly generated population in each generation before the selection is made for the next generation. The search process in GA is broadly classified as *exploration* and *exploitation*. Exploration implies searching across the entire design space. Exploitation is focusing more in the area promising a solution. Typically, crossover and mutation are ways to narrow the search to an area of design space based on the characteristics of the current population—suggesting a local minimum. To keep the search open for the global minimum, other areas must be presented for exploration. This is achieved by bringing in unbiased population and letting them compete for promotion to the next generation, along with the current set. In many variations these are allowed to breed with the current good solutions before evaluation.

The ideas presented here are basic to all of evolutionary computation and specific to GA. For each category of algorithms there are more sophisticated models that are being researched and developed, and the reader is encouraged to seek information from current literature. The specific GA algorithm implemented in the book incorporates all of these elements in a simple implementation. It is not a standard algorithm but is offered as a starting point for readers to incorporate their own ideas or revert to standard implementation.

Genetic Algorithm (A 9.4): The algorithm actually implemented is given next. In line with many other algorithms available in this text, this is provided as a way to understand and appreciate the working of the GA. It is also meant to provide a starting point for you to implement your own variation of the GA, since these techniques are sensitive to the specific problem being studied. Another important feature missing from this implementation is an encoding scheme. The design vector is handled directly to provide a direct feedback.

Step 1: Set up an initial *population*.
Generation index $t = 1$
Maximum generation index $= t_{max}$

Step 2: Genetic Operations
Two Simple crossovers
Two Arithmetic crossovers (convex)
One Directional crossover

Mutation of all seven design vectors so far
Four immigrants
Selection of the best two design vectors (from 18 vectors) for promotion to the next generation
If $t = t_{max}$, Stop
Increment generation index: $t \leftarrow t + 1$
Go to *Step 2*

Example 9.3 The example explores a Bezier curve-fitting problem. Given a set of xy data, find the *fifth*-order Bezier curve that will fit the data. The original data are obtained from the following function:

$$f(x) = y(x) = 1 + 0.25\,x + 2\,e^{-x}\cos 3x \tag{9.9}$$

The objective function is the least squared error between the original data and the data generated by the Bezier curve.

$$\text{Minimize:} \quad f(X): \quad \sum_{k=1}^{nData} \left[y_{data} - y_{B_n} \right]^2 \tag{9.10}$$

Bezier Curve: A Bezier curve is a parametric curve that is a special case of uniform B-splines. Bezier parameterization is based on the *Bernstein basis* functions. Any point P on a 2D Bezier curve (any parametric curve) is actually obtained as $P(x(v), y(v))$, where $0 \leq v \leq 1$. The actual relations are as follows:

$$P(v) = [x \quad y] = \sum_{i=0}^{n} B_i\, J_{n,i}(v), \quad 0 \leq v \leq 1 \tag{9.11a}$$

$$J_{n,i}(v) = \binom{n}{i} v^i (1-v)^{n-i}; \quad \text{is the Bernstein basis} \tag{9.11b}$$

where B_i is the vertices of the polygon that determines the curve. B_i represents a pair of values in 2D space. The order of the curve is n—the highest power in the basis functions. The actual computations of the points on the curve are easier using matrix algebra and are well explained and documented in.[14] Figure 9.12 represents a cubic Bezier curve. The following are some of the most useful properties of the curve, as observed in the figure:

- The curve is completely defined by the polygon obtained by joining the vertices in order.
- The degree of the polynomial defining the curve is one less than the number of vertices of the polygon.
- The first and last points of the curve are coincident with the first and last vertex. The remaining vertex points do not typically lie on the curve.

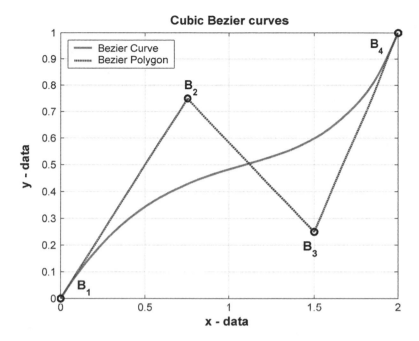

Figure 9.12 Illustration of the Bezier curve.

- The slopes at the ends of the curve have the same direction as the sides of the polygon.
- The curve is contained within the convex hull of the polygon.

In this example, the convenient matrix representations for the curve are used. They are not discussed here but are detailed in Chapter 11. They are available in the code. The source files contain code for evaluating the Bezier curve. The Bezier calculation is fairly modular. This example is not trivial. A more important reason for its inclusion is that, if understood, it signals a reasonable maturity with MATLAB programming. More importantly, it exemplifies the way to exploit the immense power of built-in MATLAB support. The author encourages the user to spend some effort tracing the progress of the calculations through the code.

Returning to Example 9.3, there are *eight* design variables associated with this problem. The design variables are the inside polygon vertex coordiantes. For $n = 5$, there are six vertices. The first and the last are known from the data points. This leaves four vertices or eight design variables representing the x and y values for the vertices. First, the global solution, if it exists, must have an objective function value of zero. Second, this is the only way to describe eight design variables in two-dimensional space (plot). The best value for the design vector in every generation is plotted to see how the curve fit improves. Since encoding is not being used in the algorithm, a reasonably long design vector is necessary to create the crossovers.

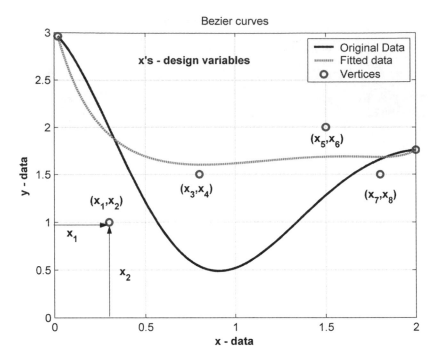

Figure 9.13 Genetic Algorithm, description of Example 9.3.

Figure 9.13 illustrates the eight design variables at arbitrary locations. They also illustrate that they are not close to the solution—which is the solid curve. There are several reasons why this is not the best example for illustration of GA. First, it is an example in continuous variables. Second, local minimums are not very distinguishable, and there are innumerable minimums. The global minimum is not substantially different from local minimum value to warrant exploration of the global solution. Third, the problem is not ill behaved. The most important reason depends not on the algorithm but on the computation of the objective function—calculating the least squared error. It uses a Newton–Raphson technique that is prone to failure if the curve folds over severely, or the vertices are very close together. One way to overcome this habit is to cause the x location of the vertices to monotonically increase (which is implemented in the population generation function). This upsets many of the crossover operations, since they are now conditionally random instead of purely random.

Initial Population: This refers to Step 1 of algorithm 9.4. Ten sets of design vectors are randomly created. From this set, the *two* best vectors are used as parents for crossover calculations. This is the pool available for propagation. This is also the minimum to perform crossovers. The fitness function is the objective function. The number of crossovers is restricted to keep the code simple. In general, the algorithm here, while including all the main features, is a little restrictive in its implementation.

MATLAB Code: There are several function m-files for this Example. A few of them relate to the determination of the Bezier curve based on the design variables. These are **coeff.m, Factorial.m, Combination.m, Bez_Sq_Error.m**. They must be available in the directory. A wrapper file is used to define Example 9.3 and is coded as **GeneticAlgorithm_Example9_3.m**. This will initiate the execution of the algorithm through the file **GeneticAlgorithm.m**. However, this is not generic, as it explicitly calls the **Bez_Sq_Error.m** function. The reader is suggested to pass the function name in the call statement, adopted in earlier codes, so that the fitness function can be generic. This would make the **GeneticAlgorithm.m** independent of the example.

The **GeneticAlgorithm_Example9_3.m** calls **GeneticAlgorithm.m** which will call additional files. The following is the list of files:

GeneticAlgorithm.m Directs the algorithm

populator.m Generates random design vector

SimpleCrossover.m Performs simple crossover between two parents

ArithmeticCrossover.m Performs convex crossover between two parents

DirectionalCrossover.m Performs one-directional crossover

Mutation.m Performs mutation of the design vectors

DrawCurve.m Draws the curve for each design vector

Figure 9.14 is the plot of the starting generation and the expected solution. There is a significant error associated with the initial choice of the design variables. The squared error, over 101 data points, is 41.8325. The design variables are plotted on the figure as vertices. Figure 9.15 is the plot of the solution after 200 generations. The error is significantly reduced to 0.2021, and the quality of the solution is visible on the plot. The figure is animated and displayed for each generation. It shows the movement of the vertices as well as the difference between the current solution and the actual solution. The best values for each generation are also written to the Command window. It can be seen that the error drops quickly, and there are large number of generations when nothing happens. The continuous nature of the problem makes it difficult to reduce the error to small values in a small number of generations.

The stochastic nature of the generation of solutions will prevent the same solution being replicated locally. However, over several executions of **GeneticAlgorithm_Example9_3.m**, a satisfactory solution to Example 9.3 should happen. It might also be mentioned that 200 generations is on the low side but appears sufficient to get close to the solution. The maximum generation value is not predictable, nor can it be established with certainty. It relies on numerical experimentation. This number appears easily to be over 1,000 in many investigations.

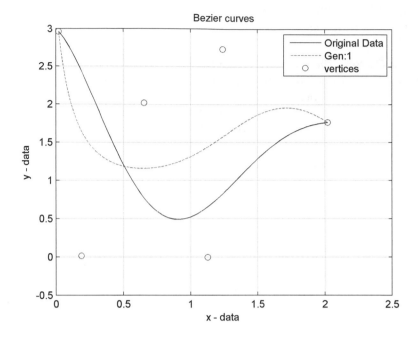

Figure 9.14 Best fit at the generation $t = 1$.

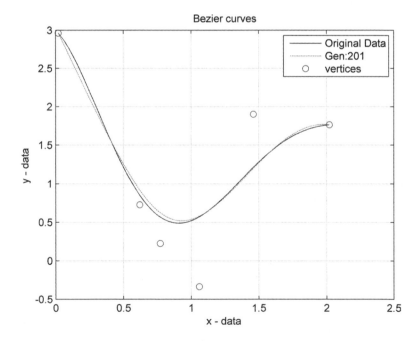

Figure 9.15 Best fit at the generation $t = 200$.

Suggested Extensions: Most of the engineering design problems involve constraints. Currently, the most popular way to handle constraints is to reformulate the problem using penalty functions (Chapter 7). There are several other alternatives to this procedure. It is possible to separately accumulate only designs that are feasible while continuing to implement the standard procedure. Second, use a strategy to drive the search process through constraint satisfaction and decrease the objective, simultaneously or separately. The algorithms indicated in this chapter can easily incorporate the needed changes.

The algorithmic procedure as implemented does not really care if the problem is continuous or discrete. It just works on a current set of values for the design variables. Discrete values will primarily require minor changes in the way the population is generated. Also, the implementation of the crossover and mutation operators will need to be changed. These changes in no way detract from the major idea behind the GA.

Closing Comments: In summary, for a stochastic technique, the GA is very impressive. The same can be said for SA. Including the fact that practical engineering problems must require partly discrete representation, the techniques in Chapters 8 and 9 are very relevant to the practice of design and development. With the enormous computational resource in desktops and laptops, an effective programming language like MATLAB, and the wide variety of algorithms that are available, it is possible now to design an individual approach to a many optimization problems. This is important to practicing engineers who can bring intuition and insight for the particular problem they have to solve in a tight time frame. In the engineering curriculum, traditional design optimization courses tend to deal only with deterministic continuous algorithms. That in itself was more than the time available for instruction. Discrete optimization or global optimization is a necessarily skill learnt by personal endeavor. The overemphasis on deterministic programming needs to be revisited in light of current developments in the area of optimization. Today in engineering departments, computer programming and its reinforcement is usually adjunct to other developments in the curriculum. New programming platforms like MATLAB offer a way to resolve this disadvantage. New algorithms also reflect a simplicity that can be easily translated and applied.

This book has been about programming in MATLAB, learning about traditional and current techniques in design optimization, and translating these optimization techniques into applications. The book is also about exploration and developing an insight to the process of design optimization. The code that is provided is only meant to be a starting point.

REFERENCES

1. Gray, P., W. Hart, L. Painton, C. Phillips, M. Trahan, and J. Wagner. *A Survey of Global Optimization Methods*. Albuquerque, NM: Sandia National Laboratories, 87185, http://www.cs.sandia.gov/opt/survey/(last accessed May 2008).

2. *Genetic Algorithm and Direct Search Toolbox 2.3.* Natick, MA: The Mathworks, 2008.

3. Arora, J. S. *Introduction to Optimal Design.* New York: McGraw-Hill, 1989.

4. Kirkpatrick, S., G. C. Gelatt, and M. P. Vecchi. "Optimization by Simulated Annealing." *Science,* 220 (4598) (May 1983).

5. Metropolis, N., A. Rosenbluth, M. Rosenbluth, A. Telle, and E. Teller. "Equations of State Calculations by Fast Computing Machines." *J. Chem. Phys.,* 21 (1953): 1087–1092.

6. Bohachevsky, I. O., M. E. Johnson, and L. S. Myron. "Generalized Simulated Annealing for Function Optimization." *Technometrics,* 28 (3) (1986).

7. Jones, A.E.W., and G. W. Forbes. "An Adaptive Simulated Annealing Algorithm for Global Optimization over Continuous Variables." *Journal of Global Optimization,* 6 (1995): 1–37.

8. Cohn, H., and M. Fielding. "Simulated Annealing: Searching for an Optimal Temperature Schedule." *SIAM Journal on Optimization,* 9 (3), (1999): 779–882.

9. Holland, J. H. "Outline of a Logical Theory of Adaptive Systems." *Journal of ACM,* 3 (1962): 297–314.

10. Dumitrescu, D., B. Lazzerini, L. C. Jain, and A. Dumitrescu. *Evolutionary Computation. International Series on Computational Intelligence.* Ed. L. C. Jain. Boca Raton, FL: CRC Press, 2000.

11. Gen, M., and R. Cheng. *Genetic Algorithms and Engineering Optimization.* Wiley Series in Engineering Design and Automation. Series ed. H. R. Parsaei. New York: Wiley Interscience, John Wiley.

12. Goldberg, D. *Genetic Algorithms in Search, Optimization, and Machine Learning.* Reading, MA: Addison-Wesley, 1989.

13. Houck, C. R., J. A. Joines, and M. G. Kay. *A Genetic Algorithm for Function Optimization.* Paper accompanying the GAOT (Genetic Algorithm Optimization Toolbox). North Carolina State University.

14. Rogers, G. F., and J. A. Adams, *Mathematical Elements for Computer Graphics,* 2nd ed. New York: McGraw-Hill, 1990.

PROBLEMS

No formal problems are suggested. You are encouraged to develop your own variants of the two algorithms, including formal ones available in the suggested literature. You are also encouraged to consider extensions to *constrained* and *discrete* problems. Use your versions to explore various examples found in the literature. In comparison to the deterministic techniques (until Chapter 8), these stochastic methods are truly fascinating and illustrate considerable robustness, and what is even more challenging to our perception, they can solve any problem with no reservations whatsoever. They are incredibly basic in construction, simple to program, and easy to deploy. They only need is a lot of time for searching the design space.

10

OPTIMIZATION TOOLBOX
FROM MATLAB

The MATLAB family of products for optimization includes the *Optimization Tool-box* and the *Genetic Algorithm and the Direct Search Toolbox*.[1] We will be referencing only the *Optimization Toolbox* in this chapter. The toolboxes are a collection of programs, m-files, and resources for optimization similar to the files you have downloaded for the book. It contains many of the methods that have been explored in this book. If you are looking to professionally engage in the practice of design optimization, then the Optimization Toolbox may prove a worthy investment. Many of the techniques and discussions in the book were kept simple to illustrate the concept, and the *m-files* in the book were also simple to provide MATLAB programming experience. Most of the m-files from the book will need to be developed further for serious practical application. There is also a very important reason for having access to the Optimization Toolbox. It allows you to take advantage of MATLAB's *open source* policy. This means you have access to the source files. You may be able to modify and extend those files to meet your special optimization requirements. Although some of the programs from the Toolbox have been used or referenced elsewhere in this book, this chapter recollects them in a single place for easy reference. As can be expected, the material of this chapter will always be dated. In fact, the current release of MATLAB is 7.6 or 2008a. The MATLAB used for developing and testing the code was 7.3 or R2006b. The Optimization Toolbox referenced in this chapter is Version 3.1 (2006b). You can expect that each new release of MATLAB and/or Optimization Toolbox will include enhancements, innovations, and extensions. By this time, you have gained enough experience to be able to take advantage of all of them. A quick idea of the new features can be gained by reading the release notes.

MATLAB 7 is a powerful development engine. In this book, we did not take advantage of the development tools that accompany MATLAB. For example, we did not take advantage of the *debugger* which could have identified the region of code that is not performing as expected. This can be useful in code creation to keep the error count low. There is also the *M-Lint Code checker* that can analyze your code and suggest significant improvements to increase performance. Then there is the MATLAB *Profiler*, which records the time spent in code execution. You can create *Directory reports* of all your code. You are encouraged to take advantage of these features as you develop your own piece of code. If you wish to package your code for sharing you might also want to consider using *GUIDE* (Graphical User Interface Development Environment) to develop a *GUI* (Graphical User Interface). Through the editor, you can also publish your code and results in HTML, Word, LaTEX, and other formats. Please use the MATLAB documentation to become more familiar with and incorporate the code enhancing activities listed here.

10.1 THE OPTIMIZATION TOOLBOX

The best place to start getting acquainted with the Optimization Toolbox is the Help browser. The following statements assume that the all of the documentation is installed on your computer. Invoking the Help browser (Full Product Help from the **Help** menu) opens the **Help Navigator** and an Internet browser where the results are displayed. Figure 10.1 displays the navigator and browser with the latter on the right. Notice, in addition to MATLAB help, you have access to help on all **Toolboxes** you are licensed for. If it does not appear it is likely that you have chosen not to install the documentation on your machine. Please do so before proceeding further.

The **help** files on the Navigator are organized as a tree with subdirectories. Clicking on *Optimization Toolbox* will open the next level of subdirectories and the browser will display several important links—see Figure 10.2. At the top, is the list of functions (m-files) that the toolbox contains organized by category or alphabetically? This gives you an overview of the optimization capability of the toolbox, and it will be easy to understand what each of the resource in the toolbox will accomplish. You can use the forward and backward arrows on the top to navigate the content that you are exposed to. There are three principal links to the documentation for the Optimization Toolbox. The first is a series of helpful exercise to get you started—**Getting Started**. The second is the **User's Guide** for details on the usage of functions. The last is a **Collection of Examples** that can effectively replace this chapter. There is also a collection of **Optimization Toolbox Demos** that run from the **Help** browser, which will give you a visual appreciation for the performance of the algorithms and the functions in the toolbox.

Most of the basic information in this chapter is from the links in the **Help** navigator of the Optimization Toolbox. Limited duplication is used for bringing attention to the more useful features. In this chapter, the emphasis is illustration of the use of the Toolbox with respect to the various examples introduced in the

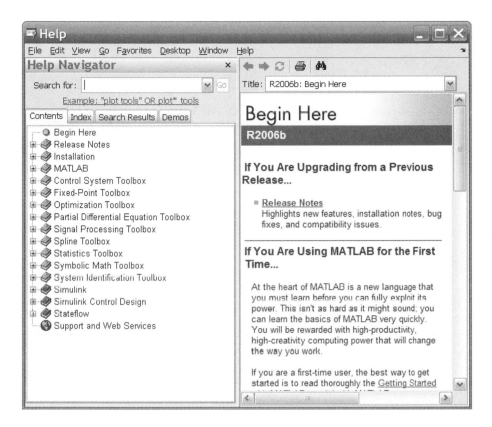

Figure 10.1 The MATLAB Help browser.

book. In point of fact, a large number of the techniques in the Toolbox are not used in the book and are not covered in this section. The experience provided in this chapter should be sufficient to understand and use the new ones effectively.

To use the *programs* or *m-files* in the toolbox, three important actions must be initiated:

1. Formulate the problem in the format expected by MATLAB with respect to the particular procedure that will be invoked. Mostly, this will involve using appropriate procedures to set the right-hand side of the constraints to zero or constant values. The Toolbox also makes it a point to distinguish between linear and nonlinear constraints. The former are introduced as coefficient elements of a matrix.

2. Parameters of the algorithm can be set or changed through the **Optimset** command. The default values can be viewed using the **Optimget** command.

3. Use the appropriate optimizer and ensure that a valid solution has been obtained. All these actions will depend on the actual mathematical model of the problem.

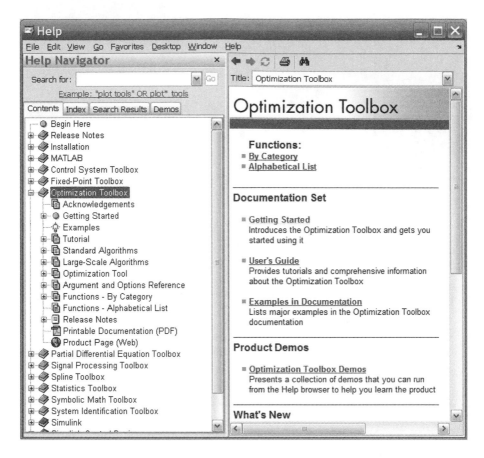

Figure 10.2 Optimization Toolbox links in Navigator.

An important element of the Optimization Toolbox is the GUI that can access all of the resources in a single page. To access it from the Command window, type

```
>> optimtool
```

Figure 10.3 displays the **Optimization Tool**. There are three panels. The first panel is the **Problem Setup and Results** panel. You can call any of the solvers, identify the m-files where the objective and constraints can be found, set the upper and lower bounds and the linear equality and inequality constraints. The second panel is where you set the algorithm control parameters. This is called the **Options** panel. Something you will use the **Optimset** command for. The thirds is the **Quick Reference** panel that will allow you access to explanations and options on the solver of your choice. This is a powerful tool and appealing and allows you to interactively explore the problem through mouse clicks. In

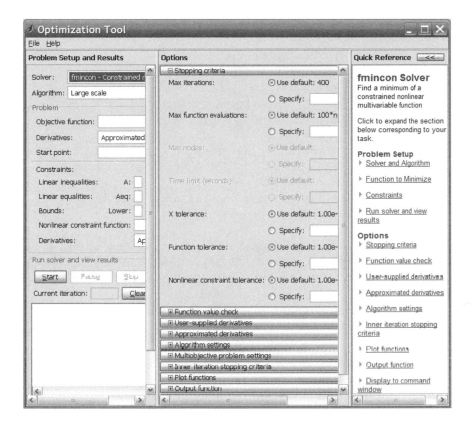

Figure 10.3 MATLAB optimization GUI: optimtool.

this chapter, we will eschew the GUI tool and run the solvers through code by planning and organizing the sequence of steps to understand the process. The reader is encouraged to use the tool once all the resources are in place.

10.1.1 Programs

The complete set of resources available in your version of the Toolbox installed on your computer can be discovered in the Command window by typing the following command:

```
>> help optim
```

The output is copied next. It includes all of the m-files and a self-explaining title. It is assigned to appropriate categories.

```
>> help optim
   Optimization Toolbox
```

Version 3.1 (R2006b) 03-Aug-2006

Nonlinear minimization of functions.
    fminbnd   -Scalar bounded nonlinear function
                minimization.
    fmincon   -Multidimensional constrained nonlinear
                minimization.
    fminsearch -Multidimensional unconstrained nonlinear
         minimization,by Nelder-Mead direct search method.
    fminunc -Multidimensional unconstrained nonlinear
                minimization.
    fseminf-Multidimensional constrained minimization,
         semi-infinite constraints.

Nonlinear minimization of multi-objective functions.
fgoalattain-Multidimensional goal attainment optimization
    fminimax   -Multidimensional minimax optimization.

Linear least squares (of matrix problems).
    lsqlin   -Linear least squares with linear
               constraints.
    lsqnonneg -Linear least squares with nonnegativity
               constraints.

Nonlinear least squares (of functions).
    lsqcurvefit-Nonlinear curvefitting via least squares
(with bounds).
    lsqnonlin -Nonlinear least squares with upper and
               lower bounds.

Nonlinear zero finding (equation solving).
    fzero -Scalar nonlinear zero finding.
    fsolve-Nonlinear system of equations solve
        (function solve).

Minimization of matrix problems.
    bintprog -Binary integer (linear) programming.
    linprog   -Linear programming.
    quadprog -Quadratic programming.

Controlling defaults and options.
    optimset -Create or alter optimization OPTIONS
               structure.
    optimget -Get optimization parameters from
               OPTIONS structure.

Demonstrations of large-scale methods.

```
circustent-Quadratic programming to find shape
 of a circus tent.
molecule-Molecule conformation solution using
unconstrained nonlinear minimization.
optdeblur  -Image deblurring using bounded linear
             least-squares.

Demonstrations of medium-scale methods.
 tutdemo -Tutorial walk-through.
 goaldemo-Goal attainment.
 dfildemo-Finite-precision filter design
 (requires Signal Processing Toolbox).
 datdemo     -Fitting data to a curve.
 officeassign-Binary integer programming to solve
      the office assignment problem.
 bandem      -Banana function minimization demonstration

Medium-scale examples from User's Guide
 objfun      -nonlinear objective
 confun      -nonlinear constraints
 objfungrad -nonlinear objective with gradient
 contungrad -nonlinear constraints with gradients
 confuneq   -nonlinear equality constraints
 optsim.mdl -Simulink model of nonlinear plant process
 optsiminit -init file for optisim.mdl
 runtracklsq-demonstrates multiobjective function
            using LSQNONLIN
 runtrackmm -demonstrates multiobjective function using
             FMINIMAX

Large-scale examples from User's Guide
 nlsf1   -nonlinear equations objective with Jacobian
 nlsf1a  -nonlinear equations objective
 nlsdat1 -MAT-file of Jacobian sparsity pattern
      (see nlsf1a)
 brownfgh-nonlinear minimization objective with gradient
and Hessian
 brownfg -nonlinear minimization objective with gradient
 brownhstr-MAT-file of Hessian sparsity pattern
      (see brownfg)
 tbroyfg -nonlinear minimization objective with gradient
 tbroyhstr-MAT-file of Hessian sparsity pattern
      (see tbroyfg)
 browneq -MAT-file of Aeq and beq sparse linear equality
  constraints
 runfleq1-demonstrates 'HessMult' option for FMINCON with
 equalities
```

```
brownvv-nonlinear minimization with dense structured
      Hessian
hmfleq1-Hessian matrix product for brownvv objective
fleq1  -MAT-file of V, Aeq, and beq for brownvv
      and hmfleq1
qpbox1  -MAT-file of quadratic objective Hessian
      sparse matrix
runqpbox4-demonstrates 'HessMult' option for QUADPROG
      with bounds
runqpbox4prec-demonstrates 'HessMult' and TolPCG options
      for QUADPROG
qpbox4 -MAT-file of quadratic programming problem
      matrices
runnls3-demonstrates 'JacobMult' option for LSQNONLIN
nlsmm3 -Jacobian multiply function for runnls3/nlsf3a
      objective
nlsdat3-MAT-file of problem matrices for runnls3/nlsf3a
 objective
runqpeq5-demonstrates 'HessMult' option for QUADPROG
      with equalities
qpeq5-MAT-file of quadratic programming matrices for
      runqpeq5
particle-MAT-file of linear least squares C and d
      sparse matrices
sc50b  -MAT-file of linear programming example
densecolumns-MAT-file of linear programming example

Graphical user interface and plot routines
  optimtool-Optimization Toolbox Graphical User Interface
  optimplotconstrviolation -Plot max. constraint violation
    at each iteration
  optimplotfirstorderopt-Plot first-order optimality
    at each iteration
  optimplotresnorm -Plot value of the norm of residuals
    at each iteration
  optimplotstepsize-Plot step size at each iteration
```

The *tutdemo* is recommended for a demonstration of the features of the Toolbox. The following items, *optimset, linprog, quadprog, fminunc*, and *fmincon*, are further explored in the chapter.

10.1.2 Using Programs

The solvers, with the same name, can be called with multiple inputs depending on the problem formulation. They can also return multiple pieces of information.

It is the user's responsibility to ensure that program is used correctly based on all the available information. For the input, the list order must be maintained. If any information in the input list is not available, then an empty matrix place holder must be used. The output is usually returned as a structure, and keys must be use to index the appropriate variable and its value. We will use the program *fmincon* to illustrate these ideas. Typing **help fmincon** in the command window returns more than the information here. It is edited to focus only on the different calling statements. Please remember that even more details are available through the **Help** browser.

```
>> help fmincon
 FMINCON finds a constrained minimum of a function of
    several variables.
    FMINCON attempts to solve problems of the form:
        min F(X)
subject to (eq. stands for equality):
A*X   <= B, Aeq*X  = Beq (linear constraints)
    C(X) <= 0, Ceq(X) = 0    (nonlinear constraints)
    LB <= X <= UB            (side constraints)

    X=FMINCON(FUN,X0,A,B)
starts at X0 and finds a minimum X to the function FUN,
    subject to the linear inequalities A*X <= B. FUN
    accepts input X and returns a scalar function value F
    evaluated at X. X0 may be a scalar, vector, or matrix.

    X=FMINCON(FUN,X0,A,B,Aeq,Beq)
(Set A=[] and B=[] if no inequalities exist.)

    X=FMINCON(FUN,X0,A,B,Aeq,Beq,LB,UB)
    if no bounds exist. Set LB(i) = -Inf if X(i) is
        unbounded below; set UB(i) = Inf if X(i) is
        unbounded above.

    X=FMINCON(FUN,X0,A,B,Aeq,Beq,LB,UB,NONLCON)
The function NONLCON accepts X and returns the vectors
        C and Ceq, representing the nonlinear inequalities
        and equalities respectively.
    (Set LB=[] and/or UB=[] if no bounds exist.)

    X=FMINCON(FUN,X0,A,B,Aeq,Beq,LB,UB,NONLCON,OPTIONS)
OPTIONS, an argument created with the OPTIMSET function.

    X = FMINCON(PROBLEM)
finds the minimum for PROBLEM. PROBLEM is a structure
```

PROBLEM.objective: with the function FUN
PROBLEM.x0: the start point
PROBLEM.Aineq: the linear inequality constraints
PROBLEM.bineq:
PROBLEM.Aeq: the linear equality constraints
PROBLEM.beq:
PROBLEM.lb: the lower bounds
PROBLEM.ub: the upper bounds
PROBLEM.nonlcon: the nonlinear constraint
PROBLEM.options : the options structure
PROBLEM.solver: solver name 'fmincon'
Use this syntax to solve at the command line a problem
    exported from OPTIMTOOL. The structure PROBLEM must have
    all the fields.

The following are the different ways to process the *output* from fmincom.

        [X,FVAL]=FMINCON(FUN,X0,..)
returns the value of the objective function FUN at
        the solution X.

        [X,FVAL,EXITFLAG]=FMINCON(FUN,X0,..)
returns an EXITFLAG that describes the exit condition
 of FMINCON. For EXITFLAG
Both medium- and large-scale:
= 1 First order optimality conditions satisfied
= 0 Maximum number of function evaluations/iterations
= -1 Optimization terminated by the output function
Large-scale only:
= 2 Change in X less than the specified tolerance.
= 3 Change in the objective function value less than the
specified tolerance.
Medium-scale only:
= 4 Magnitude of search direction smaller than the
 specified tolerance and constraint violation less than
 options.TolCon.
= 5 Magnitude of directional derivative less than
 the specified tolerance and constraint violation less than
 options.TolCon.
= -2 No feasible point found.

        [X,FVAL,EXITFLAG,OUTPUT]=FMINCON(FUN,X0,..)
returns a structure OUTPUT
OUTPUT.iterations: with the number of iterations
OUTPUT.funcCount: the number of function evaluations

```
OUTPUT.stepsize: the norm of the final step
OUTPUT.algorithm: the algorithm used
OUTPUT.firstorderopt: the first-order optimality
OUTPUT.message: the  exit message
OUTPUT.lssteplength: the final line search steplength
OUTPUT.cgiterations: the number of CG iterations
```

```
    [X,FVAL,EXITFLAG,OUTPUT,LAMBDA]=FMINCON(FUN,X0,..)
    returns the Lagrange multipliers at the solution X:
LAMBDA.lower: for LB
LAMBDA.upper: for UB
LAMBDA.ineqlin: for the linear inequalities
LAMBDA.eqlin: for the linear equalities,
LAMBDA.ineqnonlin: for the nonlinear inequalities, and
LAMBDA.eqnonlin: for the nonlinear equalities.
```

```
   [X,FVAL,EXITFLAG,OUTPUT,LAMBDA,GRAD]=FMINCON(FUN,X0,..)
returns the value of the gradient of FUN at the solution X.
```

```
   [X,FVAL,EXITFLAG,OUTPUT,LAMBDA,GRAD,HESSIAN]=
   FMINCON(FUN,X0,..)
returns the value of the HESSIAN of FUN at the solution X.
```

Three **structures** are used in this list of function calls. They are PROBLEM, OUTPUT, and LAMBDA. This list of function calls indicates 12 ways of using the fmincon function. The differences depend on the information in the mathematical model that is being solved and the post optimization information that should be displayed. In using the different function calls, any in between data that does not exist for the particular examples, must be indicated using a null vector ([ ]). Such extensive information can be obtained for each of the solvers in the Optimization Toolbox.

10.1.3 Setting Optimization Parameters

There are several parameters that are used to control the application of a particular *optimization* technique. These parameters are available through the OPTIONS structure. All of them are provided with default values. There will be occasions when it is necessary to change the default values. One example may be that your constraint tolerances are required to be lower than the default. Or EXITFLAG recommends a larger number of function evaluations to solve the problem. These parameters are set using a *name/value* pair. Generally, the *name* is usually a string, while the value is typically a *number*. The list of parameters for algorithm control can be identified by typing `optimset` at the Command prompt or typing help `optimoptions`. The list is quite long, and is not included here. The default settings for a particular solver can be found by typing optimset with the solver name as `optimset('fmincon')` —for the `fmincon` program. The options structure can be seen by typing in the Command window the following:

```
>> help optimset
  OPTIMSET Create/alter optimization OPTIONS structure.
     OPTIONS = OPTIMSET('PARAM1',VALUE1,'PARAM2',VALUE2,..)
     creates an optimization options structure OPTIONS
     in which the named parameters have the specified values.
     Any unspecified parameters are set to [] (parameters
     with value [] indicate to use the default value for
     that parameter

     OPTIONS = OPTIMSET(OLDOPTS,'PARAM1',VALUE1,..)
  creates a copy of OLDOPTS with the named parameters
     altered with the specified values.

     OPTIONS = OPTIMSET(OLDOPTS,NEWOPTS)
  combines an existing options structure OLDOPTS with a
     new options structure NEWOPTS.  Any parameters in
     NEWOPTS with non-empty values overwrite the
     corresponding old parameters in OLDOPTS.

     OPTIONS = OPTIMSET
  (with no input arguments) creates an options structure
     OPTIONS where all the fields are set to [].

     OPTIONS = OPTIMSET(OPTIMFUNCTION)
  creates an options structure with all the parameter names
     and default values relevant to the optimization function
     named in OPTIMFUNCTION.

  OPTIMSET PARAMETERS for MATLAB
  The most common parameters are
  Display-Level of display [ off | iter | notify | final ]
  MaxFunEvals-Maximum number of function evaluations
           allowed [ positive integer ]
  MaxIter-Maximum number of iterations allowed
          [ positive scalar ]
  TolFun-Termination tolerance on the function value
          [ positive scalar ]
  TolX-Termination tolerance on X [ positive scalar ]
  FunValCheck-Check for invalid values, such as NaN
           or complex, from user-supplied functions
           [ {off} | on ]
  OutputFcn-Name(s) of output function [ {[]} | function ]
```

All output functions are called by the solver after
each iteration.
  PlotFcns-Name(s) of plot function [{[]} | function]
          Function(s) used to plot various quantities in
every iteration

10.2 EXAMPLES

Four examples are illustrated in this section. They are continuous optimization
problems discussed in the book. They are from linear programming, quadratic
programming, unconstrained minimization, and constrained minimization. Only
the appropriate mathematical model is introduced prior to illustrating the use of
the Toolbox function. The details are available in the earlier chapters.

10.2.1 Linear Programming

The solver is linprog which solves the following problem

$$\text{Minimize} \quad f^T x;$$

$$\text{Subject to} \quad [A][x] \le [b]$$

$$[A_{eq}][x] \le [b_{eq}]$$

$$x^{lb} \le x \le x^{ub}$$

Example 10.1: This is same as Example 3.2. We will solve it using the Optimization Toolbox.

$$\text{Minimize} \quad f(x_1, x_2) : -1.4x_1 - x_2 + 19,120 \tag{10.1}$$

$$\text{Subject to:} \quad g_1(x_1, x_2) : x_1 + x_2 \le 2,000 \tag{10.2a}$$

$$g_2(x_1, x_2) : x_1 + x_2 \ge 640 \tag{10.2b}$$

$$0 \le x_1 \le 800; 0 \le x_2 \le 1,440 \tag{10.2c}$$

The solution in Chapter 3 was

$$x_1 = 800; \quad x_2 = 1,200, \quad f = 16,800 \text{ or } \$168.00 \tag{3.27}$$

MATLAB Code: The code is available in **Example10_1.m***. The problem
can be solved directly by typing in the Command window. The file is more
informative and can also serve as a template. The simplex method is invoked
through **linprog.m**. The default algorithm solves a large-scale problem. We
use the *option* structure to set that parameter to 'off'. The constant in (10.1)

---

*Files to be downloaded from the web site are indicated by boldface courier type.

cannot be accommodated within the program but is added to the solution output from the program. The output in the Command window:

```
Optimization terminated.
```

```
Example 10.1
```

```
x1*  =     800.000
x2*  =    1200.000
f*   =   16800.000


EXITFLAG =    1
Total time (s)=   0.0000
```

10.2.2 Quadratic Programming

The solver for QP problem is quadprog. The standard format for this program is

$$\text{Minimize} \quad f(x) = c^T x + \frac{1}{2} x^T H x$$

$$\text{Subject to:} \quad A x \leq b$$

$$A_{eq} x = b_{eq}$$

$$LB \leq x \leq UB$$

Quadratic programming refers to a problem with a quadratic objective function and linear constraints. Although such a problem can be solved using nonlinear optimization techniques of Chapter 7, special quadratic programming algorithms based on the techniques of linear programming are normally used, as they are simple and faster. If the objective function is a quadratic function in the variables, then it can be expressed in **quadratic form**. For example, the function f

$$f(x_1, x_2) = x_1 + 2x_2 + x_1^2 + x_1 x_2 + 2x_2^2$$

can be expressed as

$$f(x_1, x_2) = \begin{bmatrix} 1 & 2 \end{bmatrix} \begin{bmatrix} x_1 \\ x_2 \end{bmatrix} + \frac{1}{2} \begin{bmatrix} x_1 & x_2 \end{bmatrix} \begin{bmatrix} 2 & 1 \\ 1 & 4 \end{bmatrix} \begin{bmatrix} x_1 \\ x_2 \end{bmatrix};$$

$$f(x_1, x_2) = [c]^T \begin{bmatrix} x_1 \\ x_2 \end{bmatrix} + \frac{1}{2} \begin{bmatrix} x_1 & x_2 \end{bmatrix} [H] \begin{bmatrix} x_1 \\ x_2 \end{bmatrix}$$

$$f(x_1, x_2) = [c]^T [x] + \frac{1}{2} [x]^T [H][x]$$

Example 10.2: We have only one example of QP problem, Example 2.1, and this was extensively discussed while also being trivial. We will use a contrived example to examine the QP program.

$$\text{Minimize } f(x_1, x_2, x_3) = \begin{bmatrix} -1 & -2 & -3 \end{bmatrix}^T [x] + \frac{1}{2}[x]^T \begin{bmatrix} 3 & -1 & 1 \\ -1 & 2 & 1 \\ 1 & 1 & 4 \end{bmatrix} [x]$$

(10.3)

$$\text{Subject to: } \quad x_1 + 2x_2 - x_3 \le 2.5 \qquad (10.4a)$$

$$2x_1 - x_2 + 3x_3 = 1 \qquad (10.4b)$$

$$x_1, x_2, x_3 \ge 0 \qquad (10.4c)$$

MATLAB Code: The solution can be obtained by executing `Example10_2.m`. The default optimization parameters are used so that the options structure does not have to be changed. The solution is formatted and displayed in the Command window. It is copied from the Command window:

```
Optimization terminated.


 Example 10.2


Final Values
Optimum Design Variables


    0.5818     1.1182     0.3182
Optimum function value


   -1.9218
Lagrange Multipliers for inequality constraint


   0.0218
Inequality constraint


   2.5000
Lagrange Multipliers for equality constraint


   0.0164


Equality constraint


   1.0000


Total time (s) =   0.0156
```

10.2.3   Unconstrained Optimization

We will solve Example 6.2, which is called the Rosenbrock problem, designed to challenge unconstrained minimization algorithms. The MATLAB documentation uses the example for illustrating numerical solutions of nonlinear equations. Example 10.3 is used to illustrate the use of fminunc.m the program to be used for unconstrained minimization. For this example we will provide analytical gradients of the objective function. This is indicated in the options structure. The objective function and the gradients are calculated in a function m-file, which is referenced by the call to the solver.

Example 10.3:   The Rosenbrock problem is

$$\text{Minimize} \quad f(x_1, x_2): \ 100(x_2 - x_1^2)^2 + (1 - x_1)^2 \tag{10.5}$$

The solution to this problem is

$$x_1^* = 1.0; \quad x_2^* = 1.0; \quad f^* = 0.0$$

The problem is notorious for a large number of iterations for convergence, as well as very small changes in design as the solution is being approached.

MATLAB Code: The code for solving the example is in **Example10_3.m** while the objective function and the gradients are set up in **ObjectiveExample10_3.m**. The outputs in the Command window are copied here:

```
——————————
Example 10.3
——————————

EXITFLAG:      1

Values for the OUTPUT structure
        iterations: 21
         funcCount: 28
          stepsize: 1
     firstorderopt: 2.8177e-008
         algorithm: 'medium-scale: Quasi-Newton line search'
           message: 'Optimization terminated: relative
                    infinity-norm of gradient less than
                    options.TolFun.'

Final Values-User specified Gradients

Optimum Design Variables
————————————————————————————
    1.0000     1.0000
```

```
Optimum function value
```

```
  6.1379e-019
```

```
Gradients of the function
```

```
  1.0e-007 *
  -0.2818    0.1371
```

```
Total time (s)=  0.0313
```

For this example, the OUTPUT structure is printed to the Command window. It provides some important information about the performance to the algorithm with respect to the example.

10.2.4 Constrained Optimization

This example is about the design of an I-beam for use in a particular structural problem. It appeared in Chapter 1 as Example 1.2, in Chapter 7 as Example 7.3, and in Chapter 8 as Example 8.4. The constrained solver is `fmincon.m`.

Example 10.4: Design a cantilevered beam, of minimum mass, carrying a point load F at the end of the beam of length L. The cross section of the beam will be in the shape of the letter I (referred to as an I-beam). The beam should be sufficiently strong in bending and shear. There is also a limit on its deflection.

The mathematical model associating x_1 with d, x_2 with t_w, x_3 with b_f, and x_4 with t_f, so that the design vector is $X = [x_1, x_2, x_3, x_4]$. The problem in standard format is:

$$\text{Minimize} \quad f(X) : \gamma L A_c \tag{10.6}$$

$$\text{Subject to:} \quad g_1(X) : F L x_1/2I_c - \sigma_{\text{yield}} \quad < 0 \tag{10.7a}$$

$$g_2(X) : F Q_c/I_c x_2 - \tau_{\text{yield}} \le 0 \tag{10.7b}$$

$$g_3(X) : F L^3/3E I_c - \delta_{\text{max}} \le 0 \tag{10.7c}$$

$$0.01 \le x_1 \le 0.25; \quad 0.001 \le x_2 \le 0.05; \tag{10.7d}$$

$$0.01 \le x_3 \le 0.25; \quad 0.001 \le x_4 \le 0.05 \tag{10.7d}$$

The designer must ensure that the problem definition is also consistent with the unit system chosen for the parameters and variables. The parameters for this problem (value is given in the parenthesis) are F (10000 N); L (3m); γ (steel: $= 7860$ kg/m^3); E (210 GPa); σ_{yield} ($250E + 06N/\text{m}^2$); τ_{yield} ($145E + 06N/\text{m}^2$); and the maximum deflection δ_{max} (0.005m).

Some geometric constraints among the design variables will be needed. For example, the problem is not aware that x_1 must be greater than x_4.

$$g_4(X):x_1 - 3x_3 \leq 0 \tag{10.8a}$$

$$g_5(X):2x_3 - x_1 \leq 0 \tag{10.8b}$$

$$g_6(X):x_2 - 1.5x_4 \leq 0 \tag{10.8c}$$

$$g_7(X):0.5x_4 - x_2 \leq 0 \tag{10.8d}$$

The side constraints are brought into the problem directly in the problem formulation through linear inequality constraints from. There will be eight such constraints developed:

$$g_8(X) = -x_1 + 0.01 \leq 0 \tag{10.9a}$$

$$g_9(X) = x_1 - 0.25 \leq 0 \tag{10.9b}$$

$$g_{10}(X) = -x_2 + 0.001 \leq 0 \tag{10.9c}$$

$$g_{11}(X) = x_2 - 0.05 \leq 0 \tag{10.9d}$$

$$g_{12}(X) = -x_3 + 0.01 \leq 0 \tag{10.9e}$$

$$g_{13}(X) = x_3 - 0.25 \leq 0 \tag{10.9f}$$

$$g_{14}(X) = -x_4 + 0.001 \leq 0 \tag{10.9g}$$

$$g_{15}(X) = x_4 - 0.05 \leq 0 \tag{10.9h}$$

To complete the model, we will need the following relations based on the design variables.

$$A_c = 2x_3x_4 + x_1x_2 - 2x_2x_4$$

$$I_c = \frac{x_3x_1^3}{12} - \frac{(x_3 - x_2)(x_1 - 2x_4)^3}{12}$$

$$Q_c = 0.5x_3x_4(x_1 - x_4) + 0.5x_2(x_1 - x_4)^2$$

MATLAB Code: The problem is run through **Example10_4.m**. This file calls the solver **fmincon**. Prior to this call the parameters, linear constraints, lower and upper bound are set. The objective function is calculated in **ObjectiveExample10_4.m** and the constraints are set up in **ConstraintsExample10_4.m**. The code requires iteration information be written to the Command window by changing the default parameter in the options structure. This is not included here. The final values are printed to the Command window:

Example 10.4

EXITFLAG: 1

Values for the OUTPUT structure
         iterations: 8
          funcCount: 45
       lssteplength: 1
           stepsize: 5.1114e-010
          algorithm: 'medium-scale: SQP, Quasi-Newton,
                      line-search'
      firstorderopt: 9.0949e-013
            message: [1x143 char]

Final Values
Optimum Design Variables

    0.2500   0.0121   0.1250   0.0242

Optimum function value

  200.3002

Final Nonlinear Constraints

  1.0e+004 *
   -0.0000   -3.5357   -0.0144

Lagrange Multipliers for Nonlinear constraints

  1.0e-003 *
    0.9279   0   0

Lagrange Multipliers for Linear constraints

  1.0e+003 *
         0   0.3345   0   2.7583

Total time (s)= 0.1875

The mass of the beam is 200.3 kg. he deflection constraints are active. All of the Lagrange multipliers are zero or positive.

You are encouraged to run the example with Diagnostics on, and explore the options structure, as well as the output structure.

REFERENCES

1. Product listing from www.mathworks.com/products/product_listing/index.html. (accessed July 2005)

11

HYBRID MATHEMATICS — AN APPLICATION

This chapter is included to demonstrate the application of optimization in new directions. The term **Hybrid Mathematics** in this chapter is used to identify analytical solutions to mathematical problems using numerical techniques of optimization. There are a couple of limitations to the nature of the problems and their solutions addressed in this chapter. First, the solutions to these problems must be continuous functions. Second, the word *analytical* is used to imply closed form solutions. Third, the solutions are described by finite-length polynomials. Fourth, because the solution is obtained through numerical techniques, and the practical consideration of limiting the highest power of the polynomials, the solution will have to be considered an approximate analytical solution, even if its accuracy is sufficient for engineering purposes.

The author has been working in this area for several years.[1-6] Reference 6, created in 2007, is a Web resource that documents the work of the author in this area. It includes a collection of examples of engineering applications, including an exhaustive collection of references for those who will be interested in this area of application of optimization. It is shown that it is possible to establish analytical solutions for linear or nonlinear, ordinary or partial, single or a system of differential equations using a direct formulation coupled with the optimizer from the MATLAB Optimization Toolbox.

This approach mirrors one of the classical approaches for the solutions to mathematical problems. It involves assuming the solution as the sum of known functions multiplied by undetermined constants. The constants are then evaluated by applying sufficient number of boundary conditions leading to a system of algebraic systems of equations. Often, infinite series are used to represent such solution. Usually a recurrence relation can be generated to establish the coefficients. In this chapter, we use Bezier functions to represent the solution and use

489

optimization to determine the coefficients. The chapter uses the least squared error principle to establish the optimum solution.

Solutions to differential equations are a vast subject and can spawn much discussion in several areas. This chapter avoids existential debates and focuses only on the technique and the application. Current developments in this area can be found in other references by the author and a few in the open literature and are not included here. Two sets of examples are included in this chapter. The first set includes using Bezier curves for data representation. Here the optimization conditions are explicit and the process is non-iterative. The second set is the solution to differential equations where the solution is obtained through the Optimization Toolbox.

11.1 CENTRAL IDEA

The main thesis of the chapter is to *express the representation of the data* or *the solution to the differential equation* through a *parametric curve* or a *surface*. These curves or surfaces can be required to behave as a function by employing simple **side constraints** during their creation. These constraints are routine in applied numerical optimization. The solution therefore is the curve that will *reduce the squared error over all of the data points* or *satisfy the differential equations and the boundary conditions*. The particular curve used in this book is the Bezier curve/surface.[7]

The Bezier curve, defined through geometric construction by Bezier, can be reproduced using the Bernstein basis.[8,9] The Bernstein polynomial approximation to a continuous function mimics the gross features of the function remarkably well.[8] Furthermore, as the order of the polynomial is increased, this approximation converges uniformly to the function and its derivatives where they exist. The Bezier curve delivers, at the minimum, the same smoothness as the primitive function it is trying to emulate.[8,9]

Several kinds of parametric curves have been investigated to provide solutions to these kinds of problems. The one that appears often in connection with solving differential equations is the B-spline.[10] B-splines are essential elements in computer graphics.[11] Reference 12 uses B-splines as the basis for the finite element method. Parametric Bezier curves using the Bernstein basis functions are also a special class of the uniform B-splines using an open-knot vector. B-splines, however do not possess continuous derivatives over the whole domain, and therefore are not very useful for establishing analytical solutions.

11.1.1 Bezier Function — 2D

We defined the Bezier curve briefly in Example 9.3. We expand the definition here to a Bezier surface and also to the relations used to generate these functions using matrix representation, which allows exploitation through array programming in MATLAB. The Bezier curve is a parametric curve that is completely determined by a set of vertices or control points. The same vertices will establish a polynomial representation for the curve. The Bezier curve is very smooth, bounded, real, and easily accommodates the natural boundary conditions associated with differential systems. They can be infused with properties of a function by constraining the vertices of the independent variable. For a Bezier curve, any point (x, y) on the curve corresponds to parameter $p, 0 \leq p \leq 1$. The Bezier curve is described by

$$[x(p)y(p)] = \sum_{i=0}^{n} B_i J_{n,i}(p), \quad 0 \leq p \leq 1 \tag{11.1}$$

$$J_{n,i}(p) = \binom{n}{i} p^i (1 - p)^{n-i} \tag{11.2}$$

where $J_{n,i}(p)$ represents the Bernstein basis. $B_i's$ are the vertices of the polygon (a pair of values a_i and b_i) that determine the curve in two-dimensional space (or a triple of values for three-dimensional space). Figure 11.1 illustrates the two-dimensional Bezier curve. The vertices are shown as filled circles. The curve and all its derivatives can be obtained through simple mathematical operations using these vertices. The order of the curve in the figure is n, which is the highest order of the polynomial in the basis.

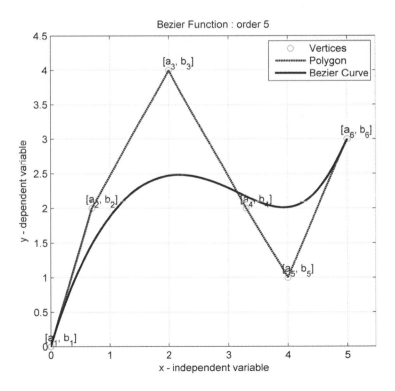

Figure 11.1 Bezier function, fifth order.

The following are some of the useful properties of these curves[9]

- The basis functions are real.
- The curve is defined analytically using the polygon obtained by joining the vertices in order.
- The degree of the polynomial defining the curve is one less than the number of vertices of the polygon. In Figure 11.1, the curves have six vertices—therefore, polynomial is of fifth order.
- The first and last points of the curve are coincident with the first and last vertex. The remaining points do not typically lie on the curve.
- The slopes at the ends of the curve have the same direction as the sides of the polygon.
- The curve is contained within the convex hull determined by the vertices.

By having the x vertex locations monotonically increasing, the curve will have the properties of a function. We will refer to these constrained Bezier curves as *Bezier functions*. Once the vertices are determined, the curve can be represented in closed or explicit polynomial form.

There are convenient matrix representations for the description of the Bezier curves, which makes computing the curves and derivatives simple. The curve in Equation 1.1 can be generated through the following matrix multiplication:[9]

$$[x(p) \, y(p)] = [P][N][B] \tag{11.3}$$

$[P]$ is the parameter vector. $[N]$ is a coefficient matrix based on the order of the curve. It is not defined here but can be found in the cited reference.[9] The coefficient matrix can be created through included code by just identifying the order of the function. $[B$ is the matrix that hold the vertices of the curve. For the fifth-order function in Figure 1.1, the various matrices are

$$[P] = [p^5 \, p^4 \, p^3 \, p^2 \, p \, 1] \tag{11.4a}$$

$$[N] = \begin{bmatrix} -1 & 5 & -10 & 10 & -5 & 1 \\ 5 & -20 & 30 & -20 & 5 & 0 \\ -10 & 30 & -30 & 10 & 0 & 0 \\ 10 & -20 & 10 & 0 & 0 & 0 \\ -5 & 5 & 0 & 0 & 0 & 0 \\ 1 & 0 & 0 & 0 & 0 & 0 \end{bmatrix} \tag{11.4b}$$

$$[B] = \begin{bmatrix} 0 & 0 \\ 0.7000 & 2.0000 \\ 2.0000 & 4.0000 \\ 3.3000 & 2.0000 \\ 4.0000 & 1.0000 \\ 5.0000 & 3.0000 \end{bmatrix} \tag{11.4c}$$

MATLAB Code: Figure 11.1 and the values in (11.4b) are obtained using the code **DrawingCurves.m**[*]. It calls **coeff.m** to set up the coefficient matrix. This, in turn, will call **Combination.m** and **Factorial.m**. Note that the parameter values and its exponents are obtained through array processing.

11.1.2 Bezier Function — 3D

The Bezier function in three dimensions or the parametric Bezier surface can be described as a vector-valued function of two parameters r and s

$$[x(r, s) \quad y(r, s) \quad u(r, s)]; \quad 0 \leq r, s \leq 1 \tag{11.5}$$

Reference 9 uses a natural Cartesian product generalization of the Bernstein operator for the bivariate functions.

$$[x(r, s) \quad y(r, s) \quad u(r, s)] = Q(r, s) = \sum_{i=0}^{m} \sum_{j=0}^{n} B_{i,j} \, J_{m,i}(r) \, K_{n,j}(s) \tag{11.6a}$$

$$J_{m,i}(r) = \binom{m}{i} r^i (1 - r)^{m-i}; \quad 0 \leq r \leq 1$$
$$\tag{11.6b}$$
$$K_{n,j}(s) = \binom{n}{j} s^j (1 - s)^{n-j}; \quad 0 \leq s \leq 1$$

Each B_{ij} represents a set of three values, defining a vertex location in E^3 (Euclidean three-dimensional space). m is the order of the surface (also the polynomial) in x-direction. n is the order of the surface (also the polynomial) in the y-direction. $J_{m,i}$ and $K_{n,j}$ are the Bernstein basis or polynomial form.

Several properties of bipolynomial product $J_{m,i}(r) \, K_{n,j}(s)$ are the same as the univariate Bernstein operator used in the previous subsection:

- The four corners of the domain can be located through the parameter combinations $(0,0)$, $(0,1)$, $(1,0)$, and $(1,1)$.
- The edges of the domain can be traced by the univariate Bernstein polynomial approximation in one parameter, as the other is set to one of the two limits of zero or one.
- The surface is within the convex hull, defined by the vertices.
- The approximation is as smooth as or smoother than the function it is trying to approximate.

It should be clear that the basic domain is rectangular. In the examples in the book, the vertex locations are defined on an intersecting net of x and y. This allows the set of vertices along a fixed x to define a Bezier function in the y-u plane. A similar advantage holds for x-u plane. Figure 11.2 defines a Bezier surface. The vertices are distinctly marked on the figure.

[*]Files to be downloaded from the web site are indicated by boldface courier type.

Bezier Surface

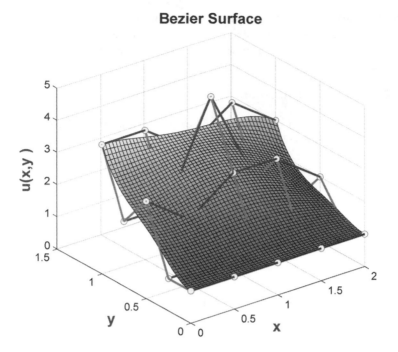

Figure 11.2 Example of a Bezier surface.

There are convenient matrix representations for the Bezier surface.[9] The particular entries for the example in Figure 11.2 are not reproduced but are indicated through the general representation:

$$[x(r, s) \quad y(r, s) \quad u(r, s)] = Q(r, s) = [R][M][B][N]^T [S] \qquad (11.7)$$

where

$$[R] = [r^m \; r^{m-1} \ldots 1] = [r^5 \; r^4 \; r^3 \; r^2 \; r^1 \; 1] \qquad (11.8a)$$

$$[M] = \begin{bmatrix} \binom{m}{0}\binom{m}{0}(-1)^m & \binom{m}{1}\binom{m-1}{m-1}(-1)^{m-1} & \cdots & \binom{m}{m}\binom{m-m}{m-m}(-1)^0 \\ \binom{m}{0}\binom{m}{m-1}(-1)^{m-1} & \binom{m}{1}\binom{m-1}{m-2}(-1)^{m-2} & \cdots & 0 \\ \cdot & \cdot & & 0 \\ \binom{m}{0}\binom{m}{1}(-1)^1 & \binom{m}{1}\binom{m-1}{0}(-1)^0 & \cdots & 0 \\ \binom{m}{0}\binom{m}{0}(-1)^0 & 0 & \cdots & 0 \end{bmatrix}$$

$$(11.8b)$$

$[N]$ is defined the same way as $[M]$ but of order n

$$[S] = [s^n \ s^{n-1} \ \ldots \ s]^T = [s^5 \ s^4 \ s^3 \ s^2 \ s \ 1]^T \qquad (11.8c)$$

MATLAB Code: Figure 11.2 is created by `DrawBezierSurface.m`. The coefficients are printed to the Command window and are not shown here. You can also explore the corresponding contour plot by uncommenting a portion of code. The code calls additional function m-files for creating coefficients listed previously.

11.1.3 Data Decoupling

Equation (11.3) can be recognized as:

$$[x(p) \ y(p)] = [P][N][B_x \ B_y]$$

leading to

$$[x(p)] = [P][N][B_x]; \quad [y(p)] = [P][N][B_y] \qquad (11.9)$$

which decouples both the x and y data. They are related through the same value of the parameter and the same order of the curve. The Bezier function, therefore, can be used to handle single-vector data (or one-dimensional data).

11.1.4 Derivatives

The second set of problems in this chapter relate to differential systems. It will be necessary to calculate the derivatives of the function in the domain. Consider a Bezier function defined by

$$[x(p) \ y(p)] = [P][N][B] \qquad (11.3)$$

The derivatives with respect to the parameter p can be computed through the matrix multiplication

$$[x' \ y'] = [P'][N][B] \qquad (11.10a)$$

$$\left[x'' \ y''\right] = \left[P''\right][N][B] \qquad (11.10b)$$

where the primes indicate derivatives with respect to p. The derivatives of y with respect to x are obtained through the chain rule. For example,

$$\frac{dy}{dx} = \frac{dy/dp}{dx/dp} \qquad (11.11)$$

We can similarly express other higher derivatives through other relations. In order for the derivative dy/dx, d^2y/dx^2, and all higher derivatives to exist, x' must be non zero between $0 \leq p \leq 1$. This can be enforced by keeping a minimum distance between two x vertex values.

The partial derivatives with respect to r and s can be established through matrix operations. For example, the first derivative of Q with respect to r can be obtained as

$$\left[\frac{\partial x}{\partial r} \quad \frac{\partial y}{\partial r} \quad \frac{\partial u}{\partial r}\right] = \frac{\partial Q}{\partial r} = \left[\frac{\partial R}{\partial r}\right][M][B]\,[N]^T[S] \qquad (11.12)$$

The partial derivatives with respect to x and y are obtained once again through chain rule. The solution to the differential equation is to identify the particular curve or surface that is sufficiently continuous and differentiable, and satisfies all of the constraints. This chapter refers to these as **Bezier solutions**. Note, unlike conventional data-fitting problem (which is referred to as curve fitting), this is a **curve fitting** fitting a curve to the boundary value problem (BVP).

Two kinds of examples are used to illustrate the technique in this chapter. The first kind deals with problems whose solutions are continuous functions but whose derivatives are not necessary for the problem. We will term them as data-handling problems. The second kind of examples will deal with differential equations. Both these sets of examples will use the least squared error (LSE) to drive the solution to the optimization problem.

11.2 DATA-HANDLING EXAMPLES

There are two examples in this category. The first deals with fitting a curve to x-y data, and we have explored one version in Example 4.2. Here, we will be working with Bezier function. The second example deals with fitting a surface to the x-y-z data. In both of these cases, the solution can be identified by explicit polynomial relations—which are what we term as the analytical solution. In these examples, we also identify the best order of the curve that will fit the data by comparing the solution for a range of the order of the Bezier curve.

11.2.1 Data Fitting with Bezier Functions

Data fitting is a process where the original data are transformed to have a different representation because of some advantages associated with the new representation. The new representation will usually be an approximation, and there will be some errors in this transformation. Traditional approach to data fitting requires three elements. The first element is the data that are being fit or reduced (given data). The second element is the family of functions that the data are considered to belong to (Bezier functions). The third is the measure that determines the acceptable approximation of the original data by the family of functions (minimum sum of the squared error over all data points).

The process used in this chapter is similar to the least squared error used in regression analysis. It is recessed within an outer cycle that is used to select the best order of the Bezier function through simple stepping of the order of the Bezier function starting from the minimum value. For this outer cycle, the least sum of the absolute error over all of the data points was used as the criteria. Therefore, there are two different measures used to arrive at the best approximation of the original data.

11.2.2   Optimum Bezier Solution

The Bezier coefficients for the approximation are determined by requiring that the squared error between the original data and the fitted data being the minimum for a given order of the Bezier function.

Problem Definition: Given a set of (m) vector data $y_{a,i}(Y)$, for a *selected order of the Bezier function* (n), find the coefficient matrix, $[B]$ so that the corresponding data set $y_{b,i}, (Y_B)$ produces the least sum of the squared error:

$$\text{Minimize} \quad E = \sum_{i}^{m} (y_{a,i} - y_{b,i})^2 = (Y - Y_B)^T (Y - Y_B)$$

$$= (Y - P_A NB)^T \ (Y - P_A NB) \tag{11.13}$$

To find the best $[B]$ we then use standard calculus to express

$$\frac{\partial E}{\partial B} = 0 \tag{11.14}$$

which leads to

$$[B] = [P_A^T P_A]^{-1} [P_A]^T Y \tag{11.15}$$

Equation (11.17) establishes the best coefficient matrix $[B]$ for a given order of the Bezier function. P_A is the parameter array based on the number of data points, m. This non iterative determination is combined with additional processing to determine the best order for a given set of data. In this section this error is the smallest sum of the absolute error over all of the data points.

$$\text{Minimize} \quad E(n) = \sum_{n=n\mathbf{min}}^{n\mathbf{max}} abs(y_{a,i} - y_{b,i}) \tag{11.16}$$

11.2.3   Example 11.1 — Smooth Data at Equidistant Intervals

The data are generated at equidistant intervals of the independent variable (x). The dependent variable (y) values are generated using a smooth function:

$$y_a(x) = \frac{\sin\left(0.1 + \sqrt{x^2}\right)}{0.1 + \sqrt{x^2}} \quad 0 \le x \le 10; \tag{11.17}$$

In this case, it is the hyperbolic sine function. Remember that this function is not known. Only the data generated using this function are input to the procedure. There are 101 data pairs. Bezier functions are fit for both the independent and dependent variables. The comparison between the original data and the data generated using the Bezier function is illustrated both graphically and through standard statistics.

Figure 11.3 shows the original data and the data fitted using the Bezier function for Example 11.1. It is evident, within the scale of the plot, that there is no error between the two curves. The best order of the Bezier curve (n) was 15. The number of data points (m) was 101. Tables 11.1a and 11.1b show that the Bezier function can reproduce the original data with identical statistics to five decimal places in both the independent and the dependent variable. This is definitely sufficient for engineering problems.

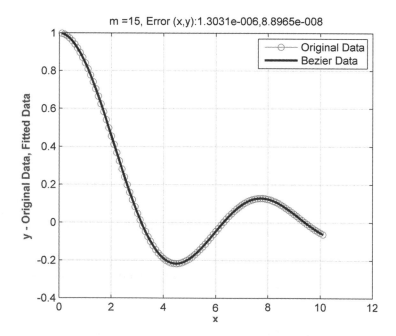

Figure 11.3   Data representation by Bezier function.

Table 11.1a   Example 11.1 Comparison of x **Data** ($m = 101, n = 15$)

Statistical Variable	Original Data	Bezier Data
Average Error	—	1.29016e-008
Sum of Avg. Error	—	1.30306e-006
Minimum	0.10000	0.10000
Maximum	10.10000	10.10000
Mean	5.10000	5.10000
Lower Quartile	2.50000	2.50000
Median	5.10000	5.10000
Upper Quartile	7.60000	7.60000
Variance	8.58500	8.58500
Std. Deviation	2.93002	2.93002
Skew	0.00000	0.00000
Kurtosis	1.78195	1.78195

Table 11.1b   Example 11.1 Comparison of y **Data** ($m = 101, n = 15$)

Statistical Variable	Original Data	Bezier Data
Average Error	—	8.80839e-010
Sum of Avg. Error	—	8.89647e-008
Minimum	−0.21723	−0.21723
Maximum	0.99833	0.99833
Mean	0.15835	0.15835
Lower Quartile	0.23939	0.23939
Median	0.06043	0.06043
Upper Quartile	7.60000	7.60000
Variance	0.12159	0.12159
Std. Deviation	0.34870	0.34870
Skew	1.20522	1.20522
Kurtosis	3.27429	3.27429

MATLAB *Code:* The code for solving Example 11.1 is available in **BezXYCurveFit.m**. The code will create Figure 11.3 and generate information for Tables 11.1 and 11.2. It calls **DataIn.m** to obtain the data statistics. The data input is obtained from **getCurveXYData.m**.

Bezier Coefficient Values: The values for the Bezier coefficients for Example 11.1 are

$$B = [B_x^T, B_y^T];$$

$$B_x = [0.1000 \quad 0.7667 \quad 1.4333 \quad 2.1000 \quad 2.7667 \quad 3.4333 \quad 4.1000 \quad 4.7667,$$

$$5.4333 \quad 6.1000 \quad 6.7667 \quad 7.4333 \quad 8.1000 \quad 8.7667 \quad 9.4333 \quad 10.1000];$$

$$B_y = [0.9983 \quad 0.9761 \quad 0.7957 \quad 0.4643 \quad 0.0501 - 0.3219 - 0.5129 - 0.4518,$$

$$- 0.1987 \quad 0.0646 \quad 0.2192 \quad 0.2392 \quad 0.1689 \quad 0.0692 - 0.0145 - 0.0619];$$

Using the expression in equation (11.3), the original 101 data for each variable can be replaced by 16 values of the Bezier coefficients. This can be considered *data reduction*.

Explicit Polynomial Description of the Data: The Bezier function in equation (11.1) for this pair of vertex values can be reduced to the polynomial functions of the parameter p. These functions can be used for further processing for additional properties of the data. The polynomial form is reduced from the actual form based on the Bernstein polynomial representation in equation (11.1):

$$x(p) = 10.000p \; - 0.0093904 \; p^{15} + 0.068148 \; p^{14}$$

$$- 0.22214p^{13} + 0.42945p^{12} - 0.54786p^{11}$$

$$+ 0.48598p^{10} - 0.30818p^9 + 0.14176p^8$$

$$- 0.047558p^7 + 0.011551p^6 - 0.0019640p^5$$

$$+ 0.00021731p^4 - 0.000014486p^3 + 0.44443 \; 10^{-6}p^2$$

$$+ 0.10000 \tag{11.18a}$$

$$y(p) = -0.33300p^{15} - 32.424p^{15} + 239.97p^{14}$$

$$- 725.02p^{13} + 1101.7p^{12} - 819.23p^{11}$$

$$+ 282.78p^{10} - 271.21p^9 + 357.81p^8$$

$$+ 1.5170p^7 - 194.05p^6 - 12.322p^5 + 83.070p^4$$

$$+ 3.3278p^3 - 16.617p^2 + 0.99833$$

$$0 \leq \; p \leq 1 \tag{11.18b}$$

11.2.4   Example 11.2 — Data Fitting Using Bezier Surface

Problem Definition: Given a set of array data $[U]$, assuming an order for each dimension (m, n), find the Bezier function coefficient matrix, $[B_U]$ so that the corresponding approximate data $[U_B]$ generates the least value for the sum of the squared error over the data array

$$\text{Minimize:} \quad E = \sum_i \sum_j (U - U_B)^2 \tag{11.19}$$

$$U_B = [R_A][M][B_U][N]^T[S_A]^T = [F_A][B_U][G_A] \tag{11.20}$$

To find the best coefficient matrix $[B_U]$ we use calculus:

$$\frac{\partial E}{\partial B_U} = 0 \tag{11.21}$$

which leads to

$$[B_U] = [G_A I F_A^T F_A]^{-1}[G_A I F_A^T U] \tag{11.22}$$

This equation establishes the best coefficient matrix $[B_U]$ for a given orders of the Bezier function. This non iterative approach allows us to add additional processing to choose the best order for a given set of data, as the order is varied between the chosen range. This error is the smallest sum of the absolute error over all of the data points.

Example 11.2: The original data is generated at equidistant points using the hat function, which was inversed in Example 9.1.

$$u = \frac{\sin(0.1 + \sqrt{x^2 + y^2})}{0.1 + \sqrt{x^2 + y^2}}; \quad -10 \le x, y \le 10 \tag{11.23}$$

Data is generated at 101×101 points. Once more, the order for the x and y dimensions was looped between 2 and 15 to determine the best order for the surface fit. For each pair of chosen order the Bezier coefficients were obtained from (11.22). Once the coefficients are known, all other information is easily obtained, including the statistics in Table 11.2. Figure 11.4 is the contour plot of the original information. Figure 11.5 is the contour plot of the corresponding Bezier data. There is a small error in the fit.

Table 11.2   Example 11.2: Data Statistics Comparison for Original Data and Bezier Data

	Original—u	Bezier—u
Minimum	−0.21723	−0.22097
Maximum	0.99833	0.97149
Mean	0.01227	0.01208
Median	0.00201	0.00054
Range	1.21557	1.19246
Mode	−0.21723	−0.22097
Variance	0.00100	0.00099
Std. Dev	0.08289	0.08281
Skew	17.92968	17.77942
Kurtosis	185.85133	183.13296

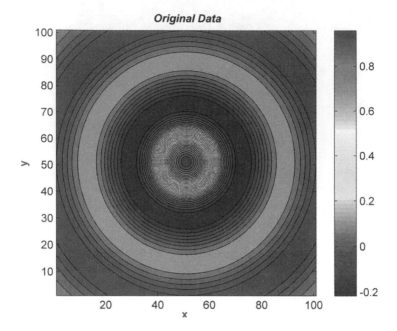

Figure 11.4 Example 11.2: Original data.

MATLAB Code: The MATLAB code for Example 11.2 is available in **BezierSurfaceFit.m**. The companion file for the data is **getSurfaceData.m**. By this arrangement you can run a different example by targeting a different example in **getSurfaceData.m** file. Table 11.2 and Figures 11.4 and 11.5 are generated by the code. There is also more information printed to the window, including additional figures that are not included here. The figures and the statistics suggest some error in the Bezier approximation. It is also true that we are at the limit of round-off error because of the large orders necessary for the fit. Remember the highest polynomial term $Cr^{13}s^{14}$.

The following information will appear in the Command window:

```
Example:Example 11.2
Problem Type::Bezier Surface fit of smooth surface data
Best order of Surface in x -dir:13
Best order of Surface in y -dir:14
No. of Parameters in x -dir:101
No. of Parameters in y -dir:101
Sum of Absolute Error          :  1.50677e+001
Maximum Absolute Error         :  7.11190e-002
Minimum Absolute Error         :  4.84507e-008
Least Squared Error            :  5.06416e-001
Average Error                  :  1.47708e-003
```

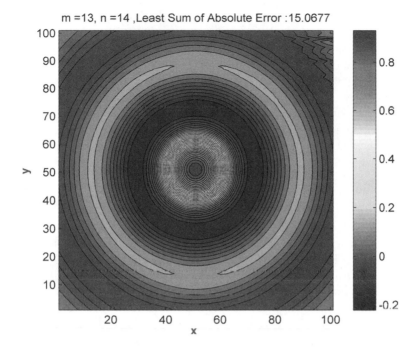

m =13, n =14 ,Least Sum of Absolute Error :15.0677

Figure 11.5 Example 11.2: Bezier function data.

```
Number of data points          :   10201
Total Time -less printing and plotting (s):10.7969
```

The polynomial coefficients and the analytical form of the solution is not reproduced here for space reasons. The figures provide a good indication of the error. The data are continuous to the thirteenth partial derivatives. Remember, the original sine function is being reproduced by polynomials. The following is not demonstrated here but the reader can easily show that the Bezier data reduction method also works on non-equidistant data, non-smooth data, and discontinuous data. The Bezier fit works as a data smoother and tries to preserve the average of the original data.

11.3 SOLUTIONS TO DIFFERENTIAL SYSTEMS

This section presents two examples to illustrate that Bezier functions can provide excellent solutions to engineering problems that are described by differential equations. They are referred to as boundary value problems and are among the most difficult to solve in engineering mathematics. Both these problems deal with multiple nonlinear equations. The first deals with ordinary differential equations and the second with partial differential equations. Before proceeding further, it deserves to be noted that analytical solutions to such types of problems are rare today. In practice, boundary value problems are solved numerically using collocation methods, or Galerkin methods or finite difference methods. Closed-form solutions, for nonlinear differential systems usually do not exist unless the problem is special. Closed-form solutions are generally investigated using series representation, perturbation methods, group theoretic methods, and special techniques that exploit particular features of the differential equations. All of the approaches require significant effort and are problem dependent. By contrast, the procedure illustrated in this section is simple, direct, universal, and problem independent.

11.3.1 Flow over a Rotating Disk

This example is a fluid flow problem. The original problem definition is available in Schlichting.[13] Figure 11.6 illustrates the features of the problem.

A disk of radius R is rotating with the angular speed ω in still fluid. The flow is steady, incompressible, has constant property, and is axisymmetric. The fluid at the disk has to satisfy the no-slip condition. The centrifugal effects cause the fluid to leave the disk near the disk. The flow above the disk must replace this airflow through a downward-spiraling flow. A cylindrical coordinate system (r, θ, z) is used for description. V_r, V_θ, V_z, are the velocity components and are shown on the figure, suggesting the boundary conditions. p is the pressure, v,

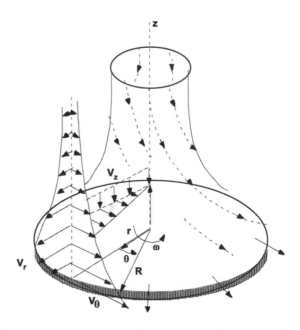

Figure 11.6   Example 11.3: Problem and variable description.

the dynamic viscosity. The continuity and the Navier–Stokes equations are

$$\frac{\partial V_r}{\partial r} + \frac{V_r}{r} + \frac{\partial V_z}{\partial z} = 0$$

$$V_r \frac{\partial V_r}{\partial r} - \frac{V_\theta^2}{r} + V_z \frac{\partial V_z}{\partial z} = -\frac{1}{\rho}\frac{\partial p}{\partial r} + \nu\left[\frac{\partial^2 V_r}{\partial r^2} + \frac{\partial}{\partial r}\left(\frac{V_r}{r}\right) + \frac{\partial^2 V_r}{\partial z^2}\right]$$

$$V_r \frac{\partial V_\theta}{\partial r} + \frac{V_r V_\theta}{r} + V_z \frac{\partial V_\theta}{\partial z} = +\nu\left[\frac{\partial^2 V_\theta}{\partial r^2} + \frac{\partial}{\partial r}\left(\frac{V_\theta}{r}\right) + \frac{\partial^2 V_\theta}{\partial z^2}\right]$$

$$V_r \frac{\partial V_z}{\partial r} + V_z \frac{\partial V_z}{\partial z} = -\frac{1}{\rho}\frac{\partial p}{\partial z} + \nu\left[\frac{\partial^2 V_z}{\partial r^2} + \frac{1}{r}\frac{\partial V_z}{\partial r} + \frac{\partial^2 V_z}{\partial z^2}\right]$$

(11.24)

The boundary conditions are

$$z = 0: \quad V_r = 0; \quad V_\theta = r\omega; \quad V_z = 0;$$
$$z = \infty: \quad V_r = 0; \quad V_\theta = 0;$$

(11.25)

Reference 13 explains that using a differential volume over the disk an approximate value for the boundary layer thickness can be found.[13] This is used for scaling the z-coordinate. Next, scaling for the velocity and pressure are introduced so that dependence on r and z can be separated. This is an essential technique for reducing the partial differential equations (PDE) to ordinary differential equations (ODE). The scaling relations for the various terms, where δ is the boundary layer, are

$$Z = z\sqrt{\frac{\omega}{\nu}} \approx \frac{z}{\delta}; \quad V_r = r\omega F(Z); \quad V_\theta = r\omega G(Z);$$
$$V_z = \sqrt{\nu\omega}H(Z); \quad p(z) = \rho\nu\omega P(Z);$$

(11.26)

Substituting in equations (11.24 to 11.25), and simplifying with Z being the independent variable and primes representing derivatives with respect to Z,

$$2F + H' = 0$$
$$F^2 + F'H - G^2 - F'' = 0$$
$$2FG + HG' - G'' = 0$$
$$P' + HH' - H'' = 0$$

(11.27)

The transformed boundary conditions are as follows:

$$Z = 0; \quad F = 0; \quad G = 1; \quad H = 0; \quad P = 0$$
$$Z = \infty \, (\approx 6); \quad F = 0; \quad G = 0$$

(11.28)

The original Navier–Stokes equations and boundary conditions are now reduced to the set of ODE and corresponding boundary conditions in equations

(11.27) and (11.28). Usually, the last differential equation for pressure is not solved with the first three. It can be obtained from the solution of F, G, and H.

The solution to (11.27), (11.28) was first obtained using a power series around $Z = 0$, and an asymptotic series for large values of Z. Numerical solutions are currently accepted as alternate analytic solutions for this problem today. MATLAB is used to generate the numerical solutions through its boundary value problem solver used for comparison with the Bezier functions. The final value of the residuals, the error in the differential equations over the range of the independent variable, should provide reasonable confidence that the Bezier functions provide an excellent solution.

Bezier Function Formulation: Three Bezier functions will be used to identify the functions F, G, and H. This is now a coupled set of nonlinear differential equations. The three functions are tied to the same parameter value and the values for the independent variable. These Bezier functions must solve the differential equation and boundary conditions in equations 11.27, 11.28. To avoid additional terminology in defining functions we will identify our three Bezier functions, as $[Z, F]$, $[Z, G]$, and $[Z, H]$. These are determined by three sets of vertices:

$$\left[P_1^F, P_2^F, \ldots, P_{m+1}^F\right], \left[P_1^G, P_2^G, \ldots, P_{m+1}^G\right], \left[P_1^H, P_2^H, \ldots, P_{m+1}^H\right].$$

The optimization problem is: Find the vertices of the Bezier functions $[Z, F], [Z, G]$, and $[Z, H]$ and their derivatives that

$$\text{Minimize} \quad f = \sum_{i=1}^{n_p} \left\{ \left[2F_i + H_i'\right]^2 + \left[F_i^2 + F_i'H_i - G_i^2 - F_i''\right]^2 \right\}$$

$$+ \sum_{i=1}^{100} \left\{ \left[2F_iG_i + H_iG_i' - G_i''\right]^2 \right\} \tag{11.29}$$

Subject to $\quad [Z_1, F_1] = [0, 0]; \quad [Z_1, G_1] = [0, 1]; \quad [Z_1, H_1] = [0, 0]$

$$[Z_{m+1}, F_{m+1}] = [0, 0]; \quad [Z_{m+1}, G_{m+1}] = [0, 0]; \tag{11.30}$$

where i represents a point on the curves and m order of the curves. The number of points, n_p on each of the curve is chosen to be 101. The objective function is the residuals of the differential equation, for the three equations, over the domain/trajectory. The problem in equations 11.29, 11.30 is solved using `fmincon` from the MATLAB Optimization Toolbox.

MATLAB Code: This is a serious optimization problem, and the accompanying code is large and split into several files. The file NCurves.m controls the solution and the comparison. The list of files used is given next.

```
%     NCurves.m-this setus up the initial guess,  linear
%        constraints (boundary conditions), calls optimizer,
%          prints, plots and compares with BVP solver
%
%        printInformation.m -Print vertices and limits
%        DrawCurves.m-Plots the multiple curves
%        set_Lin_Eq_Cons.m -Linear Equality Constraints
%        set_Lin_InEq_Cons.m-Linear Inequality constraints
%
%          fmincon.m-The optimizer
%            Obj_Rotflow_1.m-set up the objective function
%            MakeDerivatives2.m-obtain second derivatives of
%                                the curve
%
%          Compare_RotFlow_1.m-Compare solution from BVP
%                                solver and Bezier
%          MakeCurve.m    -generate x,y information from
%                                vertices
%            bvpexampledisk.m-using MATLAB BVP solver
%                bvpinit.m-define initial solution
%                bvp4c.m -boundary value solver
%                mat4bc.m-boundary conditions for solution
%                deval.m -evaluate solution at chosen
%                                locations
%                state5.m-set up state space description of
%                                diferential system
%
%     AnalyticalCurve.m: explicit polynomial representation
```

There is a lot of information written to the Command window that is not reproduced here.

The Bezier Solution: The solution to the optimization problem is presented in Figure 11.7. The design variables that are the solution to the optimization problem are the vertices, which are also displayed as markers. The Bezier functions that represent the solutions to the nonlinear boundary value problem (NLBVP) are the curves that are obtained explicitly from the vertex information and are also shown in the figure.

Figure 11.8 is the comparison of the Bezier functions with the other numerical solution obtained using the BVP solver in MATLAB. The axes are flipped to display the boundary layer picture. Figure 11.8 illustrates that the two solutions are indistinguishable within the scale of the graph. The residuals of the Bezier functions are exact. The following additional information is available with respect to the Bezier functions solving this example.

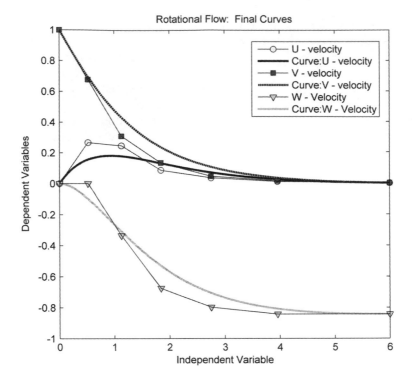

Figure 11.7 Example 11.3: Final Bezier functions.

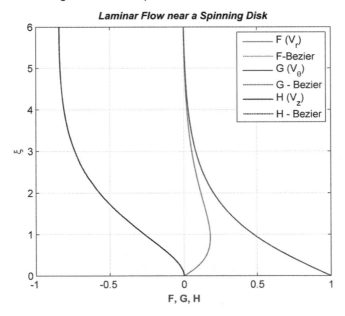

Figure 11.8 Example 11.3: Laminar flow nearing a spinning disk — Comparison of two solutions.

The order of the functions was six for all of the curves. The number of design variables was 28. The convergence criteria for the optimum were set to 1.0e-08. Iterations stopped when the objective function F could not improve further due to meeting the tolerance on the constraints (1.0e-06). The final value of f was 1.21.0e-06. This is sum of the squared error in the three differential equations over 101 points each. The target value is zero. The number of iterations of the solver was 63. Since the residuals are computed exactly, the low value of the sum of the residuals for the final iteration should provide reasonable confidence that the Bezier functions provide an excellent solution.

Analytical or Explicit Bezier Solution: The explicit solution for the independent variable and the three functions are given here:

$$Z(p) = 0.40468p^6 - 0.10827p^5 + 0.71410p^4 + 0.66665p^3 + 1.1775p^2$$
$$+ 3.1453p$$

$$F(p) = 0.83149p^6 - 2.5617p^5 + 1.4346p^4 + 3.0404p^3 - 4.3471p^2$$
$$+ 1.6024p$$

$$G(p) = 1. - 1.9360p^6 - 0.51266p^5 + 2.5747p^4 - 5.2818p^3 + 4.8806p^2$$
$$- 0.72495p$$

$$H(p) = 0.71602p^6 - 1.4886p^5 - 1.6865p^4 + 6.6602p^3 - 5.0435p^2$$

The author is not aware of a limited-term analytical solution for this problem. In fact, the last known explicit solution was obtained several decades ago and is considered too challenging to be pursued today. Here we have shown how numerical optimization in combination with a parametric function definition can easily establish solutions to the intractable problems of the past and the future. Such an approach can be a regular feature in building new mathematical model for physical process and situations.

11.3.2 Two-Dimensional Flow Entering a Channel

This is also an example from fluid mechanics where we solve the complete set of the Navier–Stokes equation for a two-dimensional problem. Obtaining closed-form solutions to such problems is not even attempted today. Such problems are solved numerically using a finite difference method or through the use of domain discretization methods like finite volume or finite element methods.

Problem Description: A steady, two-dimensional, constant property flow takes place in a two-dimensional channel as illustrated by Figure 11.9. The x-velocity (u) at the inlet is constant with the value U_0. There is no y-velocity (**v**) at the inlet. The no-slip conditions apply on both walls.

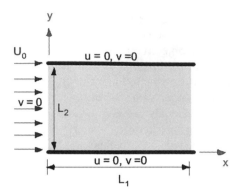

Figure 11.9   Example 11.4: Problem description.

The velocities u, v, and pressure p must satisfy the *Navier–Stokes* equation. This is a coupled system of nonlinear partial differential equations (*PDE*) and is expressed as follows:

$$\text{continuity:} \quad \frac{\partial u}{\partial x} + \frac{\partial v}{\partial y} = 0$$

$$\text{x-momentum:} \quad u\frac{\partial u}{\partial x} + v\frac{\partial u}{\partial y} = -\frac{1}{\rho}\frac{\partial p}{\partial x} + v\left[\frac{\partial^2 u}{\partial x^2} + \frac{\partial^2 u}{\partial y^2}\right] \qquad (11.31)$$

$$\text{y-momentum:} \quad u\frac{\partial v}{\partial x} + v\frac{\partial v}{\partial y} = -\frac{1}{\rho}\frac{\partial p}{\partial y} + v\left[\frac{\partial^2 v}{\partial x^2} + \frac{\partial^2 v}{\partial y^2}\right]$$

The boundary conditions are:

$$x = 0; \quad u(x, y) = u(0, y) = U_0; \quad v(x, y) = v(0, y) = 0;$$

$$p(x, y) = p(0, y) = c;$$

$$y = 0; \quad u(x, y) = u(x, 0) = 0; \quad v(x, y) = v(x, 0) = 0;$$

$$y = L_2; \quad u(x, y) = u(x, L_2) = 0; \quad v(x, y) = v(x, L_2) = 0; \qquad (11.32)$$

In the above, ρ is the fluid density and v is the fluid kinematic viscosity. L_1 is the length of the channel. L_2 is the width of the channel. The domain is called the entering region of the flow as the viscous effects through the walls will shape the velocity profile in the channel as the flow proceeds left to right. In the following development the example was solved as formulated. No simplifying assumptions or problem reductions took place, including exploiting problem symmetry.

The Bezier Function Formulation: The nonlinear BVP problem in (11.31), (11.32) is solved using Bezier functions. Here the solution will be represented by three surfaces in the solution domain. The first is the solution for the velocity in the x direction $u(x, y)$, the second is the solution for the velocity in the y direction $v(x, y)$, and the third one is the solution for the pressure $p(x, y)$.

For convenience, we collect the solution for $u(x, y)$ over the whole domain in the set of three matrices $[X, Y, U]$. Similar definitions hold for $v(x, y)$ and $p(x, y)$. The problem, therefore, is:

Find the **Bezier surfaces** $[X, Y, U], [X, Y, V], [X, Y, P]$, or the *vertices of the parametric surface* $\{A_{i,j}\}, \{B_{i,j}\}$ and $\{C_{i,j}\}$ that solve the BVP expressed in equations (11.31), (11.32) through the following optimization problem:

$$\text{Minimize} \quad F(X) = \sum_{i=1}^{p} \sum_{j=1}^{q} \left[\frac{\partial u}{\partial x}\bigg|_{i,j} + \frac{\partial v}{\partial y}\bigg|_{i,j} \right]^2$$

$$+ \sum_{i=1}^{p} \sum_{j=1}^{q} \left[u\frac{\partial u}{\partial x}\bigg|_{i,j} + v\frac{\partial u}{\partial y}\bigg|_{i,j} + \frac{1}{\rho}\frac{\partial p}{\partial x}\bigg|_{i,j} - v\left[\frac{\partial^2 u}{\partial x^2} + \frac{\partial^2 u}{\partial y^2}\right]\bigg|_{i,j} \right]^2$$

$$\tag{11.33}$$

$$+ \sum_{i=1}^{p} \sum_{j=1}^{q} \left[u\frac{\partial v}{\partial x}\bigg|_{i,j} + v\frac{\partial v}{\partial y}\bigg|_{i,j} + \frac{1}{\rho}\frac{\partial p}{\partial y}\bigg|_{i,j} - v\left[\frac{\partial^2 v}{\partial x^2} + \frac{\partial^2 v}{\partial y^2}\right]\bigg|_{i,j} \right]^2$$

$$x = 0; \quad u(0, y) = U_0; \quad v(0, y) = 0;$$

$$y = 0; \quad u(x > 0, 0) = 0; \quad v(x > 0, 0) = 0;$$

$$y = L_2; \quad u(x > 0, L_2) - 0; \quad v(x > 0, L_2) = 0; \tag{11.34}$$

The boundary conditions in equation (11.34) are modified from those in equation (11.32) for the following important reasons.

1. $u(0, 0)$ cannot be both U_0 and 0. This causes a problem in uniqueness for explicit/analytical solutions.
2. u cannot decrease from U_0 to 0 instantaneously at the start of the walls of the channel ($x = 0$) because of the no-slip condition. This causes infinite values for the partial derivatives and thereby the residuals will blow up at $(x, y) = (0, 0)$ and $(0, L_2)$.

These issues must be recognized when trying to develop analytical solutions, too. This is particularly challenging since the analytical solutions are designed to establish solution over the entire domain through continuous functions. The Bezier functions compute exact derivatives, and hence the procedure must live with large local residuals in this problem. In the solution established later, the no-slip conditions are enforced over a finite length of the walls of the channels. Such a problem is nonexistent in computational fluid dynamics (CFD) solutions, where the residuals are approximated and the continuity of the solution is not enforced beyond the degree necessary for computing residuals. The problem description here introduces large residuals at the inlet region near the walls. These large values should propagate, but damp out quickly for an acceptable solution.

The pressure boundary condition is not applied. This is true in CFD, too, because boundary conditions on both velocity pressure cannot be specified at a point in low-speed flow. The boundary conditions were used to constrain the vertices directly. The problem in equations (11.33), (11.34) is solved using fmincon from the MATLAB Optimization Toolbox. The code for this example is not included.

Problem Constants: The problem constants were selected for simplicity. The length of the plate L_1 is 45. The height of the flow domain L_2 is 1. The density ρ was 1. The kinematic viscosity ν was 0.01. The Reynolds number was 22,500. Even if this number is greater than the critical Reynolds number, the equations in (5.12) make the flow laminar. The inlet velocity in the x direction at the leading edge of the plate is constant at a value of $U_0 = 5$. The y-velocity at the inlet was 0. A consistent system of units is implied.

The Bezier Solution: The solution presented corresponds to $m = 9$ and $n = 6$. This set of orders was the same for the three Bezier functions. The total number of design variables was 210. For the solution presented, there were 75 parameter values in $r(p)$ and 75 parameter values for $s(q)$, giving a total of 5,625 points where the objective function was evaluated. Originally, the number of such points was 2,500. It was increased to 5,625 to see if the residual increased because of more points in the entry region with high residuals. It was surprising to note that the objective function decreased slightly, indicating that it is difficult to overcome the lack of definition of the problem near the walls at the entry that are the reason for the high residuals. Further increase of the parameter points did not result in any appreciable improvement in the solution. The initial guess for the solution was constant level surfaces, with a value of 5 for the u-surface, zero for v, and 10 for the p surface.

The program was stopped at the 250th iteration, as the objective decreased only in the third decimal place. Further iterations will not change the solution. The final value of the objective function was 76.2982. This is the sum of the squared residuals of the three equations, each evaluated over 5,625 points. This yields an average squared residual per equation of 0.0136. Looking at the statistics for each equation:

Maximum residual in first equation: $1.000e+000$

Maximum residual in second equation: $5.475e+000$

Maximum residual in third equation: $5.174e-003$

Sum of residual in first equation: $3.901e+001$

Sum of residual in second equation: $3.701e+001$

Sum of residual in third equation: $2.784e-001$

The residuals are shown in Figure 11.10. The figure confirms the expectation for the solution. It can be seen that the residuals are large only in the region around the wall where the flow enters. The remaining points on the domain point to a good solution.

Figure 11.11 is the solution for the u and v velocity. It is nice to see a semblance of symmetry, even though it is not enforced or built into the problem formulation. Figure 11.11 also contains the velocity profiles at 10 locations along the channel. The locations are identified in the legend. It is impressive to see the parabolic profile in the u velocity develop with the flow. The presence of symmetry is suggested in the figure. The v velocity magnitude is lower than the main flow, and as the u profile is developed, it tends to die out as the flow settles An interesting observation from the u profile is that while the differential form of the continuity equation appears satisfied, the integral form where the same mass flow is expected to cross all sections is in defect. This is an interesting contrast from traditional expectation, where typically v is ignored in most calculations.

It can be seen that the no-slip condition is only met after a significant distance from the entrance. This was identified earlier to be associated with the requirement for the uniqueness of the u velocity at the entrance. There is a practical reason, too. For this problem, the aspect ratio of the flow domain is 45 to observe the development of the flow. In previous problems, this was typically around 1. For a surface of order 9 in the x direction, there are 10 points along the length where the vertices are defined. If the vertices are at equidistant locations, then the first point where the no-slip condition can be applied is 4.5 units away from the entrance. The actual conditions should take effect beyond this location, which is supported by Figure 11.11. This problem can be controlled by a proper choice of the aspect ratio of the flow domain. For this solution, the flow is fully developed around 25 units in the flow direction. This is also the distance predicted by physical experiments.

It is the pressure solution shown in Figure 11.12 that is most astonishing. The Navier–Stokes equation handling of the pressure term is usually considered weak because it is not present directly in the continuity equation. The Bezier solution appears to make a strong statement about the pressure variation. The solution illustrates almost a linear variation of pressure in the flow direction a little after the entrance, or a constant pressure gradient. There is no pressure dependence on in the y direction. The Bezier function for pressure allowed a ninth order variation in the x direction, and a sixth-order variation in the y direction. It is impressive that the final form is linear in x and had no variation in y. Interestingly, the pressure does not appear to be affected by the high residuals at the entrance region. This suggests that the quality of the solution is good, even if the sum of the residuals is significantly higher than zero in that region. Here, the exact partial derivatives are being used unlike the other contemporary techniques. In Figure 11.12, the design values for the final pressure function are shown. It is interesting to see the layout that determines the almost-constant pressure gradient. Remember, each of these variables is determined individually by the optimizer. Their perfect alignment in rows and their constant length, is unexpected.

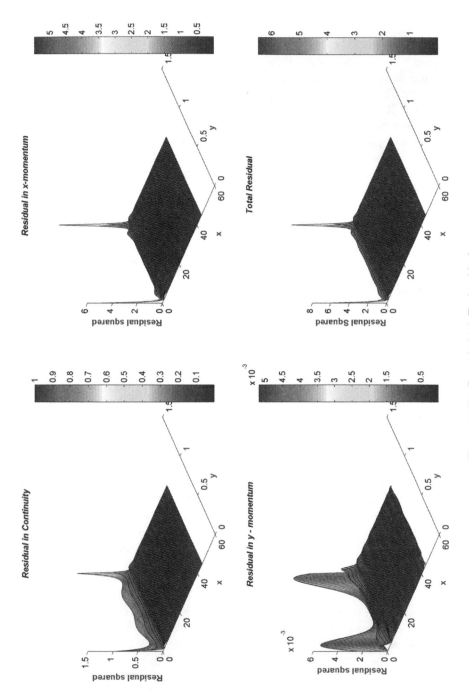

Figure 11.10 Example 11.4: Final residuals.

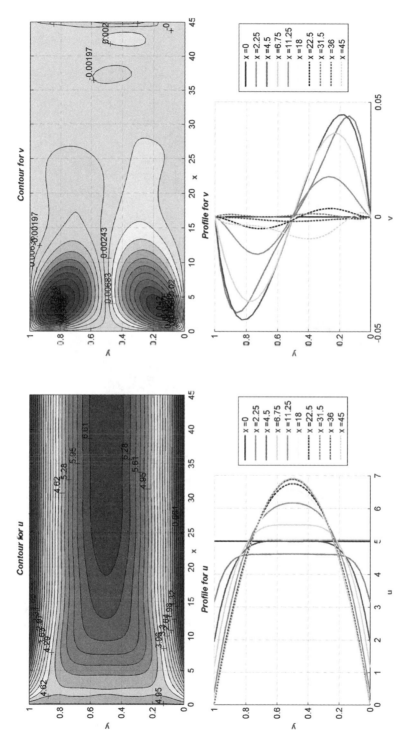

Figure 11.11 Example 11.4: u anc v velocity and profiles.

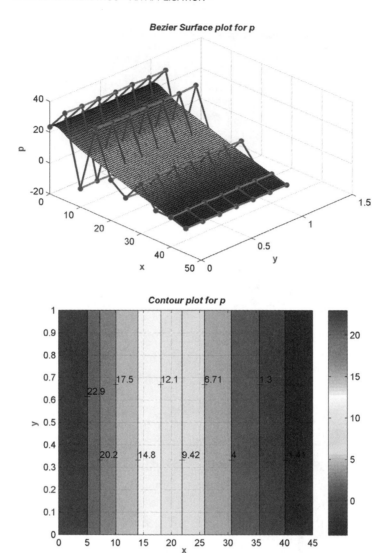

Figure 11.12 Example 11.4: p variation.

Analytical/Explicit Bezier Solution: The explicit solution in polynomial form can be easily obtained. Only the x-velocity is recorded below to indicate the nature of such solutions.

$$x(r, s) = 45r; \quad y(r, s) = s \tag{11.35a}$$

$$u(r, s) = (-8795.4r + 151260.r^2 - 964180r^3 \times +0.31783 \times 10^7 r - 0.60716$$
$$\times 10^7 r^5 - 0.46803 \times 10^7 r^7 + 0.69548 \times 10^7 r^6 + 0.16873$$

$$\times 10^7 r^8 - 246800 r^9)s^6 + (26351r - 452650 r^2 + 0.28806$$

$$\times 10^7 r^3 - 0.94755 \times 10^7 r^4 + 0.18050 \times 10^8 r^5 + 0.13795$$

$$\times 10^8 r^7 - 0.20600 \times 10^8 r^6 - 0.49387 \times 10^7 r^8 + 714780 r^9)s^5$$

$$+ (-30481r + 507190 r^2 - 0.31707 \times 10^7 r^3 + 0.10281$$

$$\times 10^8 r^4 - 0.19310 \times 10^8 r^5 - 0.14247 \times 10^8 r^7 + 0.21693$$

$$\times 10^8 r^6 + 0.49689 \times 10^7 r^8 - 691090 r^9)s^4 + (17087r - 261340 r^2$$

$$+ 0.15551 \times 10^7 r^3 - 0.48450 \times 10^7 r^4 + 0.87483 \times 10^7 r^5 + 0.58366$$

$$\times 10^7 r^7 - 0.94006 \times 10^7 r^6 - 0.18776 \times 10^7 r^8 + 227370 r^9)s^3$$

$$+ (-4744.3r + 59192 r^2 - 309050 r^3 + 855950 r^4 - 0.13559 \times 10^7 r^5$$

$$- 575990 r^7 + 0.12258 \times 10^7 r^6 + 95975 r^8 + 8706.8 r^9)s^2$$

$$+ (583.44r - 3653.7 r^2 + 8246.7 r^3 + 4947.0 r^4 - 61106 r^5$$

$$- 127900 r^7 + 127630 r^6 + 64251 r^8 - 12978 r^9)s$$

$$+ (-5r^9 + 15r^8 - 180 r^7 + 420 r^6 - 630 r^5$$

$$+ 630 r^4 - 420 r^3 + 180 r^2 - 45r + 5) \qquad (11.35b)$$

The author is not aware of a limited term analytical or explicit solution to the Navier–Stokes equation for this type of problems. Once again, we have shown how numerical optimization, in combination with a parametric function definition, can easily establish solutions to the intractable problems of the past and the future and allow development of new mathematical models.

11.4 SUMMARY

In this chapter, we have seen the application of design optimization to an area that is relatively new and novel. The optimization methods are quite mature, and you should be able to apply them to your own area of work. In the last example, we had 210 variables. Your only effort is to represent the problem in the standard format of an optimization problem. With respect to the problems covered in this chapter, or hybrid mathematics, or the solutions through Bezier functions, the examples illustrate the following:

- The formulation of the problem is simple.
- The problem set-up is direct.
- No domain discretization was necessary. This can be considered a meshless approach.

- Differential equations are handled in their original form.
- Exact derivatives are used in the residual computation.
- Standard optimization techniques are used to obtain the solution.
- The procedure is independent of the type or class of the problem.
- A single, continuous solution is defined for the entire domain.
- The solutions have a basic polynomial representation.
- Finite number of terms can span the solution space effectively.
- The procedure is adaptive and the degree of freedom of the Bezier functions (the order of the function) can be obtained by simple exploration.
- Problems with higher derivatives need not be transformed.
- Natural boundary conditions can be handled directly.

The Bezier functions have excellent blending and smoothing properties that are not available with other polynomial approximating functions. The procedure can provide good approximate solutions for numerically difficult types of BVP. It can also function as a filter to identify singular regions of the solution.

Naturally, this method may not be the right one for problems with discontinuous solutions or solutions with discontinuity in their derivatives. The work is exploratory and executed within the MATLAB interpreter environment, and hence, it does not make sense to compare execution time to any of the current methods for solving these problems—like the finite element. However, the current domain discretization techniques has seen over a billion man-hours of investment in development, compared to about 2,000 hours for the examples you can visit in Reference 6.

REFERENCES

1. Venkataraman, P. "Low Speed Multipoint Airfoil Design." Paper # 98—2402, 16th AIAA Applied Aerodynamics Conference. Albuqueque, New Mexico, June 1998.
2. Venkataraman, P. "B-Spline Based Free Form Solution of Nonlinear Systems." DETC2004-57672, 24th Computers and Information in Engineering (CIE) Conference. Salt Lake City, Utah, September 2004.
3. Venkataraman, P. "A New Class of Analytical Solutions to Nonlinear Boundary Value Problems." DETC2005-84604, 25th Computers and Information in Engineering (CIE) Conference. Long Beach, California, September 2005.
4. Venkataraman, P. "Explicit Solutions for Linear Boundary Value Problems Using Bezier Functions." DETC2006-99227, 26th Computers and Information in Engineering (CIE) Conference. Philadelphia, PA, September 2006.
5. Venkataraman, P. and J. G. Michopoulos. "Explicit Solutions for Nonlinear Partial Differential Equations." DETC2007-35439, 27th Computers and Information in Engineering (CIE) Conference. Las Vegas, Nevada, September 2007.
6. Venkataraman, P. "Explicit Solutions to Differential Equations." A Web resource. http://people.rit.edu/pnveme/ExplictSolutions2/

7. Bezier, P. *"How Renault Uses Numerical Control for Car Body Design and Tooling."* SAE Paper 680010, Society of Automotive Engineers' Congress. Detroit, Michigan, 1968.

8. Gordon, W.J. and R.F. Riesenfeld. "Bernstein-Bezier Methods for the Computer-Aided Design of Free Form Curves and Surfaces". *Journal of the Association of Computing Machinery*, vol. 21, no. 2, (April 1974): 293–310.

9. Rogers, D. F., and J.A. Adams. *Mathematical Elements for Computer Graphics. 2nd Ed.* New York: McGraw-Hill, 1990.

10. De Boor, Carl. "On Calculations with B-Splines". *Journal of Approximation Theory*, vol. 6. (1972): 50–62.

11. Farin, G. E. "NURBS for Curve and Surface Design". *SIAM* (1992).

12. Hollig, K. "Finite Elements with B-Splines". *SIAM* (2003).

13. Schlicting, H. *Boundary-Layer Theory*. New York: McGraw-Hill, 1979.

INDEX